Don G.

In the Collected Studies Series:

JACK MORRELL
Science, Culture and Politics in Britain, 1750–1870

ROY M. MacLEOD
Public Science and Public Policy in Victorian Britain

WILLIAM H. BROCK
Science for All: Studies in the History of Victorian Science and Education

WILFRED V. FARRAR
Chemistry and the Chemical Industry in the 19th Century:
The Henrys of Manchester and Other Studies

R.W. HOME
Electricity and Experimental Physics in 18th-Century Europe

A. RUPERT HALL
Newton, his Friends and his Foes

GERARD L'E. TURNER
Scientific Instruments and Experimental Philosophy, 1550–1850

MAURICE CROSLAND
Studies in the Culture of Science in France and Britian since the Enlightenment

JOHN HEDLEY BROOKE
Thinking about Matter: Studies in the History of Chemical Philosophy

A. RUPERT HALL
Science and Society: Historical Essays on the Relations of Science,
Technology and Medicine

ROBERT FOX
Science, Industry and the Social Order in Post-Revolutionary France

TREVOR H. LEVERE
Chemists and Chemistry in Nature and Society, 1770–1878

ALLAN CHAPMAN
Astronomical Instruments and Their Users: Tycho Brahe to William Lassell

COLLECTED STUDIES SERIES

Civil Engineers and Engineering in Britain, 1600–1830

Professor A.W. Skempton

A. W. Skempton

Civil Engineers and Engineering in Britain, 1600–1830

VARIORUM
1996

Published by VARIORUM
Ashgate Publishing Limited
Gower House, Croft Road,
Aldershot, Hampshire GU11 3HR
Great Britain

Ashgate Publishing Company
Old Post Road,
Brookfield, Vermont 05036-9704
USA

ISBN 0-86078-578-5

British Library CIP Data
Skempton, A.W.
Civil Engineers and Engineering in Britain, 1600–1830. — (Variorum Collected Studies Series: CS533)
1. Civil Engineering—Great Britain—History. 2. Civil Engineers—Great Britain—History.
I. Title. II. Series.
624'.0941

US Library of Congress CIP Data
Skempton, A.W.
Civil Engineers and Engineering in Britain, 1600–1830 / A.W. Skempton.
p. cm. — (Collected Studies Series: CS533)
Includes index (cloth: alk. paper)
1. Civil Engineering—Great Britain—History. 2. Civil Engineers—Great Britain—History.
I. Title. II. Series: Collected Studies: CS533.
TA57.S49 1996 96–20176
624'.0941—dc20 CIP

The paper used in this publication meets the minimum requirements of the American National Standard for Information Sciences – Permanence of Paper for Printed Library Materials, ANSI Z39.48–1984. ∞ ™

Printed by Galliard (Printers) Ltd
Great Yarmouth, Norfolk, Great Britain

COLLECTED STUDIES SERIES CS533

CONTENTS

This volume contains xxiv + 338 pages

INTRODUCTION

The papers in this volume are contributions to the history of civil engineering in Britain from about 1600 to 1830; a period preceding the railway age and the granting of a Royal Charter to the Institution of Civil Engineers in 1828. As an introduction to the subject it will be helpful to sketch the long-established version of early British civil engineering history and then to summarise the changes brought about by modern research, in which these papers play a part.

The traditional view, correct as a first approximation, is that while some important works were carried out in the 17th and early 18th century they were few and far between, and it is only from about 1760 that the profession of civil engineering came into being, in response to a rapidly increasing demand for public works such as canals, bridges and harbours. And, again as an approximation, the story of the years between 1760 and 1830 can be told essentially in terms of the achievements of five great engineers: Smeaton, Brindley, Jessop, Rennie and Telford. They, of course, were succeeded by Robert Stephenson, I.K. Brunel and other Victorians whose careers lie beyond our period.

This is the pattern followed in the excellent short book on *British Engineers* by Metius Chappell (1942) who, interestingly, emphasises the interaction between engineers and social development. But the first, and for a century the best, outline of pre-1830 history is given in the Introduction to vol. 1 of the Institution's *Transactions* (1836). In summary, the author[1] says that works of engineering had been executed before 1760, some of considerable magnitude, but they could not provide sufficient employment to support a body of men trained in work of this kind, and on occasion skilled engineers had to be brought from abroad to direct the operations. Rivers were made navigable, water was brought to London by Sir Hugh Myddleton's New River in 1613, and the Great Level of the Fens was drained in the mid 17th century by Sir Cornelius Vermuyden, from the Netherlands. Later, about 1720, John Perry, who spent his early engineering career in Russia, became famous for stopping a serious breach in the Thames bank at Dagenham; Charles Labelye led the way in bridge-building on a large scale with Westminster Bridge, begun in 1739, and 'in the unassuming character of a country mason' William Edwards had, in the bridge at Pontypridd, set an unsurpassed example of intrepidity and determination. But all these and some other undertakings, 'important though they certainly were, must, when viewed in connection with what has been effected since, be considered

[1] Known to be Michael Borthwick, assistant to James Walker, second President of the Institution.

rather as detached efforts than as the offspring of an enlarged spirit of improvement . . . giving rise to a new order of professional men'. The first of this class were James Brindley (1716–72) and John Smeaton (1724–92), and their 'worthy associates' Grundy, Henshall, Mylne, Golborne and Whitworth whose signatures are reproduced (Fig. 1). Their careers are summarised: Brindley's and Smeaton's at some length, the others more briefly. The list of works is impressive: many canals, including the Trent & Mersey and the Forth & Clyde, several fine bridges, Blackfriars among them; the Eddystone Lighthouse, harbours, river improvements and fen drainage in Lincolnshire, and so on. Then comes William Jessop (1745–1814) who, 'following Smeaton's retirement, stood at the head of his profession', with such works as the West India docks in London, the Grand Junction Canal and the harbour at Bristol.

Soon after the turn of the century, along with an ever-increasing volume of privately funded schemes, works of 'unparalleled magnitude' were now undertaken by the government. The projects, public and private, include the Caledonian Canal, roads and bridges in the Highlands, Waterloo Bridge, the London–Holyhead road, Menai suspension bridge, Plymouth Breakwater, several large harbours, great docks in the Port of London, Bell Rock Lighthouse and a dozen other important works, all completed before 1830. John Rennie (1761–1821) and Thomas Telford (1757–1834) occupied the foremost places in this period. Among their contemporaries William Chapman (1749–1832) and Ralph Walker (1749–1824) were especially notable, along with Joseph Whidbey, Hugh Baird and Alexander Nimmo (signatures also shown in Fig. 1).

Borthwick goes on to say that the profession was still young when its members first adopted the principle of union. Their *Society of Civil Engineers* was established under the auspices of the 'illustrious Smeaton' in 1771, and its founders are precisely the 'worthy associates' mentioned above. The club, for such it was, came to be known as the Smeatonian Society, and still exists.[2] However, as time went on the need was felt, particularly by the younger men, for a formal body to govern their affairs and to organise regular meetings for discussion and presentation of papers. Thus the *Institution* came into being, in 1818, with Telford as its first President from 1820.

* * *

During the next twenty years after 1836 much detail was added to Borthwick's account without changing its general thesis. Telford's (posthumous) autobiography appeared in 1838; brief biographies of Brindley, Jessop and Chapman were published in Weale's *Quarterly Papers on Engineering* (1843–44); Edward Cresy in his *Encyclopaedia of Civil Engineering* (1847) devotes some 300 pages to

[2] As it does today, with a membership of 'forty eight engineers and twelve gentlemen'. For a history of the Society *see* Watson (1989).

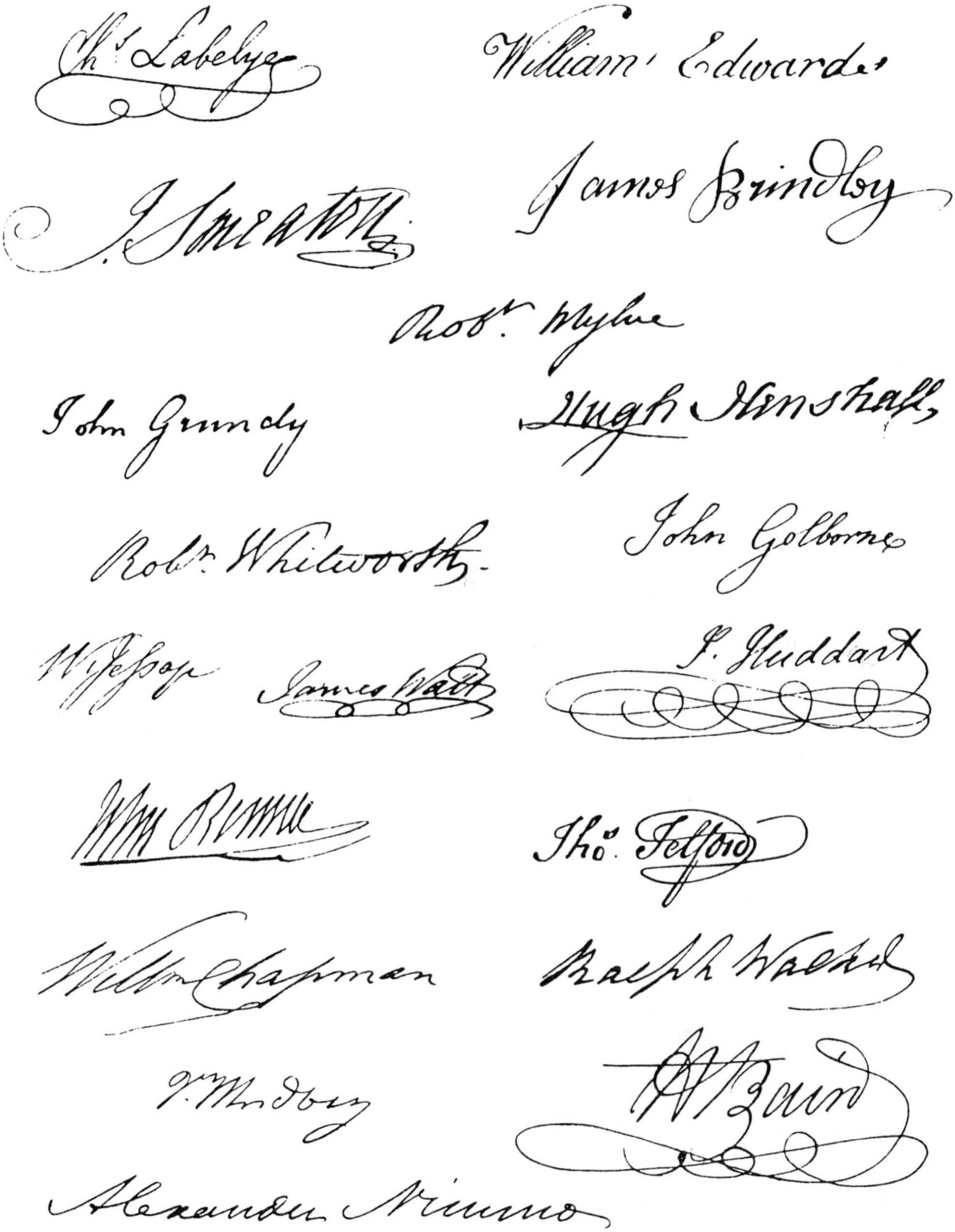
William Edwards
James Brindley
John Grundy
Hugh Henshall
Robt. Whitworth
John Golborne
James Watt
J. Huddart
Thos. Telford
William Chapman
Ralph Walker
Alexander Nimmo

illustrated descriptions of works in Britain, and Sir John Rennie's treatise on *Harbours* was published in 1854. Then in 1861 came the incomparable *Lives of the Engineers* by Samuel Smiles. In three volumes, the first two contain substantial biographies of Brindley, Smeaton, Rennie and Telford, with shorter pieces on Myddleton, Vermuyden, Edwards and the blind road-maker John Metcalf, and a few pages each on Perry, Labelye and Robert Mylne.

Behind the lucid prose and fine illustrations lies a formidable amount of inquiry, and Smiles can be forgiven for concentrating on the great names; though we may regret that the 'worthy associates' of Smeaton and Brindley are mentioned merely in passing, and even Jessop receives little more than a long footnote.

Numerous books and articles on Victorian engineers and their works were published thereafter but the need for further research on early engineering history seems not to have been felt, perhaps due in part to the authority and continuing popularity of the *Lives*. For whatever reason, however, scarcely anything was added to the subject for many years,[3] leaving unanswered questions which now appear obviously in need of investigation. Among these are:

> How were such large 17th-century works as the New River and the draining of the Fens organised, and who beside Myddleton and Vermuyden were involved?
>
> It is apparent from Priestley (1831), for long the standard history of inland navigation, that several rivers were made navigable before 1750 (most notably the Aire & Calder) including a few, such as the Wey, in the 17th century; but what was actually done in this field, and by whom?
>
> Granted that the amount of work accomplished in the early 18th century was far less than in the second half, it was nevertheless an age of expanding trade and technical innovation famously characterised by the steam engines of Thomas Newcomen. Surely more was going on than the traditional history indicates. Who were the engineers, what did they achieve and how did they operate?
>
> More needs to be known about the early Smeatonians most of whom, though recognised in the 1836 article as co-founders of the profession, had later been largely forgotten. The same applies to Chapman and a fortiori to Jessop.
>
> How were the numerous and often very large works of the 1760-1830 period organised and carried out, and who were the resident engineers and contractors?
>
> The regular publication of papers in journals began in the 1830s. What forms of publication existed before this?

---

[3] For example, the *Dictionary of National Biography* (completed 1900) adds little to the *Lives* apart from a more scholarly account (by Prosser) of Labelye. The engineer W. H. Wheeler in his *History of the Fens of South Lincolnshire* (1896), mentions reports and works by Grundy, Langley Edwards and other 18th-century engineers, but this is quite exceptional.

* * *

Among the first signs of a revival of interest were two books on the history of inland transport in Britain, by Pratt (1912) and Jackman (1916). Although excellent in their way, with a great deal on the economics and administrative aspects of roads, river navigations, canals and railways, they had little to say on engineering. However, at about this time attention again began to be given to early technical history, initially in the field of mechanical engineering and from 1930 on civil engineering. These pioneer works of modern scholarship are:

Peet (1930) on Thomas Steers (c.1672–1750), engineer of Liverpool's first dock and the Newry Canal in Ireland.

Webster (1930) on construction of the new channel in the River Dee estuary, in the 1730s, by Nathaniel Kinderley.

Sir Alexander Gibb's book (1935) on *Telford*, with extensive use of primary source material.

Williamson (1936) on George Sorocold, remembered for his waterwheels in London Bridge but of wider achievement and referred to (by Hatton 1708) as 'that great English engineer'.

Willan (1936) on *River Navigation in England 1600–1750* with notes on engineers and engineering, followed by several papers on individual navigations.

S.A. Harris (1937) on Henry Berry, Steers' successor at Liverpool and engineer of the Sankey Canal, built 1755–59.

Wright (1938) on three early Smeatonians: John Grundy, Thomas Yeoman (the first President) and John Golborne.

Darby (1940) on *The Draining of the Fens*, listing many 18th century reports.

Together these publications effected a revolution in our subject. After the war, momentum was regained. It is sufficient to mention Barker on the Sankey Canal (1948), Rolt's *Inland Waterways* (1950), L.E. Harris on *Vermuyden and the Fens* (1953) and Hamilton (1951) with new information on Perry and a facsimile of his signature, reproduced below to add to those in Fig. 1.

Inspired by Esther Wright's paper and the books of Willan and Rolt in particular, and also by attending a meeting of the Smeatonian Society in 1950, my main field of historical research was now determined: civil engineers and engineering in the 18th century, carried forward where necessary for a few decades and with a backward glance at the stranger world of the 17th century.

* * *

The first paper in this volume (presented in 1953) used the river navigations as a base for brief accounts of engineers[4] in practice before 1760. A few mistakes were perpetuated from secondary sources, but the paper remains a ready reference for this group of men. The errors are corrected in a postscript (I, 54) and a more up-to-date account of the navigations is given in paper VII along with some additional names.

It happens that none of the engineers recorded in these two papers carried out any harbour works, apart from the Liverpool docks, and indeed from Sir John Rennie's *Harbours* (1854) it seems that with the exception of major works at Dover in Elizabethan times (described in some detail with contemporary plans) not a great deal had been achieved before 1750. The examples, briefly described, in early 18th-century England are Liverpool's first dock (correctly attributed to Thomas Steers), Perry's scouring sluice at Rye, and work at Scarborough, begun in 1732. He refers to Labelye's report on Yarmouth harbour and could have added another on Sunderland where, according to Murray (1847), work started on the South Pier in 1723.

Knowledge in this field of early engineering history remained almost static for a hundred years after Sir John's book. However, in 1965 I noticed in Scarborough library an engraved plan of the harbour signed 'Willm Vincent, Engineer for Building the Pier' and dated 1747. Examination of the town minutes of meetings revealed that Vincent had been appointed harbour engineer in 1734 and, further, that he followed William Lellam whose appointment was recorded in December 1731; clearly the engineer (unnamed by Rennie) who began the work in 1732.

This opened a promising line of inquiry to which, quite independently, Professor Swan contributed (1968) by publishing references to engineers noted while researching an economics aspect of English port improvements 1660–1830. During the next few years I examined records at Rye, Yarmouth, Whitby and Sunderland. The latter seemed particularly interesting. A decision was therefore made to write up the 18th-century history of the harbour and to combine with this a study, as complete as possible, of the harbour engineers and their works which in most cases extended well beyond Sunderland itself (II).

Meanwhile I had begun what was to be a long-term study of early printed reports.

---

4 The term 'engineer' (variably spelt) in a civil, non-military, context came into use during the 1690s; a rare earlier example of 1676 is given in I 54. It became widespread by the 1720s (see Appendix), to be replaced (slowly) after 1760 by 'civil engineer'.

This culminated in a bibliographic catalogue of nearly all civil engineering reports, plans and books issued between 1640 and 1840 (Skempton 1987). It now became possible to produce a list of engineers practising from 1690 to 1750 with their principal works and publications. The list in a shortened form is presented here as an Appendix. It throws a new light on the early 18th century.

* * *

For the period after 1760 the books on canals published from 1960 onwards by Charles Hadfield and his associates are highly important contributions.[5] Dr Wright handed over her research notes on the early Smeatonians and after further investigations our paper (III) was presented in 1971, in time for the duo-centenary of the Society. It gives an account of members elected between 1771 and 1800: a representative cross section of the profession, for in those days the Society was less exclusive than it later became.

Among the early members at least four clearly required fuller treatment. John Grundy, one of the first independent consulting engineers, is the subject of a biography by Neil Wright (1983) and his works are described in paper V; a book on Smeaton himself was published in 1981, written in conjunction with several friends possessing special knowledge of varied aspects of his career; and a full-length biography of William Jessop appeared in 1979 (Hadfield & Skempton). Robert Whitworth deserves, but has not yet received, a separate study. Robert Mylne, another founder member, distinguished both as an architect and engineer, had already been given a long entry in Colvin's *Dictionary* (1954).

The authorative work on *Arch Bridges and their Builders 1735–1835* by Ruddock (1979) contains biographical material, including a chapter on William Edwards; and Boucher's book on Brindley (1968) completes the presently available studies on leading engineers of the Smeatonian period.

The career of William Jessop continued actively through the first decade of the 19th century. Otherwise the period 1800–30 is dominated by Rennie and Telford. Rennie's works are outlined in paper III; details can be found regarding his bridges (Ruddock 1979), canals (the Hadfield books), harbours (Rennie 1854) and London docks (XI and XII). Rolt's memorable biography of Telford dates from 1958, later supplemented by a symposium of specialist studies (edited by Penfold) in 1980. Of their five contemporaries listed by Borthwick (see above) Chapman alone has received a full analysis, in paper VI. The rest await proper publication, though Ralph Walker's works in the Port of London are described in papers XI and XII, and biographical notes on Alexander Nimmo (1783–1832) are given by Ruddock (1979).

Several more engineers of the Rennie–Telford period could be mentioned, including Sir Edward Banks (1770–1835) the first general civil engineering

---

5 See also Paget-Tomlinson (1978) for condensed histories of the canals and river navigations with notes on their engineers..

contractor operating on a large scale (in the firm of Jolliffe and Banks) from about 1810 (Dickinson 1931); also Robert Stevenson (1772–1850) and James Walker (1781–1862) who were still active in 1836 and therefore not included by Borthwick. Both had decisively made their mark before the 1820s. A biography of the former was published by his son David in 1878, and Walker's career is summarised in the Institution *Proceedings* for 1863. There is ample scope here for further research.

* * *

Turning now from biography to bibliography, there are about a dozen notable pre-1830 books ranging in date from Perry on Dagenham Breach (1721) to Stevenson on the Bell Rock Lighthouse (1824), and various articles in encyclopaedias, of which John Farey on canals (1806 in Rees' *Cyclopaedia*) and Telford & Nimmo on bridges (*Edinburgh Enclopaedia* 1812) are good examples. But the form uniquely characteristic of civil engineering is the printed report. A vital part of the civil engineer's task is to write reports: on a new scheme, on progress of work in hand, on the results of a consultation or simply on his own initiative. The report would be sent to the authority involved, in manuscript, and a copy kept for record. Quite often, however, the report was printed soon after receipt for circulation among the client's representatives (commissioners, board of management, and so forth), to Parliament and to landowners and other interested members of the public.

The typical printed report runs to twenty pages, often with a folding engraved plan, issued in an edition of up to 500 copies. Sometimes the plan would be issued separately and on occasion, particularly in the early days, a printed plan would appear alone, perhaps with an abbreviated report or notice in a local newspaper. The bibliographic catalogue, previously mentioned, records well over a thousand reports, and their chronological distribution reflects the growth of the profession: very few before 1720; thereafter a stream, becoming a broad river after 1760, and after 1800 a flood.

Above all others in the 18th century, Smeaton's reports were renowned for their logic and clarity and for the wide range of subject matter. These factors, together with the regard in which he was held by his fellow engineers, led to the publication of almost all his reports (some 200) in 1812. The circumstances surrounding this event are described in paper IV.

* * *

The papers on Sunderland Harbour engineers (II) and on Grundy (V) and Chapman (VI), though biographical, are nevertheless based in large part on case records: narrative and descriptive accounts of projects, derived from primary sources and field observations. The works are on the whole not well known and some were 'discovered' in the course of investigation. The second set of papers VII to XII proceed in the other direction, taking works which are well known but lacked publications on their engineering history.

The great docks in London, built in the first three decades of the 19th century, provide an example. Despite their enormous importance to trade and commerce, one looked in vain for much information on how they were built, who was involved and exactly what was achieved. Papers XI and XII bring the story to life with details of planning, organisation, progress of the works and methods of construction, and recording, as well as the chief engineers, names of the resident engineers and contractors, essential but all too often neglected contributors to the success of any large undertaking.

On a smaller scale but also of major significance are William Strutt's cotton mills in Derbyshire. They were, together with a mill at Shrewsbury by his friend Charles Bage, the first multi-storey fire-proof buildings and the first to have iron framing: progenitors of the skyscraper. Their place in structural history was therefore acknowledged, yet few details of their construction were available and even a short visit to the Derbyshire mills showed their chronology to be confused. The architect Roy Johnson, of Sheffield University, and I were fortunate to have access to the building accounts and an opportunity to examine the structures. As a result we give a definitive account, historical and technical (X). A subsequent paper (Skempton & Johnson 1962) deals with the Shrewsbury mill and its successors.

Further it transpired that Strutt and Bage had been testing full-scale structural elements and that Bage, in 1803, produced the first practical theory for the strength of cast-iron beams (Skempton 1956). As a bonus, a long forgotten portrait of Strutt was discovered in Derby[6] with the help of my assistant, and she included a reproduction of the painting in a biography (Hacker 1960).

The destruction by fire of the celebrated Albion Mill in London, in 1791, five years after completion, perhaps more than any other event prompted Strutt's initiative. Designed and built by Samuel Wyatt, another early Smeatonian, the mill had been recognised from the start as the most advanced industrial building of its day, but knowledge of its construction was vague and in part erroneous. The facts, derived from Wyatt's correspondence with Matthew Boulton and contemporary drawings, are given in paper IX. A technical analysis of the mill's innovative raft foundation can be found elsewhere(Skempton 1971).

Books and articles on the Thames abound, none better than the two volumes by Thacker (1914 and 1920). But, again, questions remained unanswered concerning the engineering history. To find how, by whom, and when, the locks and weirs were made in the 'modernisation' of the river navigation, the minute books of the Thames Commissioners and the City of London Navigation Committee were examined. They provided the basis for a full story of the developments and personalities involved over a period of seventy years from 1770 (VIII).

* * *

[6] The portrait was later restored and transferred to the Science Museum.

The final question to be considered concerns the organisation of civil engineering works. It is best to start with the system which evolved in the 1760s and is exemplified in the papers on Grundy and Chapman and on the London docks. The engineer, after the planning, reporting and Parliamentary proceedings for a new project, was responsible for detailed designs, specifications, estimates, letting contracts and general direction of the works. He charged professional fees only. Day-to-day direction was the task of the resident engineer (a term introduced by Smeaton in 1768) whose responsibilities included setting out, supervising construction and certifying work done by the contractors. They were craft-based tradesmen skilled in earthwork, masonry or carpentry, each group with its own contract. The engineer kept in touch with the resident by regular interchange of letters and occasional or frequent site visits depending upon circumstances. Land surveyors would be employed if required in the preliminary stages and at times during construction. The engineer usually had one or two assistants and a clerk, and (at least on the bigger works) so would the resident. Large canal jobs, such as the Forth & Clyde, were divided into two or three districts each with its complement of staff and contractors.

This management structure remained essentially unchanged until recent times, except for the appearance towards the end of our period of the 'general contractor'; a firm undertaking all kinds of work on a single contract, with an 'agent' on site acting as the opposite number to the resident.

Most works conducted over a long period of time, like Sunderland Harbour and the Thames navigation, had a different type of organisation. A permanent salaried engineer would be in charge, responsible for design and for supervision of what was usually a 'direct labour' workforce (paid by the authority, not on contract). This system goes back to the 1720s on several harbour works and much earlier on Bedford Level and Deeping Fen where the engineer had the old title of 'surveyor' (of the works). When a major change in plan came up for consideration, advice might be sought from outside; Jessop's 1789 report on the Thames is an example (VIII 162). But in general the engineer took full responsibility and when necessary gave evidence in Parliament.

These engineers would themselves act as consultants on other projects, on leave of absence from their normal duties; and in fact most consultancy before 1760 was carried out in this way, for the simple reason that most engineers were then in salaried positions. Perry, on six weeks leave of absence from Rye Harbour, reported on the North Level of the Fens in 1727; John Grundy senior in 1743 reported on the River Witham whilst engineer of Deeping Fen, and even Labelye, the leading engineer of the early 18th century, reported on the South Level (1745) and Sunderland Harbour (1748) on leave of absence from Westminster Bridge (II 107). His position there was transitional in its nature; he received a salary but also payment, equivalent to fees, for each day on site, and he had a full-time salaried assistant, Richard Graham, as 'Surveyor and Comptroller of the Works' acting in effect as resident engineer (Ruddock 1979).

To a limited extent another system operated in the early 18th century: that of the engineer–contractor who handled the whole job, on a 'design and build' basis, for a fixed sum. In this way John Perry made his name but lost money at Dagenham, as did Edwards at Pontypridd; John Reynolds seems to have been successful on several small harbour works (Swan 1968) as was John Smith at Laburn Lock (VII 38) and Robert Shout of Helmsley, the North Riding bridge builder (II 115).

Most if not all of these engineers had some previous technical experience: as a mason (Robert Shout), carpenter (John Smith), land surveyor (John Grundy senior), teacher of applied mathematics (Labelye) or naval captain (John Perry). The same cannot be said of the 'undertakers' of the 17th-century river navigations. Nevertheless it has to be accepted that Sir Richard Weston, William Sandys and the others recorded in paper I acted as their own engineers in the sense of planning and generally directing the works, as well as partly financing them, and indeed there is some contemporary evidence that this was the case. Clearly they would depend on skilled craftsmen for building the locks; 'water carpenters', as Labelye called them, with experience of sluices and weirs for water mills, pile-driving and timber sheeting. And on the larger navigations a land surveyor might be employed to set out the cuts.

By tradition Sir Hugh Myddleton (c.1560–1631) acted in a similar capacity on the New River. This is not altogether true, though it is certain that the responsibility for the undertaking fell on him, during construction and subsequent operation. Berry (1956), in a beautiful piece of research, shows that the original planning was done by Edmund Colthurst of Bath (d.1616) who actually began work in 1605, on a line similar to the one finally adopted, but soon had to give up for lack of adequate funding. After a pause, with Colthurst still trying to gain support, Myddleton stepped in with his own (considerable) resources and strong backing from the City. Construction began shortly afterwards, in May 1609, under Myddleton's managerial control. From the start he employed a leading mathematical practitioner[7] Edward Wright (1558–1615) as 'my Arts man' to take levels along the 40 mile length from the spring at Amwell, near Ware, to Islington in north London (the average gradient being only 5 inches per mile). Wright was succeeded by Edward Pond as 'Mathematician' (surveyor) from 1611 to completion in 1613. Colthurst, acting as 'Overseer' throughout, is the nearest equivalent to a resident engineer; the 'Master of Timber-work' William Parnell evidently played an important role, and a 'clerk' kept the accounts. The workforce totalled upwards of 200 men.

Such was the organisation of a major early 17th century civil engineering work in England, without a foreigner in sight; nor were any involved in the river navigations. It was in the draining of marshes and fenlands that Dutch engineers made their appearance, dominated by the figure of Sir Cornelius Vermuyden (c.1590–1677). His first big work, the draining of Hatfield Chase, between 1626 and 1630, was

---

[7] *See* Taylor (1954).

financed from Holland and most of the workmen were Dutch; but few details of the organisation are available (Korthals-Altes 1925). Then came the draining of the Great Level of the Fens; by far the largest undertaking in the 17th century and, in this case, financed principally by the Earl of Bedford and other Englishmen. This vast project proceeded in three stages 1631–36, 1639–41 and 1650–55. What was done in each is known (Darby 1940) and Vermuyden held the title 'Director of Works' in the second and third, but the organisation in the first two is obscure. However, regarding the third stage Dr Frances Willmoth (1993) has discovered that the mathematical practitioner Jonas Moore (1617–1679) was appointed 'Surveyor General' in 1650 with a deputy Gabriel Elliott, a 'Principal Overseer' and the 'Comptroller', up to six 'under-surveyors', three 'overseers' and a workforce of (at its maximum) 10,000 men.

Moore's duties, tersely defined by Vermuyden, were 'to sett out the works & see that they bee done'. He acted in all but name as resident engineer, constantly on site and responsible, in Vermuyden's not infrequent absence after 1653, for engineering decisions. In 1658 he produced a large-scale map of the drained Fens (a landmark in cartography) but then went on to a distinguished career in other fields.

In the 17th century some civil engineering works were carried out by military engineers, such as Sir Henry Sheres at Tangier and Dover harbours, and by Edmund Dummer, Surveyor to the Navy, at Portsmouth. But the general picture of that period remains much as outlined in the 1836 *Introduction*: despite a few great works and several smaller ones not enough was going on for the regular employment of civil engineers. What has changed is our recognition that far more was done in the early 18th century than had previously been imagined; and that a great deal more is known about the early members of the profession and their works in the years following 1760. It can now be seen that many similarities, and some contrasts, exist with the parallel development of the architectural profession, as told by Colvin (1954).

A. W. SKEMPTON

Imperial College, London
March 1996

REFERENCES

1830 Samuel Wells. *The History of the Drainage of the Great Level of the Fens, called Bedford Level.* London, Pheney.

1831 Joseph Priestley. *Historical Account of the Navigable Rivers, Canals, and Railways, throughout Great Britain.* London, Longman, Rees. Reprinted 1967. London, Frank Cass.

1836 Michael Borthwick. Introduction. *Trans. Inst. Civ. Eng. 1*, iii–xxvii.

1838 *Life of Thomas Telford, Civil Engineer* (ed. John Rickman) with folio atlas of plates. London, Payne & Foss.

1843 Samuel Hughes. Memoir of James Brindley. *Weale's Quarterly Papers on Engineering* vol. 1, pt. 1 (44 pp.).
1843 Anon. Memoir of William Chapman, *ibid* vol. 1, pt. 1 (28 pp.).
1844 Samuel Hughes. Memoir of William Jessop. *ibid* vol. 1, pt. 2 (32 pp.).
1847 Edward Cresy. *Encyclopaedia of Civil Engineering*. Vol. 1. London, Longmans.
1847 John Murray. An account of the progressive improvement of Sunderland harbour. *Min. Proc. Inst. Civ. Eng. 6*, 256–277.
1854 Sir John Rennie. *The Theory, Formation and Construction of British and Foreign Harbours*. 3 vols, London, Weale.
1861 Samuel Smiles. *Lives of the Engineers*. Vols 1 and 2. London, John Murray (numerous later editions). Reprinted 1968 with introductions by L.T.C. Rolt. Newton Abbot, David & Charles.
1863 Memoir of James Walker. *Min. Proc. Inst. Civ. Eng. 22*, 630–633.
1878 David Stevenson. *Life of Robert Stevenson, Civil Engineer*. Edinburgh, Adam & Black.
1892 R.B. Prosser. Charles Labelye. *Dictionary of National Biography. 31*, 365–366.
1896 W.H. Wheeler. *A History of the Fens of South Lincolnshire*. Boston, Newcomb; London, Simpkin & Marshall.
1912 Edwin Pratt. *A History of Inland Transport and Communications in England.* London, Kegan Paul.
1914 Fred S. Thacker. *The Thames Highway: General History*, and 1920 *The Thames Highway: Locks and Weirs*. London, Thacker. Both volumes reprinted 1968. Newton Abbot, David & Charles.
1916 W.T. Jackman. *The Development of Transportation in modern England.* Cambridge University Press. Reprinted 1962 and 1966 with introductions by W.H. Chaloner. London, Frank Cass.
1925 J. Korthals-Altes. *Sir Cornelius Vermuyden*. London, Williams & Norgate.
1930 H. Peet. Thomas Steers, the engineer of Liverpool's first dock. *Trans. Hist. Soc. Lancs. & Cheshire. 82*, 163–242.
1930 F. Webster. The River Dee reclamations and the effect upon navigation. *Trans. Liverpool Eng. Soc. 51*, 68–92.
1931 H.W. Dickinson. Jolliffe and Banks, contractors. *Trans. Newcomen Soc. 12*, 1–6.
1935 Sir Alexander Gibb. *The Story of Telford.* London, Maclehose.
1936 F. Williamson. George Sorocold of Derby. *Journ. Derbyshire Archaeol. & Nat. Hist. Soc. 57*, 43–93.
1936 T.S. Willan. *River Navigation in England 1600–1750.* Oxford University Press. Reprinted 1964. London, Frank Cass.
1937 S.A. Harris. Henry Berry, Liverpool's second dock engineer. *Trans. Hist. Soc. Lancs. & Cheshire. 89*, 91–116.
1938 Esther Wright. The early Smeatonians. *Trans. Newcomen Soc. 18*, 101–110.
1940 H.C. Darby. *The Draining of the Fens*. Cambridge University Press. 2nd edn 1956. Reprinted 1968.
1942 Metius Chappell. *British Engineers*. London, Collins.
1948 T.C. Barker. The Sankey Navigation. *Trans. Hist. Soc. Lancs. & Cheshire. 100*, 121–155.
1950 L.T.C. Rolt. *The Inland Waterways of England.* London, Allen & Unwin (and five later issues).
1951 S.B. Hamilton. Captain John Perry 1670–1732, *Trans. Newcomen Soc. 27*, 241–253.
1953 L.E. Harris. *Vermuyden and the Fens*. London, Cleaver-Hume.

1954 E.G.R. Taylor. *Mathematical Practitioners of Tudor and Stuart England.* Cambridge University Press.
1954 Howard Colvin. *A Biographical Dictionary of English Architects 1660–1840.* London, John Murray (2nd edn 1978).
1956 A.W. Skempton. The origin of iron beams. *Actes du 8e Cong. Int. d'Histoire des Sciences.* 1029–1039.
1956 G.C. Berry. Sir Hugh Myddleton and the New River. *Trans. Hon. Soc. Cymmrodorion.* 17–48. See also J.W. Gough. *Sir Hugh Myddleton, Entrepreneur and Engineer.* Oxford, Clarendon Press. 1964.
1958 L.T.C. Rolt. *Thomas Telford.* London, Longmans (and three later issues).
1960 C.L. Hacker. William Strutt of Derby (1756–1830). *Journ. Derbyshire Archaeol. & Nat. Hist. Soc. 80*, 49–70.
1960 Charles Hadfield and others. Books on canals all published at Newton Abbot by David & Charles from this date onwards (see references in papers III and VII and Paget-Tomlinson, 1978).
1962 A.W. Skempton & H.R. Johnson. The first iron frames. *Architectural Review. 181*, 176–186.
1968 D. Swan. Engineers of English port improvements 1660–1830. *Transport History. 1*, 153–168, 240–276.
1968 C.T.G. Boucher. *James Brindley, Engineer, 1716–1772.* Norwich, Goose.
1971 A.W. Skempton. The Albion Mill foundations. *Géotechnique 21*, 203–210.
1978 Edward Paget-Tomlinson. *The Complete Book of Canal & River Navigations.* Albrighton, Waine Research Publications.
1979 Ted Ruddock. *Arch Bridges and their Builders 1735–1835.* Cambridge University Press.
1979 Charles Hadfield & A.W. Skempton. *William Jessop, Engineer.* Newton Abbot, David & Charles.
1980 Alastair Penfold (ed.). *Thomas Telford, Engineer.* London, Thomas Telford Ltd.
1981 A.W. Skempton (ed.). *John Smeaton, F.R.S.* London, Thomas Telford Ltd. Reprinted 1991.
1983 Neil Wright. *John Grundy of Spalding, Engineer, 1719–1783.* Lincolnshire County Council.
1987 A.W. Skempton. *British Civil Engineering 1640–1840: a Bibliography of Contemporary Printed Reports, Plans and Books.* London, Mansell.
1989 Garth Watson. *The Smeatonians.* London, Thomas Telford Ltd.
1993 Frances Willmoth. *Sir Jonas Moore. Practical Mathematics and Restoration Science.* Woodbridge, Boydell Press.

APPENDIX

Early 18th century civil engineers, with their principal works and publications

e = designated 'engineer'. P = gives evidence in Parliament. r or b = printed report or book. p = engraved plan.

John Hadley (fl.1689–d. 1701) e, P		
Worcester waterworks		1689–91
Chester waterworks		1692–93
River Aire navigation		1699–01
George Sorocold (fl.1684–1717) e, P		
Derby waterworks		1692–93
Leeds waterworks		1694–95
London Bridge waterworks		1701–02
Derbyshire Derwent navigation	p	1703
John Perry (1670–1733) e, P		
Works in Russia (1698–1712)	b	1716
Dagenham breach		1716–20
Dover Harbour	r	1718
Dagenham breach	b+p	1721
Dublin harbour	r	1721
Rye harbour		1724–29
North Level of the Fens	r+p	1727
Deeping Fen		1729–33
Thomas Steers (c.1672–1750) e, P		
Liverpool Old Dock		1711–15
Mersey & Irwell navigation	p	1712
Liverpool docks		1717–50
Newry Canal		1737–41
Calder navigation (with John Eyes)	p	1741
William Lellam (fl.1716–d.1733) e, P		
Bridlington harbour		1716–22
Sunderland harbour		1722–31
Scarborough harbour		1731–33

John Hore (c.1690–1768) P		
Kennet navigation		1718–23
Bristol Avon navigation		1725–27
Nathaniel Kinderley (1673–1744) P		
Fenland river outfalls	r	1721
South Level of the Fens		1722–25
Eau Brink Cut	r	1730
River Dee channel		1734–37
William Palmer (fl.1722 – d.1737) e, P		
River Don navigation	r+p	1723
River Don navigation		1726–29
Yorkshire Ouse navigation	p	1725
Yorkshire Ouse navigation		1727–32
John Leaford (fl.1713–53)		
South Level		1729–53
South Level	r	1740
Denver Sluice		1746–50
Edward Rubie (c.1690–1753) e, P		
Rye harbour		1730–48
Yarmouth harbour		1750–53
John Reynolds (fl.1725–1762) e, P		
Newhaven harbour		1731–33
Littlehampton harbour		1733–36
Bridport harbour		1740–44
Southwold harbour		1747–53
John Smith (fl.1730–1767) e, P		
River Don navigation		1731–66
Naburn Lock & Dam		1753–57
Wiliam Vincent (fl.1734–d.1754) e, P		
Scarborough harbour		1734–52
Scarborough harbour	p	1747
Sunderland harbour		1752–54
John Grundy, senior (c.1696–1748) e		
River Dee channel	r	1736
Deeping Fen		1738–47
River Witham	p,r	1743, 44

Charles Labelye (1705–1762) e, P		
Sandwich harbour	p	1737
Westminster Bridge		1738–50
Westminster Bridge	r, p	1739
South Level	r	1745
Yarmouth harbour	r	1747
Sunderland harbour	r	1748
Westminster Bridge	b, p	1751
John Wooler (fl.1740–1783) e, P		
Whitby harbour		*c*. 1740–70
Whitby harbour	p	1740
Tyne Bridge failure (with John Smeaton)	r	1772
Chollerford Bridge		1772–75
Tyne Bridge (with Robert Mylne)		1774–81

ACKNOWLEDGEMENTS

The author and publisher would like to thank all the original publishers and individuals, who have given permission to reproduce the articles in this volume: L.R. Day, Chairman of the Editorial Board of the Newcomen Society (I, III, VI, VIII, X, XI, & XII); the Editors, *Industrial Archaeology Review* (II); The Royal Society of London (IV); Christopher Sturman, *Lincolnshire History and Archaeology* (V); Phillimore & Co. Ltd, Chichester (VII); Professor Andor Gomme, Society of Architectural Historians of Great Britain (IX).

The papers in the *Transactions of the Newcomen Society* are reproduced by permission of the Society.

The paper originally published in *Lincolnshire History and Archaeology* is reproduced by permission of the Society for Lincolnshire History and Archaeology.

I

The Engineers of the English River Navigations, 1620-1760

Throughout a period of 150 years before the construction of the earliest canals of the Industrial Revolution in about 1760, considerable works were carried out to improve and extend the river navigations in England. The engineers responsible for these works were living in the days before civil engineering had become fully established as a profession in Great Britain. An attempt has been made in this Paper to outline the early development of what may be called the organization of civil engineering from the work of the somewhat amateur pioneers of the 17th century to that of the professional consultants, such as John Smeaton and his contemporaries of the mid 18th century. The Paper opens with a brief survey of the river navigations, indicating the nature of the work upon which these engineers were engaged. For the economic and social history of the navigations reference should be made to the bibliographic references and particularly to the valuable monograph by Dr. T. S. Willan.[1]

A General View of the River Navigations

The significance of the river navigations and the later canals was due primarily to their functions as the most natural inland communication with the sea-ports from which all foreign trade and the very important coastal trade were handled; coupled with the high mechanical efficiency of water transport which, before the days of steam railways and motor vehicles, was the only satisfactory means of carrying heavy and bulky goods. In Table I. will be found the loads[2] which could be carried or drawn by a single horse with the forms of inland transport available before about 1830; and the vastly greater efficiency of water as compared with land carriage can immediately be seen.

Table I

Typical Loads Carried or Drawn by a Single Horse

Pack Horse	... ...	$\frac{1}{8}$ ton
Stage Waggon	on "soft" roads	$\frac{5}{8}$ ton
	on macadam roads	2 tons
Waggon on iron rails	... ...	8 tons
Barge	on river	30 tons
	on canal	50 tons

*Imperial College, University of London.

[1]Other general information will be found in Priestley (1831), Pratt (1912), Jackman (1916), Rolt (1950), and Hadfield (1950).

[2]These are average figures derived from data given in a number of authorities, including Smeaton's Reports and Telford's Pocket Memorandum Book (printed in *Life of Thomas Telford*, Ed. by Rickman 1838).

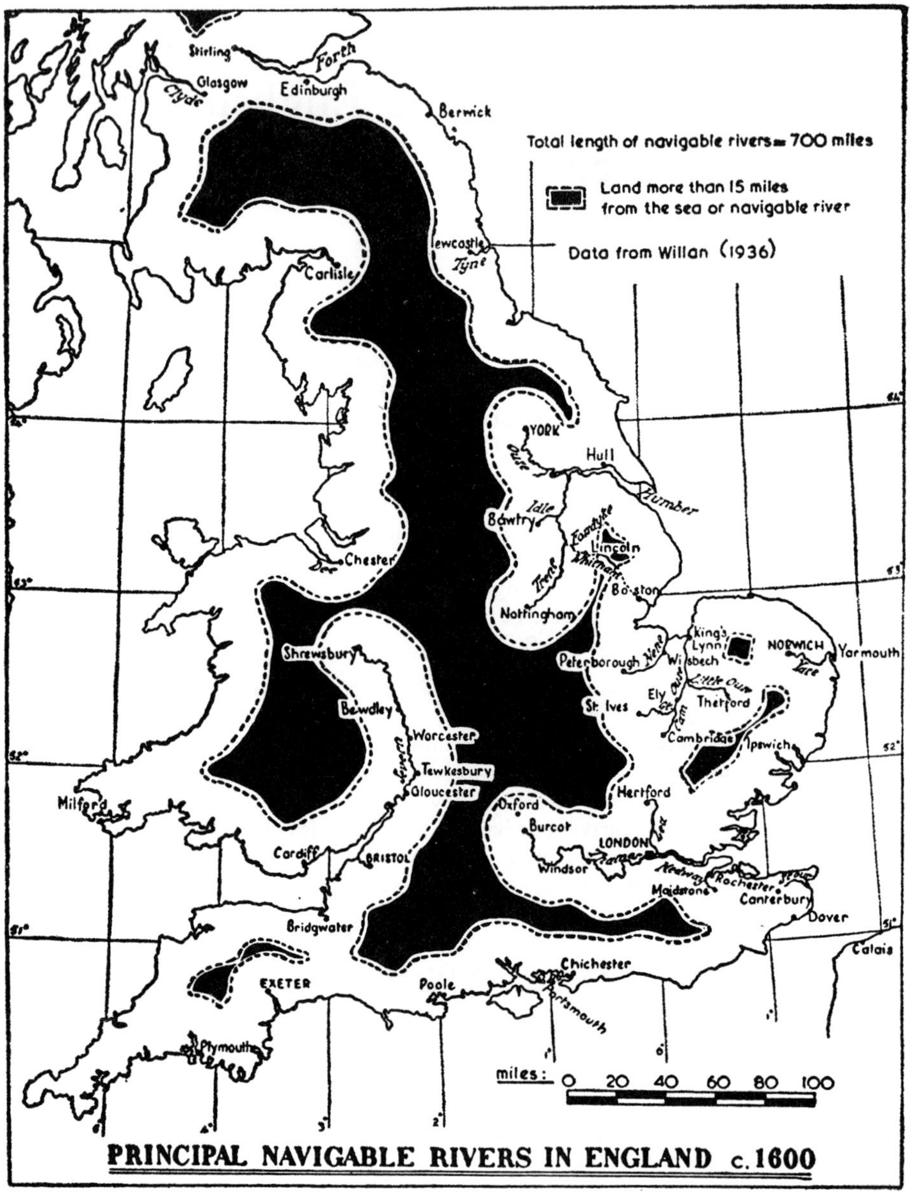

PRINCIPAL NAVIGABLE RIVERS IN ENGLAND c. 1600

FIG. 1

This efficiency was of course partially offset from the economic point of view by the greater capital expenditure required for the engineering works, amounting to over £1,000 per mile for some of the river navigations[3], and still more for the canals. Nevertheless, the average cost of carriage by inland waterways throughout the 17th and 18th centures was about 2½d. per ton per mile (ranging from ¾d to 4d), while by road the cost was about 1/- per ton per mile (ranging from 6d to 1/6d). This comparison on a mileage basis is not altogether valid, since the roads normally provided a more direct route. Even so the actual cost of water transport was between a quarter and a half of the cost by road[4]. Moreover, the roads were often impassable for waggons in winter.

At the beginning of the 17th century about 700 miles of river transport were available (Fig. 1), and on few of these rivers had engineering works of any magnitude been carried out. Thus the Severn was an 'open' river up to Shrewsbury[5]. The Trent likewise was naturally navigable to Nottingham[6], and the Ouse to York. On the Thames, the fourth and greatest of the royal rivers of England, sea going vessels could ascend to London Bridge and the 'western barges' could travel thence to Windsor without any obstructions. From Windsor to Oxford, however, there were by the 16th century 21 mill weirs across the river, through which the barges had to pass by means of stanches or flash locks. These were chiefly necessitated by the weirs. When closed they increased the depth for navigation upstream and built up a head for mill wheels. The 'flash' of water released by opening the stanch was at many sites essential for taking the barges over the shallows downstream. Flash locks are known to have existed on the Thames before 1294, and probably date back to about 1200, when navigation was established from London to Oxford, and when a number of the mills had already been long in existence[7].

A typical flash lock (Fig.2). consisted of a single gate

> "... made up of bars of wood called Rimers (K), set perpendicularly to the bottom of the passage and Lockgates (or paddles S) put down between every two of them ..which will keep a head of water..for the passage of a Barge; but must be pulled up at its arrival, (the balance beam FGH with its rimers turned aside), and the water let go till there is an abatement of the fall, before the boat may pass either down or upwards; which, with the stream, is not without violent precipitation; and against it, at many places, not without the help of a Capstan at land, and sometimes neither of them without imminent danger."[8]

In some of the steeper parts of the Thames flash locks were provided exclusively for navigation purposes[9] but, in general, milling was the more important activity in the Middle Ages, and the boats often had to wait two or three days when grinding was in progress before the miller would condescend, for a suitable toll, to open his weir.

[3]Average expenditure on all river navigation work was at least £700 per mile — based on data given by Willan (1936). Where the engineering works were simple, the cost was about £500 per mile, as on the Welland. On the Kennet, with heavy works, the cost amounted roughly to £2,000 per mile.

[4]These transport costs are summarised from extensive data given by Jackman (1916).

[5]Deepening of the channel, and 4 locks, were carried out 1824-44 between Stourport and Upton, by William Cubitt. v. Min Proc. Inst. C. E. *5*, 340-347 (1846).

[6]"The barges without the help of locks or stops (stanches) go as high as Nottingham, and farther by the help of art to Burton upon Trent" Defoe (1726) ii, p. 141.

[7]Thacker (1914) pp. 15-16, 20-22, 50-57, 113, 121, 269.

[8]Plot (1677) p. 233.

[9]In the 68 miles of river, with a fall of 120ft. between Oxford and Windsor there were at the time of John Bishop's survey of 1585, 4 stanches built for navigation only, together with 21 in the mill weirs. v. Thacker (1914) pp. 54-57.

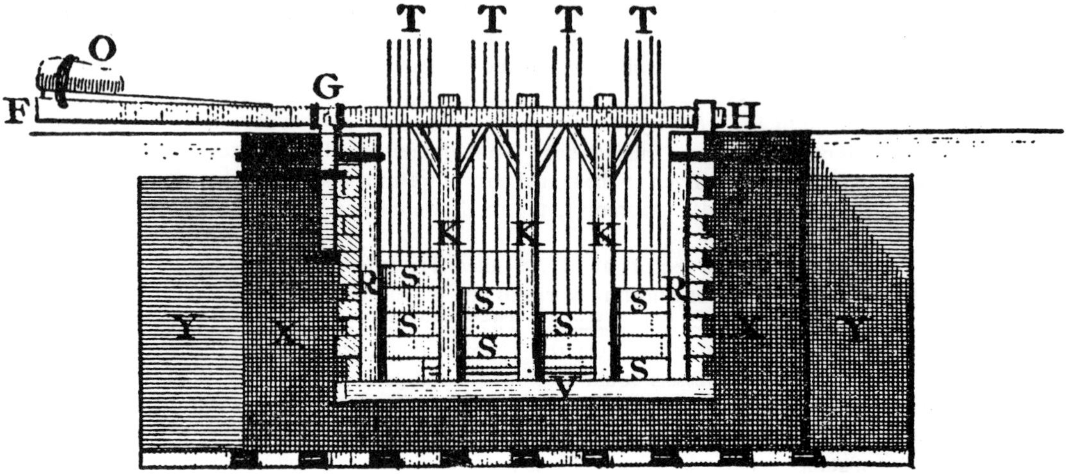

FIG. 2

Even these somewhat restricted facilities for navigation, however, were available only to a relatively small part of the country, and large areas therefore had very restricted opportunity for trade involving carriage of heavy or bulky goods. Sheffield for example had to send its exports by land to Bawtry, 20 miles away, before they could be loaded on to barges working along the rivers Idle and Trent between this town and the Humber, an almost intolerable restriction on its industry.[10] Consequently, at the end of the 17th century, Sheffield entered on a protracted Parliamentary struggle against the interest of Bawtry and the Trent, for powers to make its own river, the Don, navigable. Similarly the clothiers of Leeds, in petitioning for the navigation of the Aire and Calder, pointed out that there was "no conveniency of water carriage within 16 miles, which not only occasions a great expense, but sometimes the roads are not passable".[11] Again, before these rivers were made navigable, York received its coal not only by road from West Riding collieries 20 miles away, but also, and chiefly, from the Tyneside coalfield 200 miles distant, the coal being taken in keels down the river, by coastal vessels to the Humber, and by barge up the river Ouse to the City[12].

These disadvantages were not keenly felt in the Middle Ages, but increasing population, increasing need for coal, a gradual rise in the standard of living and a tendency towards urban life, all characteristic of the Renaissance, began to impose a strain on the existing system. Consequently, we find about the year 1600 a definite and conscious effort to improve the navigations, and, more important, an attempt to open up entirely new navigations by canalizing rivers. The engineering works involved were not spectacular as compared, for example, with those on the Briare and Languedoc canals in France, yet between 1620 and 1690 some 200 miles, and between 1700 and 1760 a further 400 miles, of navigable water were added to the river transport of

[10]Pratt (1912) p. 123.
[11]Petition of 1698, quoted by Priestley (1831).
[12]Nef (1932) Vol. 1, p. 102.

England[13]. (Fig. 3)

Almost without exception the 17th century navigation works were situated in the south, and the trade of this period consisted chiefly of coal and agricultural produce. Thus coal from the staithes at Newcastle was brought to London by sea and then up the Thames to such towns as Windsor, Reading and Oxford. Other goods carried on the river included timber, building stone, corn and malt. And from Lechlade at the head of the Thames navigation in the 17th century, where the river could carry a barge of 50 tons,

> "besides Corn of all sorts which they lade to go down stream, here comes from Severn and Avon, landed at Tewkesbury..(and brought) on horses and in Carts by land, great weights of Cheese especially...Cheshire Cheese."[14]

In the same way Tyneside coal delivered by ships at Kings Lynn was distributed into the country along the Great Ouse and its tributaries; and by 1689 this navigation extended up to Bedford. On the return journey the barges brought cargoes of wheat and other grain which was then shipped from Lynn.[15] Speaking of the Wey navigation made from the Thames to Guildford in 1651-53, Defoe emphasized the importance of this to the town of Guildford, especially for the timber trade, even from parts of Sussex and Hampshire 30 miles distant from Guildford.[16] Similarly, he pointed out that the Warwick Avon navigation, completed from the Severn up to Stratford by 1677, 'is an exceedingly advantage to all parts of the country, and also to the commerce of the City of Bristol' whence imports such as sugar, oil, wine, etc., were carried by water to Warwick.[17]

In the 18th century the amount of work increased, and the scene of activity shifted from the south to the newly awakening industrial north. One of the more notable undertakings, however, was the Kennet navigation, 1718-23, the trading on which followed the older pattern. Obviously there was an increasing trade on the southern rivers during the first half of the 18th century in consequence of the increasing population and prosperity of the country as a whole, and of London in particular. But this period is chiefly remarkable for the opening up for commercial transport of the northern counties, and especially of Yorkshire and Lancashire. Indeed, by 1760 few places in a broad belt running from Hull across to Liverpool were inaccessible to navigable water. (Fig. 3.)

The 18th century work began with the Aire and Calder navigations, one of the most important of its day, and still, in the 20th century, providing what is said to be the most economical means of handling mineral traffic in the world.[18] By 1703 these rivers had been made navigable to Leeds and Wakefield, and Defoe, writing 20 years after the opening of the navigation, having described the great cloth market at Leeds, went on to emphasize the advantages conferred by the navigation on that city and other towns in Yorkshire.[19]

---

[13]By 1800 the total length of the English waterways amounted to 2,000 miles: approximately one-third consisting of canals built since 1760, one-third river navigations constructed 1600-1760 and one-third rivers already navigable in 1600.

[14]Thomas Baskerville "An account of Bridges over the famous River Tems" Harleian M.S 1692, quoted by Thacker (1920) p. 41.

[15]Willan (1946) p. 19.

[16]Defoe i, p. 145.

[17]*Ibid.* ii p.42.

[18]Rolt (1950) p. 132

[19]Defoe ii, pp. 207-8, The navigation was let out by the undertakers at rents which rose, by stages, from £800 a year in 1704-5, to £4,400 in 1751-58 and to £8,500 a year in 1772-74 (Ramsden 1953).

I

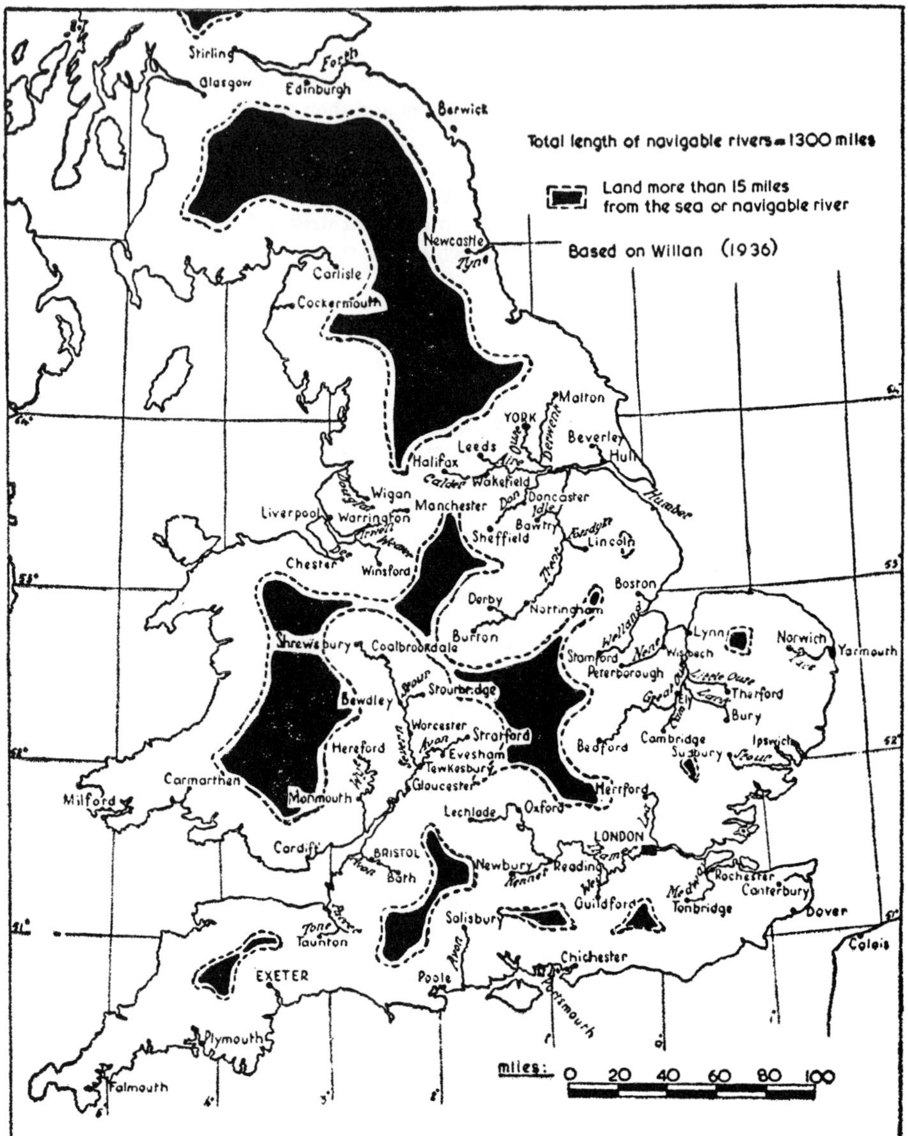

PRINCIPAL NAVIGABLE RIVERS IN ENGLAND c.1760

FIG. 3

By 1735 the Don had been made naviagable almost to Sheffield, and the Yorkshire Ouse had been greatly improved.

In the north west, in the same period, works on the Mersey and Irwell had brought Manchester on to navigable water, and the Weaver navigation had been completed to Winsford. The trade on the Weaver consisted primarily of salt brought down from Cheshire and of coal from south Lancashire brought upstream to the salt works. In 1743, ten years after the navigation had been opened, the total tonnage carried was 31,500 tons and by 1760 the trade had increased to 60,000 tons,[20] a figure equal to the tonnage passing Marlow, on the Thames, at the same period.[21] By then indeed the river navigations formed the foundation of a system of water transport which the canals completed, and without which many of the canals would have been impossible.

Of the technical features of the navigations in the 17th and 18th centuries the pound lock and the lateral cut were the most important. The use of stanches or flash locks on the rivers continued from medieval times until the 19th century, although to a limited extent, and chiefly owing to the value of the 'flash' of water which helped to take the barges over the shallows and there by reduced or eliminated the need for repeated and costly dredging operations. But the stanches had been built usually in mill weirs, and more as an adjunct to the mills than as works designed to improve navigation. When navigation interests began to dominate, the pound lock (Fig. 4), was "almost as important for economic development as the discovery of steam power".[22]

With the single-door stanch a period of an hour or two elapsed before there was a sufficient "abatement of the fall" for boats to pass, and a vast quantity of water (the "flash") escaped from the river above. This loss of water was so great as to prohibit the use of stanches on canals, where the supply of water was often the most critical problem. When the pound locks were used, only the small volume contained within the lock chamber, between the two doors, was lost; and the time required to operate a pound lock was a few minutes instead of hours.

Locks were therefore a great aid to navigation; although more costly than stanches, and requiring, for their full effect, a channel dredged and embanked where necessary, with a sufficient depth throughout its entire length. Thus a river properly equipped with locks implied a single navigation authority, in contrast with the medieval practice of private ownership of individual weirs and stanches.

Locks were first constructed in Holland where the canals of that country pass through the dykes and enter tidal water in estuaries or rivers. They had vertical lifting gates enclosing a large basin, and the oldest example securely dated is that built in 1373 at Vreeswijk.[23] The earliest locks overcoming differences in land, as well as water. level were built on the Stecknitz Canal 1391-98, near Lubeck.[24] These also had a large basin, and were opened for navigation every second or third day. The modern form of masonry-walled lock, holding only one or two barges at a time, was developed in Lombardy by Bertola da Novate, on the Bereguardo Canal[25] 1452-58, and particularly by Leonardo da Vinci, who by 1497 had used his wonderful invention of

[20]Willan (1951) pp. 39, 208, 228.

[21]Thacker (1920) p. 235. In 1760 barges of 170 tons burden could travel up to Wallingford: 100 ton barges to Oxford and 65 ton barges to Lechlade (ibid. 1914 p. 120).

[22]Willan (1936) p. 94.

[23]Telford (1830) pp. 221 & 229, quoting Hedae "Historia Episcopum Ultrajecti".

[24]R. Wreden (1919) "Vorlaufer und Entstehen der Kammerschleuse" Jahrbuch des V.D.I., p. 130, Vol. 9, (Berlin).

[25]Parsons (1939) pp. 370-371 & 377.

I

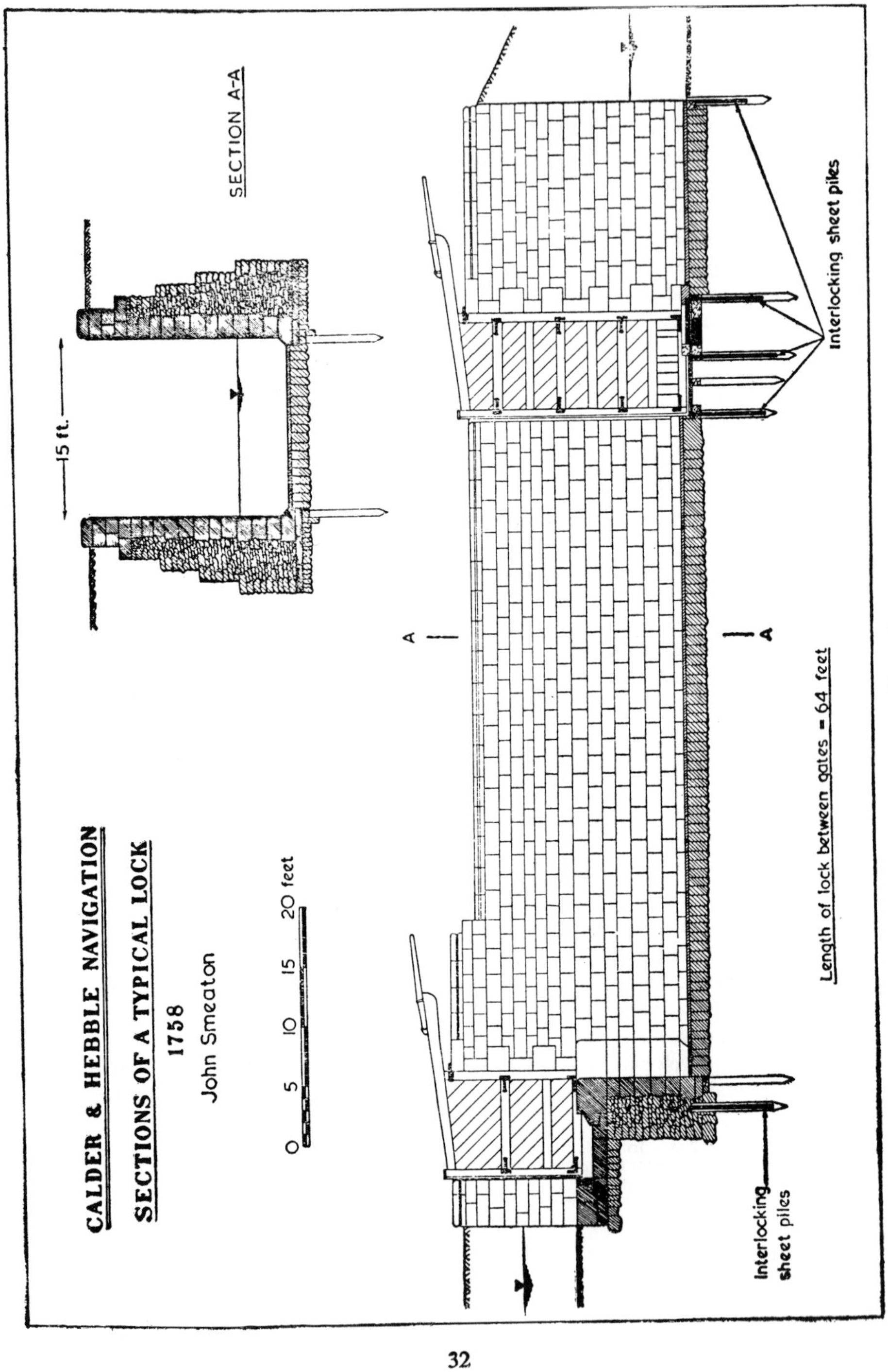
CALDER & HEBBLE NAVIGATION
SECTIONS OF A TYPICAL LOCK
1758
John Smeaton
0 5 10 15 20 feet
SECTION A-A
15 ft.
A
A
Interlocking sheet piles
Interlocking sheet piles
Length of lock between gates = 64 feet

FIG. 4

mitre-gates in the locks on the Milan Canal.[26] Mitre-gate pound locks were introduced to France in 1550,[27] and four were built, with ground sluices and basins 200ft. long and 70ft. wide, by Jean de Locquenghien on the famous Brussels Canal 1550-61.[28]

In England the first pound locks were constructed on a lateral cut beside the R. Exe, known as the Exeter Canal 1564-67 (Fig. 5). The engineer "John Trew of Glamorganshire, gentleman",

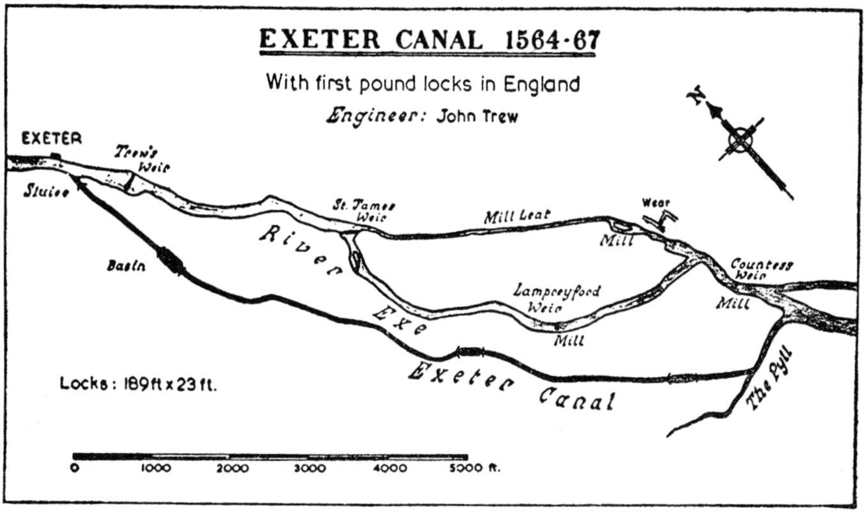

FIG. 5

was paid a fee of £200 and granted tolls from boats using the cut. The works, which cost about £3,000, included a wharf at Exeter, a diversion weir across the river, a regulating sluice at the intake of the cut and three locks each with a basin, like those on the Stecknitz and Brussels canals, of sufficient size to hold several boats: this being a necessary feature since navigation in the river below the cut was possible only at or near high tide.[29] A sketch in a manuscript in the Bodleian Library[30] shows three sluice paddles in each mitre-gate at the upstream end of the locks. The lower end, however, was closed by a single gate. But both upper and lower mitre-gates were provided in the next English lock, built 1571-74 on the R. Lea at Waltham Abbey;[31] and this became standard practice.

The locks at Exeter and Waltham Abbey, and the first survey in 1585, of the Thames between Oxford and Maidenhead by John Bishop[32] are the signs of a new epoch in English river navigations which developed with increasing momentum after 1620. So far as locks are concerned, the earliest in the 17th century were a set of three on the Thames at Iffley, Sandford and Abingdon

[26]Parsons loc. cit. pp. 379-382.

[27]Ibid pp. 383-384, 428-429.

[28]H. Engels (1834) "Notice historique et descriptive du Canal de Bruxelles au Rupel" *Ann. Trav. Publ. Belg*. Vol. 1, pp. 120-173.

[29]De la Garde (1845).

[30]Twyne-Langbain MSS, 1.fol.76 (c. 1620).

[31]Sieveking (1915).

[32]set out in detail by Thacker (1914) pp. 49-57.

(1624-35).[33] The 15 mile reach, with a 30ft. fall, between Oxford and Burcot had always been difficult and, in fact, was not navigable in the late 16th century for the largest barges which were almost double the burden of those in use a hundred years earlier. The channel was deepened; and the locks had a length of 75ft. between the gates and a width of 20ft. between the masonry walls of the lock chamber. The Thames locks were followed by those on the Warwick Avon (1635-39) and by ten on the Wey (1651-53) (Fig. 6). These latter were, and still are, turfsided locks in which the sides of the chamber are formed by earth slopes, masonry walls being used only at the abutments of the gates. By the 18th century the use of locks on the English rivers was widespread. Thus, fourteen were built on the Aire and Calder(1699-1703), the Kennet navigation (1718-23) had eighteen locks in as many miles, and on the Weaver (1730-32) eleven locks were constructed in a length of 17 miles. On the Calder twenty-six locks had to be provided in 24 miles when this navigation was extended above Wakefield (1759-64).

Lateral cuts were excavated in order to eliminate bends or particularly awkward stretches, and on some rivers involved heavy engineering works. Thus, of the total length of 15 miles of the Wey navigation, no less than seven miles were in cut, and on the Kennet (Fig. 7) 11 miles of cut 54ft. wide were made in the total length of 18½ miles. As mentioned above, the Exeter 'canal' was actually a lateral cut, and in contemporary accounts of the navigations the cuts are often referred to by the term "canals".

With these constructions, and with various ancillary works such as waste-weirs, floodgates, bridges and wharves, the engineers of the river navigations not only provided an important system of commercial transport for their own day, but also created the protoypes for much of the canal engineering of the later 18th century. Too often the work of Brindley and his contemporaries is thought to have been without precedent in England. Actually, the canals continued and developed traditions already established by the river navigations.

Civil Engineers in England Before 1771

In the early 18th century a considerable number of men were practising as civil engineers, but it cannot be said that the profession of civil engineering in England became established until the time of Smeaton. The formation of the Society of Civil Engineers[34] in 1771 may conveniently be taken as marking the beginning of this new era. The distinction between consultant and contractor was by then reasonably clear and the Society, in fact, consisted chiefly of consultants who met from time to time in London. The Industrial Revolution was in progress and, with a general expansion of trade investments and industry, the number and magnitude of public works was such as to demand specialised technical knowledge both in design and construction.

Before about 1600, in contrast, few public works had been carried out in Britain and, although the designers of the cathedrals were not only great architects but also skilled engineers, there was little civil engineering in the usual sense of the term, even in the 16th century. In this respect England was less advanced than Italy, France and Holland.

In order properly to understand the work of the navigation engineers it is helpful to distinguish between the various ways in which civil engineering was practised during the period under

[33]Thacker (1914) pp. 62-82 and (1920) pp. 125-155. The three pound locks replaced flash locks; but two other flash locks in this reach of 15 miles were still retained.

[34]Apart from a break between May 1792 and April 1793 when a re-organization took place, the Society has now been in continuous existence for nearly two centuries. Its constitution is 48 engineers and 12 honorary members. The closely equivalent Architects Club was founded in 1791. (See *Trans.* XVIII).

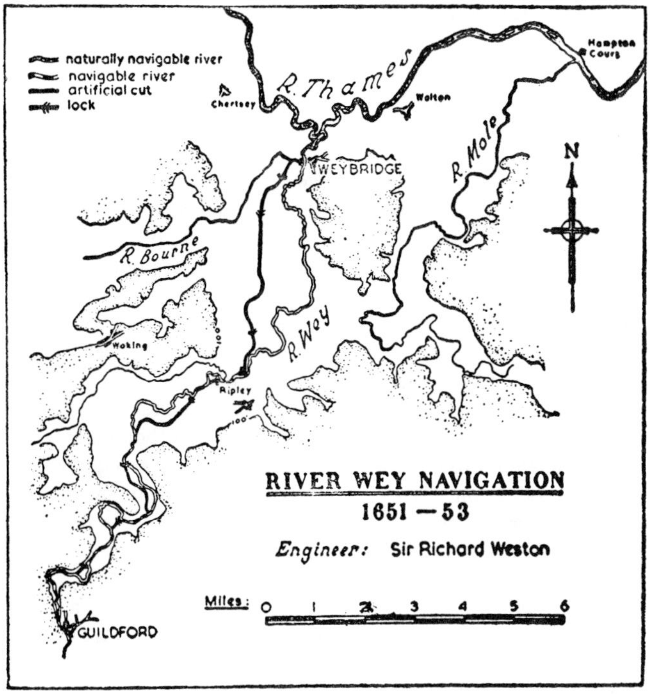
naturally navigable river
navigable river
artificial cut
lock
R. Thames
Chertsey
Walton
Hampton Court
WEYBRIDGE
R. Mole
N
R. Bourne
Woking
R. Wey
Ripley
GUILDFORD
RIVER WEY NAVIGATION
1651 — 53
Engineer: Sir Richard Weston
Miles: 0 1 2 3 4 5 6

FIG. 6

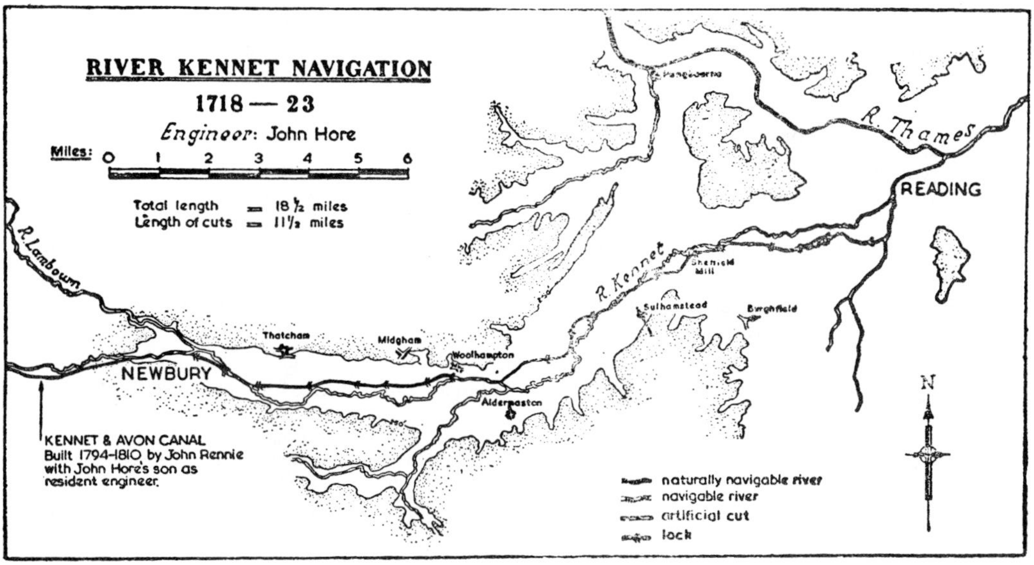
RIVER KENNET NAVIGATION
1718 — 23
Engineer: John Hore
Miles: 0 1 2 3 4 5 6
Total length = 18½ miles
Length of cuts = 11½ miles
R. Lambourn
R. Thames
READING
R. Kennet
Sheffield Mill
Sulhamstead
Burghfield
Thatcham
Midgham
Woolhampton
NEWBURY
Aldermaston
KENNET & AVON CANAL
Built 1794-1810, by John Rennie
with John Hore's son as
resident engineer.
naturally navigable river
navigable river
artificial cut
lock
N

FIG. 7

I

consideration. No classification should be interpreted too literally, but it seems reasonable to separate the engineers into four groups, and may be briefly described as follows:

1. *Undertaker-engineers* who financed a scheme personally (or with a few subsidiary partners), prepared the plans and supervised construction — acting therefore as client, contractor and consultant. In general they carried out one work only, employing a surveyor and such carpenters, masons and other craftsmen as were required. Typically they were moderately wealthy men who foresaw the benefits arising from the undertakings, who had a taste for civil engineering and who hoped for an adequate financial return for their expenditure of time, skill and capital. Sir Hugh Myddleton is the best known[35] in this group, for his achievement in bringing the New River water supply to London in 1613. Scarcely less worthy of note is *Sir Richard Weston*[36] who carried out the River Wey navigation.

2. *Part-time engineers* who were for a portion of their lives employed as civil engineers, either in a salaried position or as a contractor, but who at other times earned their living in various ways not directly related to engineering. Several in this group were, so far as is known, employed solely on one navigation, acting, for example, as clerk or manager when not engaged in engineering work. *Robert Pownall* of the Weaver navigation is one concerning whom a certain amount of information is available, but for the others we know little beyond the record of their appointment as "surveyor", "engineer" or "inspector" of the works at a time when construction was in hand. More important are the men employed on a number of works and surveys, usually in one region of the country. For an appreciable part of their lives they acted as professional civil engineers, and lack of opportunity was perhaps the only reason why they did not spend their whole time in that capacity. *John Hore* is an example. He was engineer for the Kennet and the Bristol Avon navigations, and he prepared plans for the Chelmer and the Stroudwater schemes. Yet from the age of about forty he made his living chiefly as a wharf owner at Newbury and found no engineering employment other than the maintenance of the Kennet.

3. *Full-time engineers* who were able to derive the whole or most of their livelihood from civil engineering work; at times acting as consultant, at times as contractor or, on occasion, becoming one of the joint proprietors of a scheme for which they were also the engineer. In the early 18th century important public works were often carried out by men of this type. Thus, apart from his river navigation and canal works, *Thomas Steers* built the first docks at Liverpool whilst the heroic and ultimately successful effort of John Perry[37] to stem the breach in the Thames embankment at Dagenham has always been regarded as an epic of civil engineering. On this job Perry was both designer and contractor, but later he held a position approximating to that of consulting engineer for the draining of Deeping Fen.

Of this group of engineers Sir Cornelius Vermuyden is the most celebrated, and his work in draining the Great Level has recently been described in an authoritative form by Mr. L. E. Harris[38].

---

[35] D. N. B., Smiles (1861) Vol. 1. pp. 94-152. From recent research published by Sir Jonathan Davidson (Presidental Address, *Journ. Inst. Civ. Eng.* 1948. Vol. 31, pp. 1-30), it appears that much of the technical work, and certainly the surveying, of the New River was the responsibility of Edward Wright (1558-1615) who is celebrated as one of the first to present a correct mathematical treatment for Mercator's projection.

[36] Biographical notes of engineers whose names are printed in italics will be found in the next section of the paper.

[37] D. N. B.: Smiles (1861) Vol. 1. pp. 73-82; Hamilton (1951) in *Trans* XXVII.

[38] Harris (1953)

4. *Full-time consulting engineers* who derived the whole or greater part of their livelihood from civil engineering as consultants or as salaried engineers. In 1771 when the Society of Civil Engineers was founded John Smeaton[39] and Robert Mylne[40] were the most distinguished, while John Grundy Jr., John Golborne and other early members of the Society,[41] although less well known, were also in practice as consulting engineers. From that time onwards an ever increasing number of such men entered the profession; but before 1771 there were several civil engineers working full-time, either as consultants or in a salaried position, apart from the early members of the Society. Of these *Henry Berry* is notable as the engineer of the Sankey canal and of a number of docks at Liverpool, and Charles Labelye[42] holds an honoured place as engineer of Westminster Bridge, and as the inventor of the caisson method of building sub-aqueous foundations. Henry Mill should also be noted as the first engineer of the New River Company of whom there is definite knowledge. He was appointed to this post in 1720, and was succeeded in 1771 by Robert Mylne who had for some years acted as his assistant.[43] "That great Genius Mr. Brindley"[44] should, perhaps, strictly be included in group (2) since he turned to civil engineering only after twenty years a millwright; yet he then accomplished so much, and would have accomplished still more but for his untimely death, that he must be placed with his equally great contemporary John Smeaton. Finally there are few men of an earlier generation, for example Charles Kinderley, John Grundy Sr. and *William Palmer*, concerning whom not a great deal is known, but who seem to have supplemented their work as professionally employed engineers by surveying, and even, as in the case of Grundy, by teaching "mathematicks". But surveying is closely allied to civil engineering and they may be classified tentatively in the present group.

In the evolution of the profession three periods can be recognized. Firstly, there is the 17th century when the number and magnitude of works were insufficient to provide for more than a few men, and schemes were carried out by "undertaker-engineers". Secondly, there is the early 18th century, when the demand for public works increased considerably and the system of financing them changed from that of the individual venturing his own capital to that of groups of men who jointly raised the necessary money, often by loan, and engaged engineers for the design and construction of the job. Some of these engineers were employed only for a part of their lives. Others, by acting occasionally as contractors or undertakers, as well as consultants or salaried engineers, were able to spend all their lives in civil engineering, while a few, not necessarily more important, practised full-time in the latter capacities. Thirdly there is the period after about 1760 or 1770 by which time, as previously mentioned, the demand for specialised knowledge had become sufficient for the profession of civil engineering to be established and the function of consultant and contractor to be distinguished.

The Engineers of the Navigations

Having sketched the economic and technical significance of the river navigations, and the contemporary developments in the organization of civil engineering, attention can be turned to the lives of the engineers who carried out these works.

---

[39] D. N. B.; Smiles (1861) Vol. 2 pp. 2-89.
[40] D. N. B.
[41] Esther Wright (1938) in *Trans.* XVIII.
[42] D. N. B.
[43] Dickinson (1948) "Water Supply of Greater London" *Engineer*, Vol. 186, pp. 207-208.
[44] John Goldborne in a report dated 1777 quoted by Wright (1938). For Brindley see D. N. B. & Smiles (1861) Vol. 1. pp. 307-476.

THE ENGINEERS OF THE ENGLISH RIVER NAVIGATIONS, 1620-1760

ARNOLD SPENCER[45] (1587-1655) born at Cople, Bedfordshire; a landowner of some affluence who became one of the pioneers in river navigation when in 1618 he undertook the improvement of the Great Ouse from St. Ives to St. Neots. In this stretch of 16 miles he built six sluices[46] and deepened parts of the river by the use of an "engine" of his own design, which was some form of dredge. The work seems to have been completed in three or four years, but Spencer wished to extend the navigation, and soon after 1628, under a Patent granted by the King, he built a sluice above Eaton Mills and succeeded in scouring the river up to Great Barford. From 1633 to 1635 he made further improvements, including two short lateral cuts, and although he never carried the navigation to Bedford, which had been his ambition, it is clear that for a period of about seven years before the Civil War an appreciable trading was carried on to St. Neots and to Great Barford. Spencer, however, had been unable to repay all his expenditure, and with the decline in trade during the War he could not meet his creditors, nor could he properly maintain the river. In 1647 he was compelled to mortgage the navigation, and in 1652 it was assigned in trust on the condition that Spencer should be paid an annuity of £20. He lived only another three years and, in the words of a contemporary, "he died poor (and) his being concerned in the navigation did much to waste his estate". Nevertheless in later and more peaceful times the navigation proved beneficial both to its proprietors and the country.

WILLIAM SANDYS[47] (c.1600-c.1670), the son of a wealthy and famous family, was keenly interested in river navigation and, indeed, was "called Water-works Sandys from his taste for improvements of that kind". Between the years 1636 and 1639, at an expenditure said to exceed £20,000, he made the Warwick Avon navigable from Tewksbury up to Bidford, a village some 8 miles below Stratford. Opposition from local landowners forced Sandys to purchase "with excessive charge mills, meadows and other grounds to cut in some places a course for this watery work" but he also built a considerable number of stanches[48] and locks. Trading flourished on the Avon; but on the Teme, which Sandys also proposed to improve, work was checked by the Civil War. Shortly after the Restoration in 1660, Sandys, in partnership with two other members of the family, undertook the navigation of the Wye up to Hereford. The chief difficulties were the swiftness of the stream and the large weirs which supplied power for mills and iron forges along the river. Sandys, who "was undoubtedly responsible for the work",[49] made new cuts with locks to provide a passage for the boats past the weirs. But the scheme was not really a success and by the end of the century the locks had fallen into decay. Later efforts were made at considerable cost, but the navigation was never perfected.

JOHN MALET[50] a country gentleman of Somerset, undertook the navigation of the River Parret and Tone, financing the work from the resources of his estate. It is not known whether he was responsible for the technical aspects.

SIR RICHARD WESTON[51] (1591-1652) is memorable for his introduction to England of the clover rotation, and he also carried out the finest navigation work of the 17th century, that of the River Wey. He succeeded to his father's estate of Sutton, Surrey, in 1613, and to his title in

---

45 Willan (1946).

46 In connection with this navigation both 'sluices' and 'stanches' were built (see under Henry Ashley Sr.) but the technical difference is not clear.

47 Willan (1936) and (1937a).

48 Rolt (1950) p. 71 who states that the two stanches remaining on the Avon are probably the last examples in England still in use.

49 Willan (1937a).

50 Willan (1936).

51 D. N. B.

1622. He first proposed a scheme for the Wey in 1635 but largely owing to the difficulties and unrest of the time little was accomplished before the outbreak of the Civil War. Sir Richard was a royalist and a catholic; his possessions were sequestered, he fled the country and in 1644 he was in Flanders. By 1649 he had returned home and was actively forwarding his proposals for the Wey navigation. He obtained the necessary Act in 1651 and immediately commenced operations. Financially he was in partnership with three others, but he put up half the capital of £6,000 and also supplied timber to the value of £2,000. There is contemporary evidence that Weston was the designer of the scheme and that he personally directed the labour force of 200 men employed on the work. The length of the navigation was only 15 miles, with a fall of 86ft., but 7 miles of new cuts were made (Fig.6) together with ten locks, four weirs, twelve bridges, and wharves at Guildford and Weybridge. The work was completed exactly three centuries ago in 1653, being supervised after Sir Richard's death in 1652 by his son George Weston.

ANDREW YARRANTON[52] (1616-1684), is an interesting but somewhat enigmatic figure in the late 17th century scene who may perhaps be considered as one of the first Englishmen to have spent a good proportion of his working life in civil engineering[53]. By the age of 45 he had served throughout the Civil War in the Parliamentary army, had established a small iron works at Astley, close to the farm in which he had been born, had surveyed the Stour and the Salwarpe with a view to their navigation and, at the Restoration, had been imprisoned twice, and eventually tried at Worcester assizes and acquitted. This was in 1662 when an Act had been obtained for the Stour navigation, with Lord Windsor as principal undertaker. Little progress was made, however, until Yarranton offered to complete the work in return for a third share of the profits, and in 1665 the navigation was opened from Stourbridge to Kidderminster. In the same year he surveyed the Trent and it is probably at about this date that he canalised the Dick Brook from its junction with the Severn up to his iron works at Astley. Fragments of the massively constructed masonry locks still remain, and it has been suggested[54] that "had the times been more propitious Yarranton and not Brindley might be remembered as the progenitor of the English canal system".

He then visited Saxony and Holland and on his return he "seems to have settled down as a kind of consulting engineer".[55] Certainly from about 1670 he was busy preparing plans and reports, among others, on the formation of a harbour at Newhaven, with two sluices in the river to "let down flashes of water" to scour the silt. With his son Robert he produced plans for improving the Thames navigation, and also for docks in London. In 1674 he was consulted about the serious state of the Dee, prepared a plan for a new cut from Chester to the estuary, and coupled this with the reclamation of some of the salt marshes — a scheme put into effect in a modified form by Nathaniel Kinderley 60 years later. He made proposals for improving the harbour of Dublin and surveyed the Wiltshire Avon. In 1677 he published his famous book *England's Improvement by Sea and Land.* During this period he was also engaged by Lord Windsor as engineer to extend the Warwick Avon navigation up to Stratford.

DANIEL WIGMORE[56] was undertaker for making the Welland navigable from Deeping to Stamford, on which he spent £5,000. Nothing else appears to be known of him except that he drew an income of £400 to £500 p.a. from tolls paid by the boats using the navigation.

[52] D. N. B., Smiles (1863).
[53] William Dodson, who was employed on the Great Level, is a still earlier example (Harris 1953).
[54] Rolt (1950) p. 33.
[55] D. N. B.
[56] Willan (1936)

THE ENGINEERS OF THE ENGLISH RIVER NAVIGATIONS, 1620-1760

SAMUEL FORTREY[57](1622-81). was the son of a London merchant and held a minor appointment at the Court of Charles II as well as being, at one time, a clerk of the deliveries of ordnance in the Tower, and author of *England's Interest and Improvement* published in 1663. But he was also "well versed and expert" in affairs of navigation and, in 1672, entered into a contract with the Corporation of Lincoln to improve the Foss Dyke. After the Corporation had spent £500, Fortrey was to bear one-third of the costs, and would receive one-third of the profits, and he was throughout to "afford his best advice, direction and assistance". The work was duly completed, and shortly afterwards, in 1675, Fortrey was appointed to carry out the Wiltshire Avon navigation from Christchurch to Salisbury. Here he held only a salaried position. By agreement he was to receive £50 a quarter for five quarters and £250 on the opening of the navigation[58]. The subsequent history of this work is uncertain. The Bishop of Salisbury cut the first spit and construction was still in progress two years later in 1677. It is said that a heavy flood destroyed the works shortly after their completion, possibly about 1680, but subsequent efforts were made to repair the damage and Defoe records the navigation as extending within two miles of the City. It seems, however, never to have been fully successful, technically or financially. In addition to these various activities Fortrey was a Bailiff in the Great Level and to his pen is attributed *The History or Narrative of the Great Level of the Fenns . . . with a Map of the Said Level as Drained* (1685). If it is by Fortrey, the publication was posthumous.

HENRY ASHLEY, Senior[59] (c.1630-1700) started life as a tanner in Huntingdon but in 1674 "having for some years lived near the river (Great Ouse) and the sluices, and being skilful in such work"[60] he took a lease on the navigation which Arnold Spencer had built between 1618 and 1635. This had been allowed to fall into partial decay and, owing to the improved drainage of the Fenland part of the river resulting from Vermuyden's scheme (completed in 1655), the depth of water in the neighbourhood of St. Ives was often too small for the passage of boats. Ashley therefore constructed a stanch just below St. Ives, and by 1687 had restored and improved the river up to Great Barford, with two new sluices, three stanches and a small lateral cut. In that year he became undertaker for the whole length of river above St. Neots, and by 1689 he had extended the navigations a further 7 mile upstream to Bedford, building five new sluices in this length, and thus completing the original intention of Spencer. Unlike his predecessor, Ashley was successful financially and the profits from the undertaking were at least £400 p.a.

HENRY ASHLEY, Junior[61] (1654-c.1730), probably worked with his father on the Great Ouse from about 1680 and was certainly very active in maintaining their financial interest in the navigation during the last decade of the 17th century. At this time a bitter feud existed between the Ashleys and the Jemmatt family, to whom Arnold Spencer's Patent had descended, and on several occasions the young Ashley threatened Jemmatt's agent with the sword. No blood seems to have been shed, and the argument was finally settled in the more peaceful surroundings

---

[57]D.N.B., Willan (1936).

[58]This was a high rate of payment. Some idea of the annual incomes of various classes can be gained from the following data (Gregory King's Tables 1688) : Eminent merchants £400, Persons in the law £154, Eminent clergymen £72, Persons in liberal arts and sciences £60, Artisans and Handcrafts £38, Labouring people £15.

[59]Willan (1946).

[60]These are his own words; but Joseph Hunt, a carpenter of Godmanchester who may have built some of the sluices, stated in 1689 that Ashley was "skilled in works of navigation."

[61]Willan (1946).

of the Court of Chancery, ultimately to Ashley's advantage. In 1700 he was appointed undertaker of the River Lark navigation from Mildenhall to Bury St. Edmunds, but he appears to have assigned his rights to that town and the gentlemen of the district, and acted only as the engineer for the work.[62] In 1719 he rebuilt the stanch below St. Ives, on the Great Ouse, but presumably in this latter part of his life he derived his income primarily from the profits of that navigation.

DANIEL DUNNELL[63] was engaged in 1699 as engineer, at a salary of £3 a week, for the Exeter Canal improvements, involving an extension to Topsham and the construction of a new lock with upper and lower mitre-gates set 350ft. apart, in masonry abutments, enclosing a basin with a width of 75ft. at the water surface. The depth of water was 14ft., the entrance width 26ft. and the abutment walls of the upper gates were 19ft. high, 12ft. thick at the base and 6ft. at the top, whilst those of the lower gates were 25ft. high. The estimated cost was over £2,000.

JOHN HADLEY is now remembered only as the inventor of a mechanism for raising and lowering water-wheels in accordance with the level of the stream, for which he obtained a patent in March 1693, a device used by George Sorocold (q.v.) in the new machinery for the London Bridge water-works. Referring to these water-wheels Maitland[64] speaks of Hadley as "that great Master of Hydraulicks" and, indeed, he was one of the leading engineers in the time of William III. His invention was first used at Worcester where between 1689 and 1691 "John Hadley of West Bromwich"[65] carried out a piped water supply scheme, raising water from the Severn by a water-wheel and pumps to a cistern from which it could be distributed to the houses in the city. In 1692 he undertook a similar scheme in Chester[66]. He was joint engineer with Sorocold for the Marchant's Waterworks in Westminster, concerning which Sir Godfrey Copley wrote in a letter dated 4th June, 1696,

> "I have been this day and am to meet Mr. Sorocole [sic] and Mr. Hadley . . . I do think his engine (water-wheels and pumps) the best piece of work which I have seen and I find Hadley to be a man of Mathemat. and Bookish".[67]

He appears to have had a considerable practice in this line of work as, on one occasion, he was compelled to write to the London Gazette stating that, having found certain of his workmen offering to set up pumps in his name, without his knowledge, he could in future accept responsibility for any design or work only if under a written communication from himself.[68]

But John Hadley was also the engineer for the Aire and Calder navigation, and further research may well show that he was responsible for a number of other works of the period.[69] In his diary for 3rd December 1697, Ralph Thoresby writes,

> "Proposals were now first made for making the rivers Aire and Calder navigable. I accompanied the Mayor (of Leeds) and Mr. Hadley to view the river. . I followed the wind-

[62]Defoe *Tour* Vol. 1, p. 53.
[63]Willan (1936) pp. 81 and 92. De la Garde (1845).
[64]William Maitland *The History of London* edn. of 1756.
[65]John Noake *Worcester Relics* (1877), Desaguliers (1744).
[66]Private Comm. from the Secretary Chester Waterworks Co.
[67]Armytage (1953).
[68]Rhys Jenkins (1918).
[69]In the first edition of D.N.B. it is assumed that the engineer is the same man as the famous John Hadley (1682-1744), inventor of the sextant and Vice-President of the Royal Society; but this error is corrected in the edition of 1908. That John Hadley who carried out the Aire and Calder navigation is identical with the engineer of the Chester water supply may be safely presumed, since Ralph Thoresby refers to both works as being by "Mr. Hadley", whom he met in 1697.

ings of the river and measured it with his surveying wheel till wearied. . .viewed the several mills and shallows, and with much ado finished as far as Ferry Bridge against night".

For the next day Thoresby records,

"Down the river to Weland, observed the sands etc., and upon the whole, the ingenious Mr. Hadley questions not its being done, and with less charge than expected, affirming it the noblest river he ever saw not already navigable".

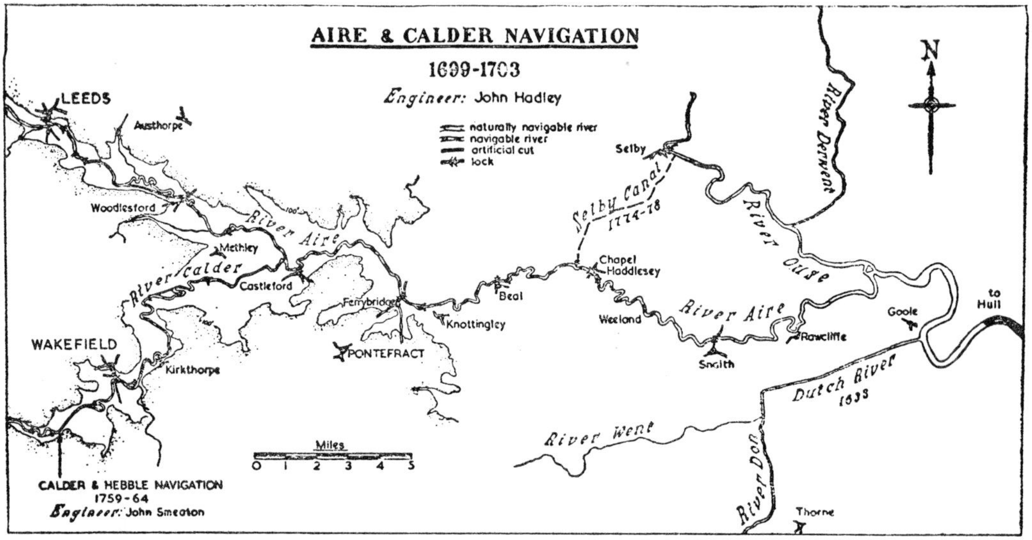

FIG. 8

Hadley attended the House of Commons Committee on the Bill in 1698, and the Royal Assent was received on 4 May 1699. Five days later Hadley was appointed in charge of the works at 400 guineas.[70] Drawings preserved among Smeaton's Designs in the Royal Society library show that the locks were fourteen in number and that five lateral cuts were made,[71] (Fig. 8). The river was naturally navigable from its junction with the Ouse up to Weeland. From thence to Leeds a distance of 30 miles, only ten locks were required, as the total fall was not more than 68ft. On the Calder from Castleford to Wakefield, four locks were built in a length of 12 miles, the fall in this distance being 28ft. The locks were 56ft. by 19ft. with masonry walls, and with 3ft. 6in. depth of water over the sills. The works were completed in 1703, at a cost of £28,000.[72] A view of the navigation near Leeds, (Plate VIII) published in 1715, shows a lock, with a weir extending across the river, and two barges sailing upstream. In the course of time trading increased and plans were made for improvements, but little was done until 1774 when a scheme was carried out, based on a report made in December 1771 by Smeaton.[73] This involved five additional lateral cuts, and all locks were rebuilt to give a depth of 5ft. on the sills. At the same time a canal was made from Haddlesey to Selby.

[70]Private communication from Dr. T. S. Willan.

[71]The largest of these was the 12 mile Coyer Cut, near Woodlesford, in land rented by the Undertakers of the navigation from the Urwin family of Temple Newsam (Ramsden, 1952).

[72]Priestley (1831).

[73]Smeaton's Reports Vol. 2. pp. 131-140.

GEORGE SOROCOLD[74] (c.1655-?1717) was well known to such notable figures in the scientific world as Sir Godfrey Copley and Ralph Thoresby.[75] He was celebrated particularly for his water supply works, including the famous water wheels built in 1701-02 at London Bridge.

Sorocold was responsible for the installation of similar schemes, all including a piped supply system, at Derby[76](1692-93), Leeds[77](1694-95), Westminster (1696), Norwich (1700) and elsewhere. A few years later he put forward proposals for an Upper Pond at New River Head,[78] and these were put into effect in 1708 with subsequent increases in the pumping capacity. For some of these water-works Sorocold acted as undertaker-engineer, for others he was designer and contractor, and it is probable that in a few cases he was consultant only, the work being carried out by a local contractor.

That Sorocold's reputation in this field was deservedly high is proved by the waterwheel and pumps at Bridgnorth which he built in 1706 and which remained in use until 1857 — a continuous life of 150 years. But it is generally accepted that he was also engineer for the Howland Dock at Rotherhithe, constructed 1696-1700 and the first great wet dock in England. Whilst the evidence for his part in this work is not definite, it is certain that in 1708 Sorocold was consulted with regard to the first dock at Liverpool. In March 1709 the Freedom of the town was conferred upon him, and his plans were completed by September. The Act was obtained in the following year, but the work was actually carried out to a revised design by Thomas Steers (q.v.).

Sorocold's contribution to river navigation was less important than the water supply and dock works. After five unsuccessful attempts to obtain powers for the navigation of the Derwent from Derby to the Trent, the Mayor and Burgesses of the town presented a Bill which passed the Commons in January 1703, and was considered but rejected by the Lords in the following month. The scheme had evidently been prepared by "Mr. Sorocold the Engineer"[79] since he attended the Parliamentary committees and submitted a map of his designs, showing four new cuts and nine locks, taking a total fall of 50ft. in a length of 10 miles. (Plates IX and X). No further action was taken until 1717 when the Corporation of Derby again petitioned for a Bill. In a letter accompanying the Petition the Mayor enclosed a map of the county, showing the lead-smelting mills in relation to the facilities for land and river transport, prepared "by the ingenious unfortunate mathematician Mr. Sorocold".

This curious phrase presumably refers to the accident that befell Sorocold at the Lombe's silkmill in Derby, which was built during the years 1717-1721 to his designs.[80] When showing some visitors around the works he slipped into the river and was carried under the 23ft. diameter waterwheel which, "like Jonah's whale, spewed him out . . . to the mill tail, where he was taken up and received no hurt at all". This is part of Defoe's account of what he called Sorocold's "odd experiment", but whether Sorocold lived to see the mill completed, and to carry out the

[74]Williamson (1937). Peet (1930). Rhys Jenkins (1918).

[75]Armytage (1953), Thoresby (1715) and Thoresby's Diary.

[76]When visiting Derby in 1698 Celia Fiennes noted that at the waterwork, "they can set the engine higher or lower just as the water is." Since Hadley obtained his patent for adjusting the level of waterwheels before the Derby work was completed, Sorocold may have based his design on Hadley's wheel at Worcester (the Chester wheel was built at about the same time as that at Derby). The two engineers were certainly acquainted and worked together at Westminster in 1696.

[77]The Leeds waterworks are described in detail by Williamson and Crump (1941).

[78]Private Communication from G. C. Berry of the Metropolitan Water Board.

[79]Rhys Jenkins (1918). This is one of the first instances in England of the term "engineer" as applied to men practising in the civil and not in the military sphere.

[80]Chaloner (1953) "Sir Thomas Lombe and the British Silk Industry" *History To-day* Vol. 3, pp 778-785.

Derwent navigation when the Act was eventually obtained in 1720, or whether the Mayor's letter indicates that he suffered more from the after-effects of his immersion and died soon after, there is yet no evidence.[81] Unfortunately there is also no certain evidence as to his parentage and date of birth, yet in the contract for the Derby water-works of 1692 he is described as "George Sorocold, of Derby gentleman" and it is probable that he was not born into the artisan class. If this is so then, of all the civil engineers practising before 1760, Brindley is the only one who is known to have graduated, as it were, from millwright to civil engineer.

THOMAS STEERS[82](1672-1750), is entitled to a place among the notable civil engineers of England. At the age of 18 he enlisted in the 4th Regiment of Foot and served in the Battle of the Boyne and for four years in Holland. After the Peace of Ryswick in 1697 he returned to London and lived at Rotherhithe where he almost certainly worked on the Howland Dock, presumably as George Sorocold's assistant. In 1698 he married, and before leaving Rotherhithe in 1709 six children were born and baptised there. In March 1710 the Act for the "Old Dock" at Liverpool had been passed with designs by Sorocold, but two months later it was resolved to start construction to a plan by Mr. Thomas Steers. Presumably Sorocold, had recommended his former assistant should act as resident engineer, and Steers, a man of ability and initiative, had at once revised and improved the designs. The dock, completed in 1715, was built out in the tidal estuary of the Pool, and it is must be reckoned as a considerable engineering feat for the period. By 1718 Steers had reclaimed much of the surrounding land[83] and had laid out the new streets as well as building some houses, in one of which he lived with his family from 1719. Two years earlier he had been appointed Dock Master, and in 1717-1718 he built a graving dock leading off the basin of the main dock, together with a long timber jetty.

Meanwhile, in 1712, he had surveyed the rivers Mersey and Irwell with a view to a navigation from Bank Quay, Warrington,[84] for 15 miles up to Manchester, and his map shows eight locks (total fall=52ft.) and two short cuts. The Act for this navigation was obtained in 1721, Steers was one of the undertakers of the scheme;[85] probably also the engineer.The work was completed about 1725. The cost of the navigation was about £15,000. In 1721 Steers also became an undertaker for the River Douglas navigation, but the work was not carried out until 1738-1742.

At various times in his career Steers acted as an architect.[86] In 1740 for example, he built a notable town house in Hanover Street for Thomas Seel, as well as the Old Ropery Theatre. But his principal architectural work was St. George's Church which was erected between 1726 and 1734 to his designs, with James Shaw as contractor. This was a typical early Georgian structure of some merit, but not without technical fault as the foundations of the tower proved inadequate and extensive rebuilding became necessary about 80 years later.

---

[81]His wife died in 1728 and was buried at All Saints', Derby. But she is described in the register simply as Mrs. Mary Sorocold, neither wife nor widow.

[82]Peet (1930). Mr Peet gives a full account of Steer's life and work at Liverpool and consequently only an outline is required here.

[83]Stewart-Brown (1930) "The Pool of Liverpool" *Tr. Hist. Soc. Lancs. Ches.* Vol. 82, pp. 88-135.

[84]The Mersey had been made navigable up to Warrington in 1698, Thomas Pattern being the undetaker (in the business sense only; the engineer is not known). He was also an undertaker for the Weaver in 1730 (Willan 1951).

[85]For Steer's activities as undertaker for these navigations see Willan (1936) and Wadsworth and Mann (1931).

[86]Other 18th century examples of architectural work by civil engineers include St. Mary's Church, Bridgnorth (1792-94) by Telford; and the City of London Lying-in Hospital (1770-73) Wick House, Richmond (1775) and several other country houses by Mylne. Similarly several bridges were constructed by architects, as for example those at Richmond (1774-77) and Chertsey (1780-85) by James Paine, and the beautiful Maidenhead Bridge (1772-77) by Sir Robert Taylor.

Soon after the completion of St. George's, Steers was engaged as consultant for the Newry Canal in Northern Ireland.[87] This was not only the first canal of major importance in the British Isles, but also our first summit-level canal. In 1736 Steers was paid 50 guineas for a survey, and from the following year, when construction began, until 1741 he was on the site for a total of 25 months and received fees amounting to £1200. The work was completed in 1745 under the direction of Acheson Johnson. From Newry, at the head of Carlingford estuary, the canal rose 65ft. in a length of 9 miles to Poyntz Pass at the southern end of the summit-level. This was 3 miles long, mostly in cutting up to 18ft. deep, and received its water supply from a small reservoir fed by streams in the locality. From the northern end of the summit the canal descended, in a length of 7 miles, to meet the R. Bann at Portadown, below which the river was navigable, for 9 miles to Lough Neagh, at an elevation 22ft. lower than the summit. There were 15 locks on the canal, 60ft. by 15ft. in area with a depth of 5ft. on the sills, and two sea locks 120ft. by 22ft. with a depth of 10ft. of water in the tideway, one at Newry and the other 2½ miles nearer the sea. The canal was built chiefly for the transport of coal to Dublin from the Tyrone Collieries, lying a few miles south west of Lough Neagh; and it involved an expenditure of approximately £50,000.

Steers was elected Mayor of Liverpool in 1739. In 1738 he had also begun construction of the Salthouse Dock and the "Dry Basin" with a graving dock at Liverpool.[88] These were large works and required 15 years for their completion, but in spite of being now nearly 70 years old, in 1740 Steers carried out a survey of the Calder.[89]

JOHN HORE[90] (c.1690-1762), came from a family of maltsters and men of property living in Newbury. His father was one of the undertakers of the River Kennet Navigation, the Act for which was obtained in 1715, and John Hore was appointed surveyor and engineer to the navigation in 1718. Whether he had any previous experience of this type of work is not known, although it is possible that he had been employed on the survey of the Bristol Avon a few years earlier. The works (Fig.7) were carried out between 1718 and 1723 and consisted of 11½ miles of new cuts 54ft. wide, eighteen locks 122ft. long and 19ft. wide, a dock basin at Newbury to hold ten large barges, a stone bridge at Burghfield and a number of timber swing bridges. Chief among the natural difficulties that had to be overcome was the comparatively steep gradient of the river, amounting to a fall of 138ft. in the length of 18½ miles from Newbury to Reading, and in 1721 several of the locks and weirs then under construction were destroyed by floods.[91] But the Engineer had also to contend with the active hostility of Reading, in which even the Mayor and Recorder took part. Hitherto the town had been the port on the Thames for all its inland district, and it now feared to lose this advantage.

The navigation was completed in 1723 and by 1725 John Hore had left the Kennet. During his work there he had received a salary of £60 per annum, and had also been appointed surveyor

[87] Peet (1930), Telford (1830) pp. 268 & 271, Journ. Irish House of Commons 1750, Vol. 5, Appendix pp. xxvi-xxviii.

[88] "Dry" basin because it was dry, or practically so, at low tide. By 1765 three more graving docks had been built leading off this "Dry Basin".

[89] Clegg (1922). Steers was associated in this work with John Eyes, a well known surveyor in the north of England.

[90] Thacker (1932), Willan (1936).

[91] The average gradient of this part of the Kennet is about 85 inches per mile and, of the rivers mentioned in this Paper, only the Calder above Wakefield is steeper, 90 inches per mile; here, also, troubles were experienced with floods. These figures may be compared with the following gradients (inches per mile): Severn, from Bridgnorth to Gloucester, 15 ins. Thames, Oxford to Windsor, 22 ins; Aire, Leeds to Weeland, 27 ins.; Mersey and Irwell, Manchester to Warrington, 42 ins.; Wey, Guildford to Weybridge, 65 ins.

I

for life at the same remuneration, and granted the rights of a wharf-owner at Newbury. Some trouble seems to have arisen over payment of expenses incurred during the work, and in 1724 Hore quarrelled with the proprietors who immediately replied by displacing him from the two life appointments. Fortunately in the following year he found employment as engineer in charge of the Bristol Avon navigation, which was carried out during the years 1725-27. His next task was the drawing up of plans for the Stroudwater scheme for which he proposed the construction of a canal 8 miles long with its own artificial water supply. An Act was obtained in 1730 for this work but nothing was done at the time to carry it into effect. In 1732 he surveyed the Chelmer with a view to its navigation but this too was not carried out until many years later.

Meanwhile he had continued in his efforts to recover the debts incurred on the Kennet, but without success, and was now reduced to "low Circumstances". In a petition he complains that "The governing Proprietors held all their meetings in London, and have acted there in an arbitary Manner" yet he finally won his case and in 1734 was reinstated on the navigation which he had built. After this date he seems to have been engaged solely in his wharf business at Newbury and as maintenance engineer for the river. John Hore's work on the Kennet places him in the first rank among the navigation engineers. His son, born in 1730, was resident engineer on the Kennet and Avon canal under John Rennie, and thus continued a family tradition over three quarters of a century.

WILLIAM PALMER[92]of York was working on the rivers in his native county during the period 1720-40, and although little is known of his life it seems probable that, as Willan says, he "had a substantial and well-paid position" as consulting engineer and surveyor. In November 1722 he was engaged to survey the Don and, at intervals during the next two years, he received a total of at least £45 for his services. Early in 1726 he accompanied a leading promoter of the scheme to London to attend the Parliamentary proceedings for the Bill, which was obtained in May. Works on the Don were put in hand immediately after the Act had been passed and by November his fees amounted to not less than £100. In 1727 details of payment to Palmer cease, but there is no reason to suppose that he did not continue to supervise the work until the navigation was completed, in about 1729, from Tinsley (3 miles from Sheffield) to Barnby.

In 1725 Palmer submitted plans for the improvement of the Yorkshire Ouse, which were accepted, and he was put in charge of the work. John Perry had also been consulted, and suggested "flood-gates and sluices" (probably stanches) for overcoming the shallows, but his advice was rejected for Palmer's proposal to contract the river into a channel 90ft. wide and so cause it to deepen itself. For his survey of the Ouse Palmer received £50 and in the next year he received 10 guineas for giving evidence in Parliament for the Bill. Work started in 1727 and by 1732 sufficient improvement had been made for the transport of some 20,000 tons a year. Drake, in *Eboracum* (1736), was rather critical of the scheme as carried out by "Mr. Palmer an engineer of our own growth, as I may call him", and advocated the making of six new cuts to improve the tidal flow and hence the scouring action of the river. But the City had already spent about £6000 and felt that the navigation was reasonably satisfactory. Thus further work was not carried out until 1757 when a lock was built at Naburn. While directing operations on the Don and Ouse, Palmer was consulted on proposed improvements of Beverley Beck and

[92]The account of Palmer and the Don has been derived from entries in the "Records of the Burgery of Sheffield" Leader (1897). For his other work see Willan (1936) and (1937b.) also Drake (1736).

in 1736, he was asked "to view the works" on the Aire and Calder.[93] Under an Act of 1740 the Don navigation was improved from Barnby to the head of the Dutch River and it is possible that Palmer was responsible for this continuation of his earlier work on the river.

THOMAS ROBINSON[94] was appointed "surveyor general of the works" on the Weaver in 1730 when construction started on that river. He had a salary of £40 a year and acted as resident engineer, supervising the building of the locks and weirs by the contractors, a payment to one of whom is recorded in the words "to Peter Walley engineer in part of £300 for one lock". The navigation was opened in 1732, eleven locks and weirs having been constructed, but no lateral cuts. The length of the river was 20 miles, with a fall of 42ft. Probably various minor works were carried out during the next three years, since Robinson's appointment did not cease until 1735, when he was granted a gratuity of £63 for his faithful services as "inspector of the navigation". He maintained a connection with the Weaver at least until 1753, being paid £10 a year to attend meetings "in the stateing and settling of the undertakers' accounts", but what other employment he found is not known.

THOMAS YEOMAN[95] (c.1700-1781) merits notice as the first President of the Society of Civil Engineers, from its inauguration in 1771 until 1780. In 1735 he published "A plan for making the river Chelmer navigable". In 1764 he was elected F.R.S. and in the following year he and Smeaton prepared a joint report for improving the Lea navigation; both being employed later to carry out their scheme. From 1766-69 he was directing the works on the Stort navigation and during this time he was also consulted on several other rivers, notably on the Clyde, for which in 1768 he prepared a joint report with Brindley and Golborne . Like most engineers of the mid 18th century Yeoman was called upon to prepare schemes for Fen drainage, and he published a report in 1769 on the North Level and the outfall of the Wisbech River. His last engagement of which there is record was the preparation of plans and estimates for the Stroudwater canal in 1775. The canal was built chiefly between 1776 and 1779 but it is probable that owing to his advanced years Yeoman did not himself direct the work.

NATHANIEL KINDERLEY[96] was the son of Charles Kinderley who in 1720 proposed the improvement of the Nene outfall by training walls extending 7 miles below Wisbech, and also suggested a cut on the Great Ouse immediately upstream of Kings Lynn. These works were not carried out until 1773 and 1821, when they were named "Kinderley's Cut" and the "Eau Brink Cut" respectively. One of the strongest advocates of their construction (amply justified in the event) was his son, in "The Ancient and Present State of the Navigation of the Towns of Lyn, Wisbech, Spalding and Boston", (1751). The Kinderleys are therefore well known in the story of the Fens, but Nathaniel's largest single work was on the Dee. In an Act of 1733 he was appointed undertaker for a scheme providing a new channel for the Dee from Chester to the estuary, coupled with the reclamation of a large area of salt marshes. Kinderley obtained the Act in trust, and nominated forty men as his assignees to finance the work. The cut, completed in 1737, was about 8 miles long and 350ft. wide, and resulted in an increase in the maximum depth of water at Chester from 9ft. to 15ft. A bank was built on the estuary side of the cut for the whole length down from the city, together with a wall on the opposite side for a distance of 3½ miles from the seaward end. The cost had been £48,000 but still further capital was

---

[93] Private communication from Dr. Willan.
[94] Willan (1951), where plans of the Weaver and of several locks and cuts are given.
[95] Esther Wright (1938).
[96] Darby (1940), Willan (1936), Jackman (1916).

raised in 1741, probably in connection with the reclamation works. By 1754 about 1400 acres of land had been reclaimed and technically the undertaking was at least a partial success. But Chester was fighting a losing battle against the sand and the greater natural advantages of Liverpool, and by the mid 19th century the older city ceased to be a seaport.

ROBERT POWNALL[97] (c.1715-1780) worked on the Weaver, in various capacities for 45 years. In 1735 he became a clerk at Winsford, with £40 p.a. Seven years later he investigated a complaint that slips in the river banks were more frequent since the navigation had been opened, but he found as many slips above the navigable limit as below, and his report was accepted as disproving the claims. In 1757 improvements in the river were considered desirable. Pownall with three others made a detailed survey of the existing locks. Early in the following year Henry Berry (q.v.) was engaged to prepare plans for the new works required. Construction started soon afterwards and Pownall was promoted to be "inspector and superintendent of the navigation" at £50 a year, as well as cashier at an extra £20 a year. Berry was in charge of two new cuts and locks to be built at Saltersford and Pickerings, while Pownall was to "overlook" the rest of the work. In this capacity he was responsible for a new lock at Northwich and, after Berry had encountered foundation trouble at Pickerings, Pownall rebuilt the lock and weir there, with John Golborne as consulting engineer. In 1761, his ability having been proved, his salary was raised to £130 p.a. although he was required to act as collector of dues at Northwich in addition to his other duties. Three years later he prepared plans for a short branch canal, which were approved by Brindley, and then carried out under Pownall's direction. In 1765 he and Hugh Henshall made surveys for a possible connection between the Weaver and the proposed Trent and Mersey canal, while in the following year, with John Golborne, he gave evidence in Parliament during the passage of the Bill for that canal. Before his death in 1780 Pownall built four more locks and lateral cuts on the Weaver, and throughout the preceding twenty years he had continued his work as collector of tonnage dues and as manager of the navigation. His salary of £130 p.a. incidentally was equivalent to about five times the pay of the master carpenter employed on the navigation in 1774 who received 2/- per day.

HENRY BERRY[98] (1720-1812) is important as the engineer of the Sankey Canal, the earliest canal of the Industrial Revolution in England, and as Liverpool's second Dock Engineer. Born of yeoman stock at Parr, near St. Helens, he worked as assistant to Thomas Steers on the Salthouse Dock, in Liverpool, and succeeded him as "Engineer to the Docks" in 1751, with a salary of 52 guineas a year. The dock was completed in 1753 and in the following year Berry and William Taylor made the first survey of the Sankey Brook at a charge of £66. In March 1755 the Act for the Sankey navigation was passed, but Berry immediately suggested to the principal proprietor of the undertaking that far better results would be obtained by constructing a canal,[99] than by making the Brook itself navigable. Berry's revised plan was accepted and the canal was opened in 1757. During this period it seems that he continued to hold his appointment as Dock Engineer and it was ordered in May 1755 "that liberty be given to Mr. Berry for 2 days a week to attend the making of Sankey Brook navigable, he providing and paying a skilful person to supervise the work of the Dock in his absence". It is said that Berry received 16/6d.

[97] Willan (1951).

[98] Harris (1937), Willan (1951) and Barker (1948) who gives a plan and many details of the Sankey Canal.

[99] Berry would have realised the desirability for a canal from his first survey, and Barker (1948) has suggested that the less ambitious scheme of a navigable river was put forward as being more readily acceptable in Parliament. The Act contained the usual clause permitting the making of "such new cuts as are proper and requisite" and, in fact, this was invoked for making the entire waterway a cut.

a day as his consulting fees on the Sankey Canal. The canal was 10 miles in length, with a fall of 78ft. taken in nine locks, and the works included several weirs and sluices for flood control. Technically it may be regarded as the link between the canalised rivers and the canals of Brindley and his associates, with heavy works of embankments, aqueducts and tunnels. Berry's position as assistant to Thomas Steers, the engineer of the Newry Canal as well as several river navigations, is significant.

In January 1758 Berry was engaged to prepare designs for improvements in the Weaver navigation. Operations commenced later that year but he soon ran into trouble. A freshet in March 1759 had forced a breach in the bank of the new cut at Pickerings and, although this was easily repaired, a few months later the weir blew up immediately after construction owing to a piping failure in an underlying bed of fine sand,[100] while in March 1760 one of the walls in the adjacent lock "was bulged owing to the foundation giving way". As a consequence Berry's services were no longer retained, and Robert Pownall (q.v.) took charge of all the works, with John Golborne acting as consulting engineer from time to time.

Unfortunate as was this experience on the Weaver, Berry never lost the confidence of Liverpool. By 1765 he had built three graving docks leading off the "Dry Basin" and next year he started the construction of George's Dock which was completed in 1771. During the next thirteen years he was employed on various engineering jobs in the Liverpool area such as repairing a bridge, superintending new pavements and roads, and making a survey of the shore of the Mersey estuary, and in 1774 he was consulted on the new dock at Hull.[101] John Grundy was engineer for this work, but Berry's expert knowledge on dock engineering must have been valuable at a time when so few men in England had experience in this field. In 1785 an Act was obtained for two new docks at Liverpool and Berry produced the designs for these. The King's Dock was built by him, and was opened in 1788, but the Queen's Dock had not been completed at the time of his retirement in 1789 at the age of 70.

JOHN SMEATON (1724-1792) the greatest English civil engineer of the 18th century was on several occasions concerned with river navigation, but only one comes within the period before 1760. In 1757, when he had been engaged for little more than a year on the Eddystone lighthouse, he was appointed engineer for the Calder navigation from Wakefield to Salter Hebble near Halifax. He wrote from Plymouth on July 15th saying that he would come to Halifax in October, when work on the rock had to stop for the winter storms, and asked for any previous plans of the Calder to be sent to him meantime. Accordingly he received the scheme prepared in 1740 by Steers and Eyes.[102] Smeaton arrived in Halifax and started his survey on 24 October. His report and plan (Plates XI and XII) were produced in November 1757, he came up from Plymouth in April 1758 to attend the House of Commons, and the Act received the Royal Assent on 9 June of that year. Work seems to have commenced in 1759, the first lock was built in 1760,[103] and the navigation was opened in 1764.[104] In the total length of 24 miles there were $5\frac{3}{4}$ miles of new cut, and the fall of 178ft. was taken in 26 locks (Fig. 4); 12 weirs and a number of bridges

[100]From Golborne's report (Willan 1951, p. 81) it seems highly probable that the failure was due to the partial washing out of a fine sand layer, an all too common phenomenon in river works with inadequate cut-off piling beneath their foundations. The weir at Pickerings was rebuilt with deeper foundations extending to a gravel statum.

[101]Sir William Wright "The Hull Docks" *Min. Proc. Inst. C.E.* Vol. 41 (1875) 83-91.

[102]For Smeaton's letter to Halifax see Clegg (1922).

[103]Smeaton (1791) p. 118.

[104]ibid pp. 65, 114.

were built, and dredging was carried out at some 14 places in the river. It is interesting to note that there were 12 mills in this 24 mile length of river. Smeaton specified one of his newly discovered hydraulic limes as mortar for the masonry of the lock walls and where the composition of the mortar used in the Calder works is given[105] he mentions that Joseph Nickalls was resident engineer on the navigation. Nickalls later became one of the founder members of the Society of Civil Engineers and its President from 1784 to 1792.[106]

In October 1767 a flood, greater than any known in the previous forty years, overtopped the banks and damaged several of the locks. Smeaton reported on the state of the navigation in November and advised the immediate repair of the banks and the building of eight new flood gates.[107] These measures were duly carried out, but further consideration of the Calder navigation and of John Smeaton's career lie beyond the scope of the present enquiry.

ACKNOWLEDGMENTS

I am greatly indebted in many ways to Dr. T. S. Willan, without whose help this Paper could not have been written. During the time when the material was being collected and prepared for publication several friends were generous in sustaining discussions and providing facts and references, particularly Dr. S. B. Hamilton, Mr. L. E. Harris, Mr. R. Glossop and Dr. T. C. Barker. Miss Patricia Mimmack traced and lettered all the maps.

DISCUSSION

MR. L. E. HARRIS thought that Dr. Skempton had been bold in his attempt to classify the various types of engineers but he had undoubtedly succeeded admirably. He had no difference of opinion to voice except with regard to Myddleton. It was his opinion that Myddleton could not be classed as an "Undertaker engineer". No evidence had been advanced to prove that he was more than the financier providing the money for the New River undertaking. Granting that Myddleton was an 'engineer', equally well Francis, 4th Earl of Bedford, and his thirteen co-adventurers in the drainage of the Great Level might be so classified.

The only difference which he could see between the two cases was that Myddleton employed engineers, whose social status was such that they remained anonymous while the Earl of Bedford was saddled with Cornelius Vermuyden, a man of vivid personality who refused to remain anonymous. Though Dr. Willan in his book asserted that Myddleton, Forde and Weston "must all bear the title of engineer", he personally could see no justification for the use of the word "must" in relation to Myddleton.

With regard to Samuel Fortrey, Mr. Harris pointed out that there were three men of this name; grandfather, father and son. The father, Samuel the second, was one of the original Bailiffs of the Bedford Level Corporation and subsequently Conservator under the Act of Incorporation of 1663. In 1663 also he published his book or pamphlet entitled *England's Interest and Improvement.* In 1680, the year before he died, the minutes of the Corporation record that "Samuel Fortrey, junior, Gent." was elected Conservator in the room of John Holman and it was this Samuel Fortrey (d.1688) who wrote *The History or Narrative of the Great Level* and not his father.

William Dodson was a Director of Works with the Bedford Level Corporation after Sir

[105]ibid p. 118.

[106]Esther Wright (1938).

[107]Smeaton's Reports Vol. 2, pp. 125-130. It has been said in several modern publications that the locks were destroyed by this flood. Smeaton's report of November 30th, states, however, that three were damaged and of these only one was destroyed.

Cornelius Vermuyden. He travelled on the Continent and while in Holland took out several patents for drainage mills.

There was little known about John Grundy, except that he was engineer to the Deeping Fen Adventures in succession to John Perry and that he taught mathematics at Spalding Grammar School. However, in a manuscript volume in the library of the Spalding Gentlemen's Society at Spalding (compiled in 1796 by Thomas Hawkes who was Treasurer to the Society at that time) Grundy is mentioned in a transcript of *A Historical Account of the Proceedings of the works of Draining Deeping Fen &c with remarks thereon*, which was written in 1763 by Joseph Featherstone, who himself was engaged on these works. Featherstone relates that at a meeting of the Adventurers held at Spalding in April 1738, "the late Mr. Grundy then being in Wales, Humphrey Smith Esq. attended the meeting and made an Agreement with the Proprietors for self and Mr. Grundy to execute the Scheme at the yearly sum of £300 per annum, that is £150 a year each — to expire in April 1744". He then goes on to say that on Monday the 21st day of May Mr. Grundy left Iscoyd [sic] for Spalding. Was Grundy, when in Wales, engaged on work on the navigation of the Dee and was he associated with Nathaniel Kinderley?

John Grundy in 1745 wrote a reply to a pamphlet of the same year written by Daniel Coppin adversely criticising Grundy's scheme for the improvement of the navigation of the Witham. In the Appendix to his pamphlet Grundy includes a general criticism of sluices on rivers and goes on to assert that the use of sluices had been condemned by such authorities as Guglielmini, Castelli, Westerdyke and others. This is important as showing that Grundy was acquainted with two of the most significant books of that time on the behaviour of rivers, viz. Domenico Guglielmini's *Della Natura di Fumi* of 1697 and Benedetto Castelli's *Delli Misure dell'Acque Correnti* of 1628, both published, of course, in Italy. This, Mr. Harris thought, tended to show that Grundy, though probably self-trained, was not only a teacher of mathematics and a surveyor, but also, within the limits of the training and qualification of those days, a scientifically trained engineer. He was indeed one of the early forerunners of the true engineers of whom Charles Labelye (a contemporary of Grundy) was perhaps, the first. Labelye was probably the last of the foreigners who were able, or permitted, to show the English what engineering in the true sense was, and to help to establish in the social scale the profession of engineering.

A general discussion then ensued in which many members took part.

Dr. A. W. Skempton, in reply, thanked the various speakers for their remarks and particularly Mr. Harris for his valuable contribution. With regard to Myddleton, he certainly did not carry out detailed engineering work on the New River but it seemed probable that he was in charge of the operations in more than a financial sense. The information on Fortrey, Dodson and Grundy was welcome, and it was greatly to be hoped that Mr. Harris would continue and extend his research on the engineers of the early 18th century working in the Fens. Presumably by the phrase "true engineers", as applied to Grundy and Labelye, Mr. Harris implied that they viewed their work as well as a science as an art. It would be generally agreed that they were among the first in England to have this outlook. But Dr. Skempton felt that men of an earlier generation, and especially Thomas Steers and John Hadley, were also scientifically-minded as the age in which they were working permitted—witness Sir Godfrey Copley's opinion that Hadley was "a man of Mathemat. and Bookish". But it must be remembered that one of the first text-books containing scientific matter of any real engineering value[108] was published

[108] Belidor *La Science des Igenieurs*, Paris 1729.

in 1729, and that the theoretical principles of hydraulics, soil mechanics and structures were not correctly formulated until the latter half of the century by men such as Coulomb, Chezy and Euler.

Principal References

(*Published in London unless otherwise stated*)

Armytage, W. H. G. 'George Sorocold and Sir Godfrey Copley', *Jour. Derby Arch. Nat. Hist. Soc.*, Derby, 1953.

Barker, T. C. 'The Sankey Navigation', *Trans. Hist. Soc. Lancs. Cheshire*, Vol. 100, pp. 121-155, Liverpool, 1948

Clegg, C. *Our Local Canals, Halifax Antiq. Soc.* Halifax, 1922.

Darby, H. C. *The Draining of the Fens*, Cambridge, 1940.

Defoe, Daniel. *A Tour through the Whole Island of Great Britain*, ed. G. D. H. Cole, 1928.

Desaguliers, J. T. *A Course of Experimental Philosophy*, Vol. 2, 1744.

Dictionary of National Biography.

Drake, F. *Eboracum*, 1736.

Fiennes, Celia. *The Journeys of Celia Fiennes* ed. Christopher Morris, 1947.

Jackman, W. T. *The Development of Transportation in Modern England*, Cambridge, 1916.

Jenkins, Rhys. 'George Sorocold', *The Engineer*, Vol. 126, pp. 323-4, 1918.

Garde, P. C. de la. 'Memoir of the Canal of Exeter from 1563 to 1724'. *Min. Proc. I.C.E.*, Vol. 4, pp. 90-102, 1845.

Hadfield, C. *British Canals*, 1950.

Hamilton, S. B. 'Captain John Perry 1670-1732'. *Trans Newcomen Soc.*, Vol. 27, 1951.

Harris, L. E. *Vermuyden and the Fens*, 1953.

Harris, S. A. 'Henry Berry, Liverpool's Second Dock Engineer', *Trans. Hist. Soc. Lancs, Cheshire* , Vol. 89, pp. 91-116, Liverpool, 1937.

Leader, J. D. *Records of the Burgery of Sheffield*, 1897.

Nef, J. U. *The Rise of the British Coal Industry*, 1932.

Parsons, W. Barclay. *Engineers and Engineering of the Renaissance*, Baltimore, 1939.

Peet, H. 'Thomas Steers, The Engineer of Liverpool's First Dock', *Trans. Hist. Soc. Lancs, and Ches.*, Vol. 82, pp. 163-242, Liverpool, 1930. (pub. in book form 1932).

Plot, Robert. *The Natural History of Oxfordshire*, Oxford, 1677.

Pratt, E. A. *A History of Inland Transport and Communication in England*, 1912.

Priestley, J. *Historical Account of the Navigable Rivers, Canals and Railways throughout Great Britain*, Wakefield, 1831.

Ramsden, G. 'Two Notes on the History of the Aire and Calder Navigation', *Thoresby Soc. Miscellany*, Vol. 12 (4), pp. 383-395, 1953.

Rolt, L. T. C. *The Inland Waterways of England*, 1950.

Salis, H. R. de. *Bradshaw's Canals and Navigable Rivers of England and Wales*, 1918.

Sieveking, A. F. 'Origin and early history of locks', *The Field*, Vol. 125, pp. 648-650, 1915.

Smeaton, John. *Narrative of the Building and a Description of the Construction of the Eddystone Lighthouse*, 1791.

ibid. *Reports of the late John Smeaton, F.R.S.*, 1812.

Smiles, Samuel. *Lives of the Engineers*, 1861.

ibid. *Industrial Biography*, 1863.

Telford, Thomas. 'Inland Navigation', Article in *Edinburgh Encyclopaedia*, Vol. 15, pp. 209-315, 1830.

Thacker, F. S. *The Thames Highway: General History*, 1914.

Thacker, F. S. *The Thames Highway: Locks and Weirs*, 1920.

Thoresby, R. *Ducatus Leodiensis*, 1715.

ibid. *The Diary of Ralph Thoresby, F.R.S.*, ed. Joseph Hunter, 1830.

Wadsworth, A. P. and Mann, J. L. *The Cotton Trade and Industrial Lancashire, 1600-1780* Manchester, 1931.

Willan, T. S. *River Navigation in England 1600-1750*, Oxford, 1936.

ibid. (a). 'The River Navigation and Trade of the Severn Valley', *Econ. Hist. Rev.*, Vol. 3, pp. 68-79, 1937.

ibid. (b). 'Yorkshire River Navigations', *Geography*, Vol. 22, pp. 189-199, 1937.

ibid. 'The Navigation of the Great Ouse between St. Ives and Bedford in the Seventeenth Century', *Bedford Hist. Rec. Soc.*, Vol. 24, Streatley, 1946.

ibid. 'The Navigation of the River Weaver in the Eighteenth Century', *Chetham Soc.*, Vol. 3, (3rd Series), Manchester, 1951.

Williamson, F. 'George Sorocold of Derby', *Jour. Derby Arch. Nat. Hist. Soc.*, Vol. 57, pp. 49-93, Derby, 1936.

Williamson, F. and Crump, W. B. 'Sorocold's Waterworks at Leeds, 1694', *Thoresby Soc.*, Vol. 37, pp. 166-182, Leeds, 1941.

Wright, Esther. 'The Early Smeatonians', *Trans. Newcomen Soc.*, Vol. 18, pp. 101-110, 1938.

Yarranton, Andrew. *England's Improvement By Sea and Air*. 1677.

TABLE II. PRINCIPAL NAVIGATIONS, 17th Century.

Date	River	Extent of Navigation	Length (miles)	Engineer
1618—35	Great Ouse	St. Ives to Great Barford	24	Arnold Spencer
1624—35	Thames	Burcot to Oxford	15	
1636—39	Warwick Avon	Tewkesbury to Bidford	32	William Sandys
1638—	Parret and Tone	Bridgwater to Ham Mills	12	
1651—53	Wey	Weybridge to Guildford	15	Richard Weston
1662—	Wye	Chepstow to Monmouth	14	William Sandys
1662—65	Worcester Stour	Kidderminster to Stourbridge	12	Andrew Yarranton
1665—*73*	Welland	Deeping to Stamford	10	
1672—*74*	Foss Dyke	Lincoln to R, Trent	11	Samuel Fortrey
1675—77	Warwick Avon	Bidford to Stratford	8	Andrew Yarranton
1675—*80*	Wiltshire Avon	Christchurch to Salisbury	30	Samuel Fortrey
1674—89	Great Ouse	St. Neots to Bidford	15	Henry Ashley Sr.

Dates in italics are approximate.

TABLE III. PRINCIPAL NAVIGATIONS, 1700 — 1760.

Date	River or Canal	Extent of Navigation	Length (miles)	Engineer
1699—1703	Aire and Calder	Weeland to Leeds and Wakefield	42	John Hadley
1699—*1710*	Trent	Wilden Ferry to Burton	23	
1700—	Lark	Mildenhall to Bury St. Edmunds	14	Henry Ashley Jr.
1705—	Essex Stour	Manningtree to Sudbury	25	
1718—23	Kennet	Reading to Newbury	19	John Hore
1720—*22*	Derby Derwent	R. Trent to Derby	10	George Sorocold
1722—25	Mersey and Irwell	Warrington to Manchester	15	Thomas Steers
1725—27	Bristol Avon	Bristol to Bath	16	John Hore
1726—29	Don	Barnby to Tinsley	23	William Palmer
1727—32	Yorkshire Ouse	Humber to York (impr.)	35	William Palmer
1730—32	Weaver	Frodsham to Winsford	20	Thomas Robinson
1735—37	Dee	Estuary to Chester	8	Nathaniel Kinderley
1738—42	Douglas	Ribble to Wigan	16	Thomas Steers
1740—	Medway	Maidstone to Tonbridge	16	
1755—57	Sankey Canal	Mersey to St. Helens	10	Henry Berry
1758—78	Weaver	Frodsham to Winsford (impr.)	20	Robert Pownall
1759—64	Calder	Wakefield to Halifax	24	John Smeaton
1759—72	Bridgewater Canal	Worsley to Runcorn	34	James Brindley

(impr.) — improvements to existing navigation
Dates in italics are approximate.

THE ENGINEERS OF THE ENGLISH RIVER NAVIGATIONS, 1620-1760

Postscript: Corrections and Additions

ARNOLD SPENCER (1587–1655). It is now known that the six 'sluices' on the Great Ouse were pound locks (see below VII 27).

SAMUEL FORTREY (1622–1681) was, additionally, one of the 'Adventurers' of Deeping Fen and 'Director of Works' there from 1669 (Deeping Fen Minute Book, Lincolnshire Archive Office). *The History ... of the Great Level* (1685) is attributed to his son SAMUEL FORTREY junior (c. 1650–88); it accompanies the second issue (1684) of Jonas Moore's map of the Fens originally published in 1658 (see above Introduction). He held the post of 'Surveyor' of the Middle Level from 1684 and, like his father and younger brothers William and James, was a Conservator of Bedford Level. (Samuel Wells *The History ... of Bedford Level* [1830], L. Gaches *Fenland Notes & Queries* 4 [1898–1900] 352–358.)

HENRY ASHLEY (c. 1630–1700). Here, again 'sluices' on the Great Ouse were pound locks.

DANIEL DUNNELL. The extension of Exeter Canal was carried out by RICHARD HURD, appointed 'ingineere' in 1676. Dunnell enlarged the canal and built the new lock (see below VII 30).

JOHN HADLEY (d. 1701). The patent of March 1693 was granted to 'John Hadley of Worcester, engineer'; another early use of the term in a non-military sense.

GEORGE SOROCOLD (fl. 1684–1717). The suggested birth date of c. 1655 is plausible but nothing is known of his early life before his marriage at Derby in December 1684. Nor, it seems, have the date and place of his death been discovered. No evidence exists that he was involved in building the Howland Dock at Rotherhithe which was designed by the shipwright JOHN WELLS (1662–1702). See: D. Swan *Transport History 1* (1968) p. 157 and Philip Banbury *Shipbuilders of the Thames and Medway* (1971 p. 139).

NATHANIEL KINDERLEY (1673–1742). His christian name is not Charles: a mistake originated by Samuel Wells (*op.cit* p. 628) and repeated many times up to and including Darby's *Draining of the Fens* (1940 and 1956). It is he who provided a new channel for the River Dee at Chester, was a 'surveyor' of the South Level 1723–25 and proposed, in a report of 1721, improvements of the Fenland river outfalls. His son NATHANIEL KINDERLEY junior published in 1751 a much enlarged version of the 1721 report. An account of the Kinderley family is given by W.H. Wheeler *Fenland Notes & Queries 3* (1895–97) 113–118.

THOMAS YEOMAN (c. 1709–1781). His age at death in January 1781 has recently been found to be 71 years; so, at last, a reliable estimate of his date of birth can be stated. The 1735 date for his plan of the River Chelmier comes from a rare misprint in Richard Gough *British Topography* (1781 *1*, 368); it should be 1765. The earliest reference to him so far discovered is in 1741 when he was a foreman (millwright) at Abbey Mills, Bromley-by-Bow. He moved to Northampton in 1742. For a fuller account of his subsequent career see III; also VIII, 159 for his work on the Thames.

PLATE VIII

VIEW OF PART OF THE CALDER NAVIGATION.

I

PLATE IX

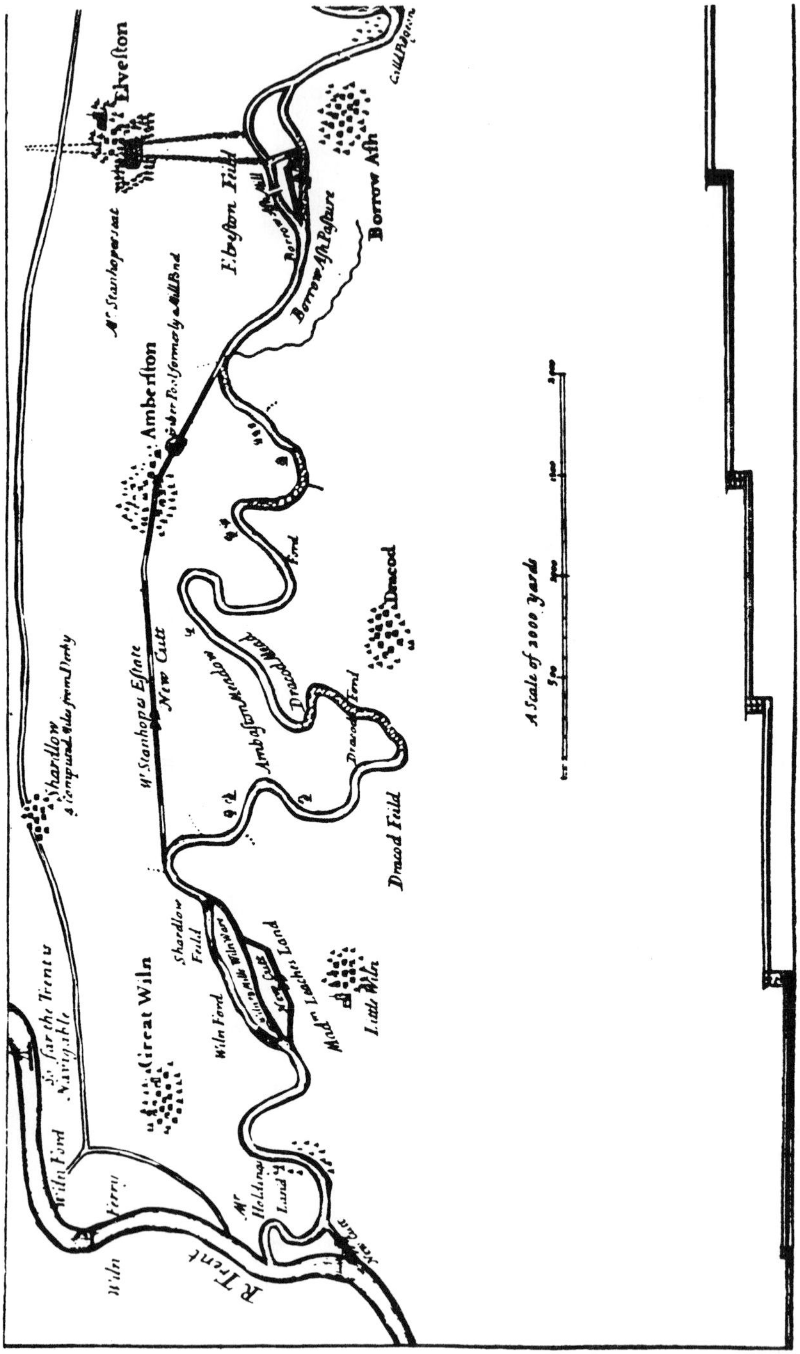

Survey of the River Derwent (i).

PLATE X

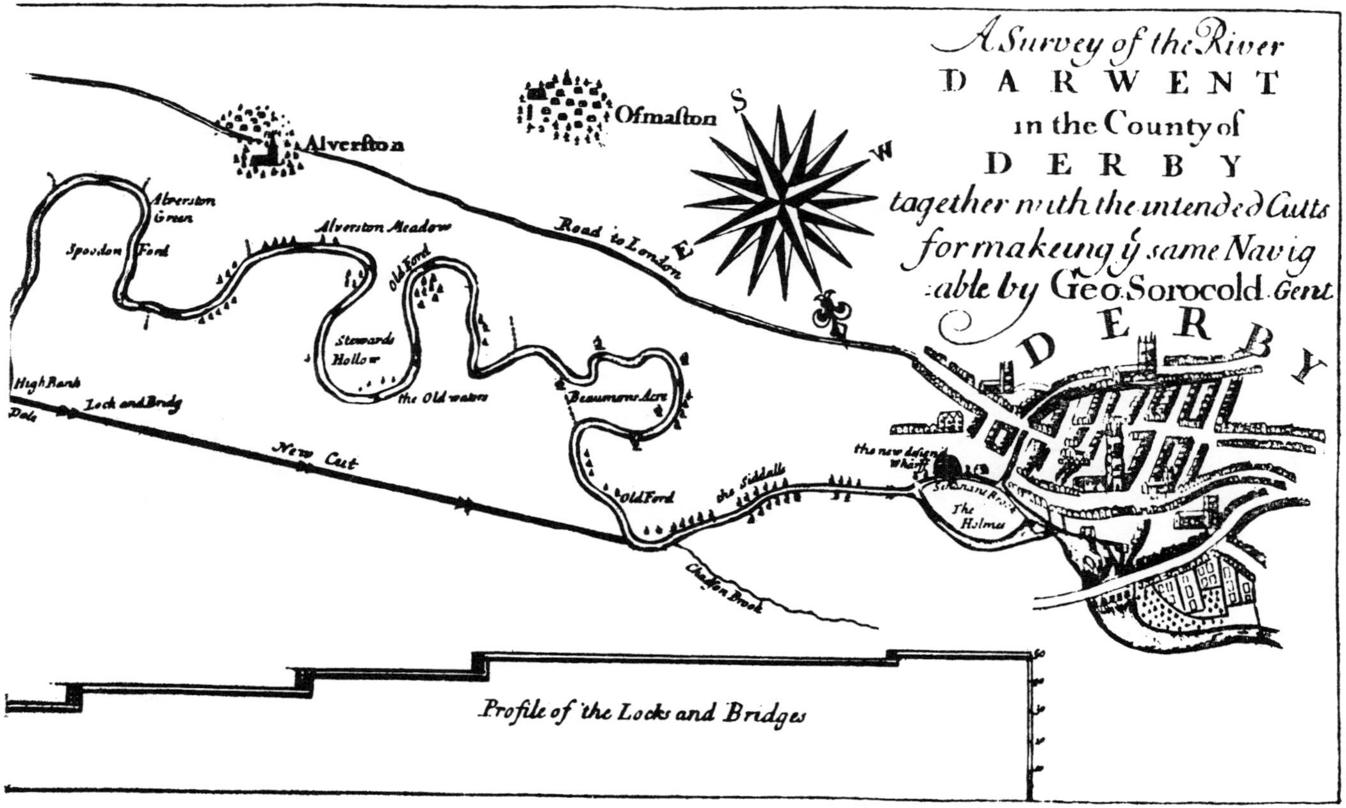

SURVEY OF THE RIVER DERWENT (ii).

PLATE XI

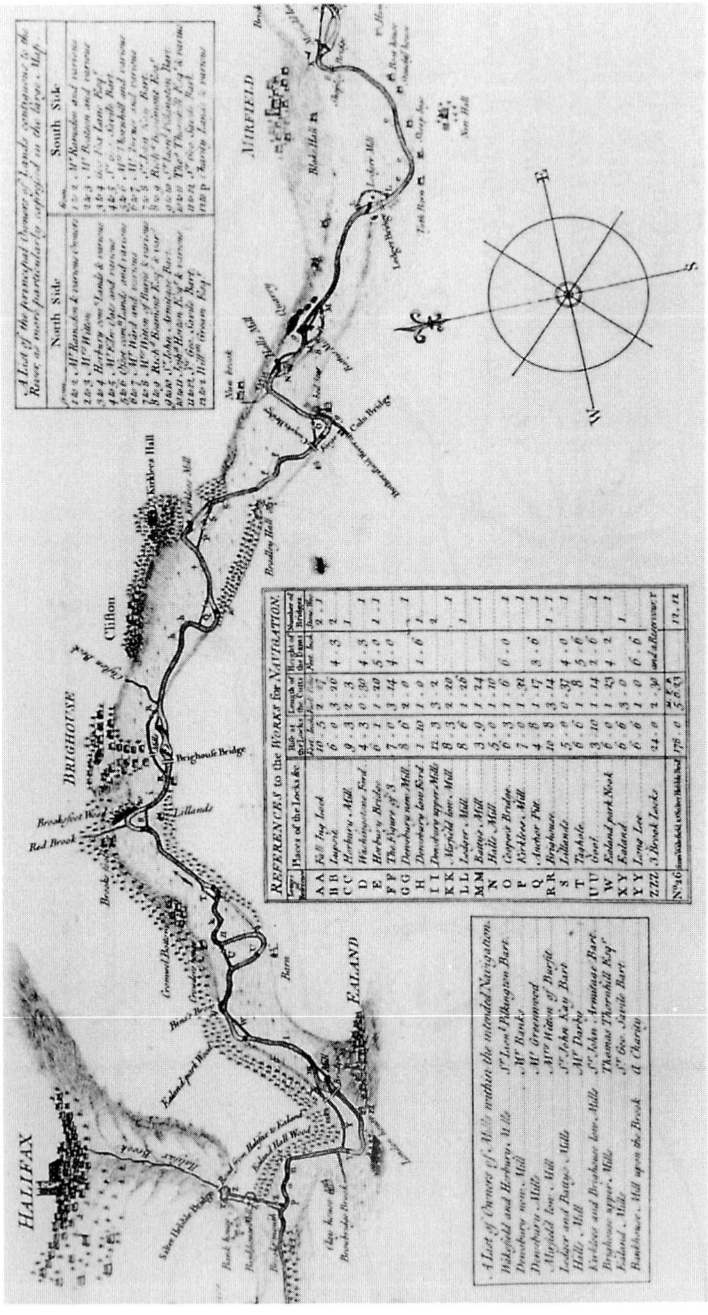

PLAN OF THE RIVER CALDER (i)

PLATE XII

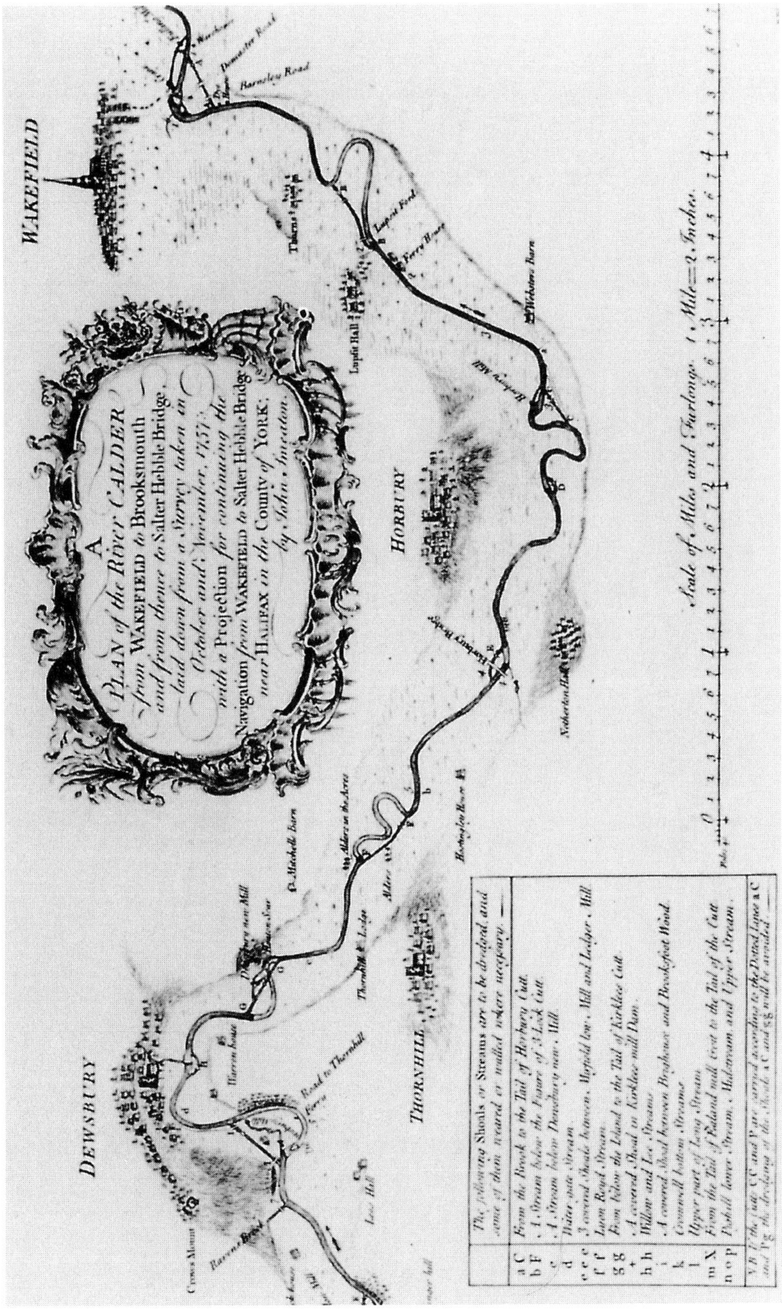

Plan of the River Calder (ii).

The Rolt Memorial Lecture, 1975*
The Engineers of Sunderland Harbour, 1718-1817

* The Council of the Association for Industrial Archaeology decided to establish a special lecture at Annual Conferences in memory of L. T. C. (Tom) Rolt, the first President of the Association.

Introduction

In the choice of a suitable topic for the first Rolt Memorial Lecture I decided without hesitation to present the results of research on some aspect of engineering biography, a field in which Rolt himself made such outstanding contributions. And as the lecture was to be given in Durham there seemed to be much in favour of devoting my attention to the early engineers of Sunderland harbour—one of the great civil engineering works in the North East. Very little reliable information has hitherto been published about these men as individuals, nor has much regard been paid in general to engineers of public authorities in the eighteenth century, a class to which they belong.

In order to keep the lecture within reasonable bounds I shall cover only the first hundred years of the harbour's history, the period extending from the first engineering report in 1718 to the death of Matthew Shout in 1817. By the latter date the sands and rocks on an exposed coast at the mouth of the River Wear had given way to a safe and commodious harbour with its entrance piers and lighthouse and a deep channel maintained by steam dredging. Much of course still remained to be done, but 1817 marks an important stage in the evolution of the modern harbour.

Organization of the Harbour Works

The works were carried out under the 'Commissioners of the River Wear and Port and Haven of Sunderland' (the 'Wear Commissioners' for short), established by an Act of 1717. This and subsequent Acts, as listed below, gave the Commissioners power to raise the necessary funds for harbour improvements and maintenance by a levy on all coals shipped from the River Wear downstream of Newbridge (near Chester-le-Street) or, from 1759, downstream of Biddick Ford.[1]

3 Geo. I, 22 June 1717
13 Geo. I, 24 March 1727
20 Geo. II, 17 June 1747
32 Geo. II, 2 June 1759
25 Geo. III, 25 April 1785
49 Geo. III, 12 May 1809

From 1752, without any significant break in sequence, the Commissioners always had as one of their principal officers an engineer to design and supervise construction of the harbour works; and it is very probable that the same arrangement applied in the earliest stage of development in the 1720s. These harbour engineers, as I shall call them, were full-time salaried officials, though they were able occasionally, and with approval of the Commissioners, to undertake for a suitable fee consulting work for other authorities.

In addition, the Wear Commissioners from time to time called in outside consultants for advice. But it must be emphasized that the harbour engineers had full responsibility for the works and indeed for the planning. This system, though by no means uncommon, differs from the more familiar type of organization in which a consulting or chief engineer is in charge, making regular visits to the site, with a resident engineer carrying out his designs. At Sunderland the works were executed almost entirely by direct labour, not by contractors, and the harbour engineers supervised this labour force, sending in regular returns of expenditure to the Commissioners. I shall of course say something about the consultant's reports and a good deal about the history of the harbour, relying to a considerable extent on the excellent paper published in 1847 by John Murray;[2] but my main concern is with the harbour engineers: their work at Sunderland and elsewhere, their careers, and any biographical details that can be discovered.

104

James Fawcett (*d.*1721)

Soon after the 1717 Act had been obtained, the Commissioners asked James Fawcett to survey the harbour and put forward a scheme for its improvement. His manuscript report is entitled:

A Scheme and Proposal for the Improvement of Sunderland Harbour with Estimates of the Charge for Building Piers and other works therein proposed; by James Fawcett.

It is dated, Sunderland, 19 August 1718. On the same day he also submitted a description of the harbour, accompanied by a map (Fig. 1).[3]

Except for some earlier attempts to deepen the North Channel, known as the Stell, the harbour in 1718 was still in its natural state. Fawcett proposed to build a masonry pier 700 ft long, projecting in a NNW direction from the south shore at point A, and also a short pier from the north shore, near the West Beacon, leaving an entrance 300 ft wide between the pier heads. His estimate for the piers amounted to £10,600. In addition he recommended the construction of a masonry quay wall along the south shore from the pier to the Custom House, and another from the north pier across the Ham Sand to the foot of the Ballast Hills. These walls would cost about the same as the piers, making a total expenditure of rather more than £20,000. The area so enclosed would be twenty acres in extent. Fawcett also emphasized the need for deepening the channel, but wisely refrained from guessing the cost involved. In preparing the estimates for the piers and walls he remarks:

I have carefully compared the prices allowed in other parts of the Kingdom where I have been concerned in the Publick works (or that have otherwise come to my knowledge) with the Observations and Informations I have received from the best workmen in these parts

and he goes on to say:

If the Commissioners shall think me in any way qualified to Serve them, I am willing to be retained on as moderate and reasonable Terms as they can Expect, and shall Act therein with the Utmost Diligence and Fidelity, to which I hope 'twill please God to add Success, and I should Count it a peculiar happiness to be an Instrument of advantage to this Place

It is known that work began on the harbour improvements in 1719 and that during the years 1719-20 at least £1,020 was spent.[4] This could very well represent the cost of building the substantial quay wall running about 250 ft south-west from

Fig. 1

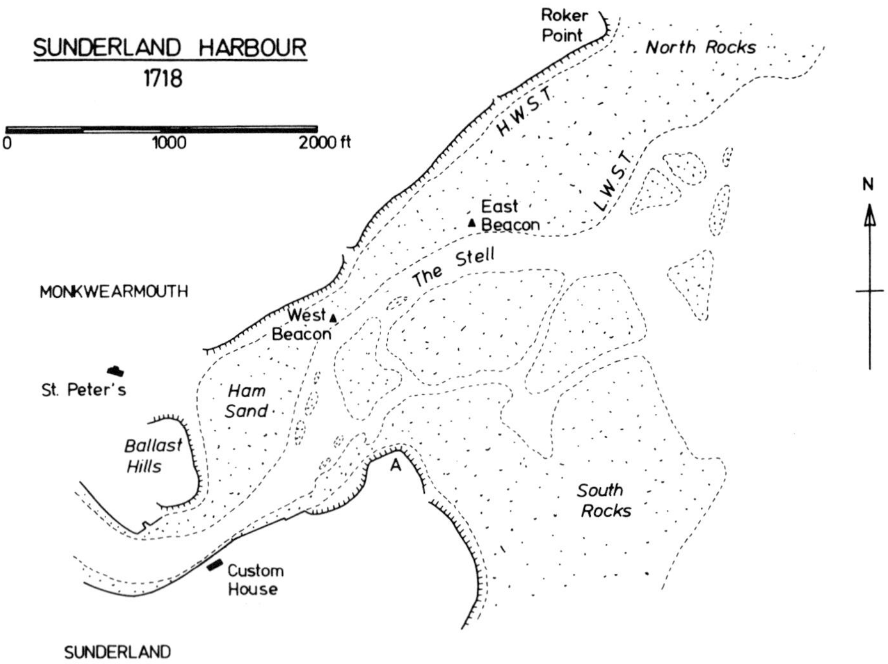

Fig. 2

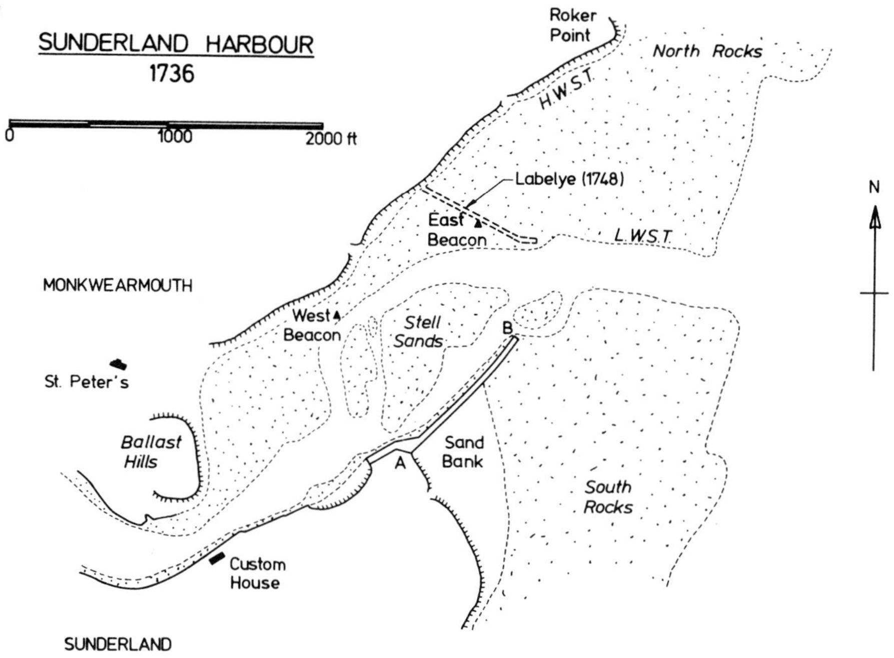

point A (Fig. 2), and I like to think that it was erected under the direction of Fawcett.

Nothing else has so far been discovered concerning Fawcett's career, except that from his will it appears he spent some time in Barbados.[5] Clearly he was a skilled surveyor and, from the quotation given above, he had previous experience of public works. He died in March 1721.[6]

William Lellam (*d.*1733) Harbour engineer probably from 1722 to 1731

When work began on the South Pier about 1723[7] it was on a very different plan from that proposed by Fawcett, and in all probability the pier as built (Fig. 2) was designed and executed under the direction of the engineer William Lellam. Certainly it was he who gave evidence to the Parliamentary Committee on the Bill for the 1727 Act at a time when construction of the pier was under way.

We first hear of Lellam in 1717 at Bridlington harbour, when he presented accounts for the repair work which had been carried out on the North Pier in the preceding year, together with estimates for completing these repairs and for rebuilding the South Pier.[8] The latter was to be about 670 ft in length, built in timber framing with stone infilling; and there is no reason to doubt that Lellam directed its construction, beginning in 1719.[9] What seems very likely is that he worked at Bridlington until 1722 and then, having got the new pier well under way, he moved to Sunderland, produced a design for the South Pier there, and began its construction in 1723.

By 1727 the South Pier at Sunderland had been carried 800 ft out from the shore, the total expenditure on the harbour had risen to just over £18,000, and Lellam estimated a further £6,000 as being necessary for completion of the work then in hand.[10] Powers to borrow additional funds were granted to the Commissioners in the Act of 1727, and by 1733 the total expenditure amounted to £28,000 of which no less than £19,800 had been spent on the South Pier.[11] The pier, which probably reached completion about 1730, now extended for 1,000 ft in a north-easterly direction from point A to B (Fig. 2). Built in solid masonry, and having a top width of 30 ft, with a parapet, it was founded on rock or, where the overlying gravel was too deep, on timber piles driven to rock.[12] Its chief purpose,

Plate 1 Stockton Bridge, designed by Joseph Robson in 1763 and built 1764-69. Engraver unknown.

apart from providing shelter, was to direct the river current and ebb tide towards the main approach channel. By the standards of the period the South Pier at Sunderland ranks as a major piece of maritime construction.

Lellam spent the last two years of his life at Scarborough, where he became harbour engineer in December 1731.[13] Scarborough Corporation had recently submitted a petition for a Bill granting powers to enlarge their harbour, and obviously needed the services of an experienced engineer. Lellam rapidly mastered the problems involved. In February 1732 he produced before the Parliamentary Committee his survey of the harbour, a plan for extending the Old Pier, and an estimate (£11,500) of the cost involved. The extension was to be 440 ft long, built in solid masonry 30 ft wide at the top, 35 ft high, with sides battering out to give a width of 60 ft at the base, and founded at a depth of at least 6 ft below low water.[14]

The Act for Scarborough harbour improvements having been obtained in June 1732, Lellam at once began preparations for building the pier extension, and work started under his direction in December.[15] But he died quite suddenly at the end of May 1733 and it was left to William Vincent (q.v.) to carry on and complete the structure.[16]

The pronunciation of his name is revealed by the phonetic spelling 'Lealam' which occurs occasionally in the Scarborough records. The correct spelling is given by his clear and well-formed signature 'Wm: Lellam' on a document, dated 28 December 1732, in which he and Mark Burleigh stipulate their fees and expenses for going to London and attendance there on a petition to Parliament on behalf of the Wear Commissioners. This shows, incidentally, that the Commissioners had no engineer of their own at that time. The individual fees were to be £1.05 per day; 50p per day for expenses in London and £2 for the return journey plus the hire of horses, and wages of a servant at 11½p per day.[17]

John Thomas

The 1717 Act granted powers to the Commissioners for a term of 21 years, i.e. to June 1738. By 1733, having completed the South Pier and, in doing so, having mortgaged the future receipts to some extent, the Commissioners could expect a net income during the next five years of only about £5,000. This sum was clearly insufficient to build a north pier, badly needed to provide protection for north-easterly gales, and to make other necessary improvements. Consequently, they petitioned for a Bill for additional powers,[18] but without success.

After another four years the Commissioners petitioned Parliament again, and called in as their chief witness the engineer John Thomas. He produced a plan of the harbour and stated that, in addition to the pier already built, 'another on the North Side was absolutely necessary, ships being frequently stranded on the sands there'. He also laid before the committee an estimate, amounting to £12,000, for building a north pier.

For reasons which are quite obscure, this second petition also failed, as did a third in February 1738, which never reached even the committee stage. In consequence, the Commissioners were presumably disbanded in June; by which date the total expenditure on the harbour (which almost exactly balanced the total receipts) amounted to £33,000.[20]

During the preceding four-and-a-half years the sum available for harbour maintenance and improvements had been about £5,000, as expected. Whether Thomas was employed at Sunderland during most of this period, or whether he had simply been called in as a consultant in 1737, I do not know. It can be assumed that he is the same John Thomas, 'an engineer', who gave evidence in 1721 on the Bill for Bridport harbour.[21]

Isaac Thompson (*c.*1703-1776)

In 1736 Isaac Thompson of Newcastle made a survey of the harbour and of the River Wear for a distance of eleven miles upstream to Newbridge. This may well have been carried out as part of the preparations for the Parliamentary petition of March 1737, together with John Thomas's estimate for a north pier. In association with Mark Burleigh, a Sunderland merchant, Thompson published the survey in the form of a large map, entitled

A Plan of the River Wear from Newbridge to Sunderland Bar as it Appeared at Low Water [and] A plan of the Mouth of the River Wear, Harbour & part of the Town of Sunderland

The map, on a scale of 600 ft to an inch, was engraved by John Tinney and is dated 1737. It forms the basis of Fig. 2. Thompson includes soundings taken along the river and the Stell (the north channel) and for some distance out to sea.

Why the harbour was neglected for so long after 1738 is not clear, but not until January 1747 did the inhabitants of Sunderland and the local landowners and gentry submit a petition for a new Act.[22] Evidence at the committee stage was given only by Burleigh and Thompson. They quoted figures of expenditure taken from the pre-1738 Accounts—a Mr. White, former Agent to the Commissioners, being unable to attend owing to his great age—and they described the bad state of the harbour, saying that the only proper method to improve it was by lengthening the pier already built on the south side and erecting a pier on the north side.[23] The petition was successful and led to the Act passed in June 1747.

It may be noted that Isaac Thompson, land surveyor and printer and 'a person of considerable literary attainments', gave public lectures on scientific subjects from 1739—later joined by William Elstob and by the mathematician Robert Harrison.[24] He published *A Description of the Orrery* (Newcastle, *c.*1750) and, jointly with Harrison, *A short Account of a Course of Natural and Experimental Philosophy* (Newcastle, 1757). He gave further evidence on the River Wear to a Parliamentary committee in 1759.[25]

Charles Labelye (1705-1762)

A week after the Act of 1747 had been obtained, the newly appointed Commissioners held their first meeting, on 24 June, and from that date the Minutes of the Wear Commissioners have survived. It was immediately resolved 'That a proper person be found to give his opinion on such Works as may be proper for the preservation and improvement of the River, Port and Haven'; and in September 1747 the Commissioners ordered:[26]

That a letter be wrote to Mr. Labillee to know if he will please come to Sunderland and give the Commissioners his opinion in Regard to building a North Pier & making other improvements.

At this time Labelye had just completed a detailed examination of Yarmouth harbour and was about to become involved in problems arising from the subsidence of one of the piers of Westminster Bridge, and so it was not until June 1748 that he felt free to ask the Bridge Commissioners for leave of absence to tackle the Sunderland job.[27] On his journey north he took the opportunity of calling at Whitby to see the harbour there.

Labelye arrived at Sunderland in July. Full preparations had been made for his visit. He carried out the necessary investigations with his usual thoroughness and clarity, and set out his recommendations and findings in

A Report relating to the Improvement of the River Wear, & port of Sunderland. Made after a View thereof taken in July 1748.

The report is signed and dated 16 September 1748, addressed from Crown Court near King Street, Westminster.

This document, running to forty-one closely written foolscap sheets, is one of the most comprehensive engineering reports of the early eighteenth century. An abstract was prepared, at the request of the Wear Commissioners, by Isaac Thompson and printed in December 1748. Thompson also made two copies of the plan showing the various proposed works. Labelye's fees and expenses for his visit to Sunderland came to £78.75 and he received additionally £52.50 'for his plan and report'.

Labelye began by giving a description of the river and harbour, with frequent references to Mr. Thompson's map. He expressed strong approval of the South Pier, both with regard to the solidity of its construction and also its position, but he thought that much further work should be carried out. His main recommendations were as follows:

(i) To remove the shoals from the river, partly by contracting its width in certain places by means of stone groynes and wharfing, and partly by harrowing the bed or, if this was not sufficient, by dredging.

(ii) To build stone wharf walls along the shore at the foot of the Ballast Hills, the erosion of which was a cause of much silting in the harbour, and to carry the wharfing at least as far as the West Beacon (Fig. 2). Sand and gravel dredged from the harbour should be placed behind the wharfing to create valuable new land.

(iii) In order to provide shelter from NE storms a new North Pier should be built in the position shown (by dotted lines in Fig. 2), leaving a width of about 600 ft between the new and the existing (south) pier heads. The new pier was to be about 25 ft wide at the top, with a parapet, and sides battered at 1:6, and with a 'foreland' of rock fill on the seaward face rising at a gentle slope to 4 ft above high water springs.

(iv) Foundations of the wharfing and of the pier should be taken to rock or if this was not possible they should be piled, the heads of the piles being cut off below low water so that they were permanently submerged; and the outer row of piles to be driven as close together as the nature of the ground permitted. The core of the pier should be built with stone blocks, not loose rock fill.

(v) It was essential to concentrate the flow into one channel. The channel south of Stell Sands was preferred as it provided a more direct course to the outer or approach channel. This south channel could be improved by dredging, by extending the wharfing previously mentioned from the West Beacon to the new North Pier head near the East Beacon, and if necessary by blocking off the Stell with stone groynes.

(vi) The works described above would result in a deepening of the channel as far as the pier heads, but would have little or no effect on the outer bar located at the mouth of the approach channel. The piers could be extended in an easterly direction, though at great cost, and this would provide some benefit; but the best way of increasing the depth of water over the bar was by dredging.

(vii) For this purpose, and for the dredging required elsewhere, 'ballast engines' had to be provided such as those used in preparing the foundations for Westminster Bridge:

> In these instead of pulling up the Tool or Bag (which brings the Ballast up) by Hand, which goes but very slowly, there is in these Engines, a purchase of an Iron Wheel and Pinion which 3 Stout Men turn round by means of a Whinch, and they rise near a Hogshead or about 4 Hundred weight of Sand or Ballast at every Pull; Besides the 3 Men at the Whinch there is one to guide the Tool, and another that Trims and Stowes in the Barge what is brought up.

Here we have the earliest descriptions of the 'bag and spoon' dredgers which were now (i.e. in 1748) coming into general use on the Thames.

(viii) Finally, Labelye emphasizes the need for the Commissioners to have an engineer to direct and supervise the works.

The importance of these recommendations, in determining the overall strategy of the subsequent improvements, will become obvious as the history of the harbour is described. John Murray, writing nearly a century later, refers to 'the judicious views of Labelye', and the Abstract of his 1748 report was thought to be of sufficient interest to merit reprinting in the Tidal Harbours Commission of 1846.

Jacob Spencely (*c.*1712-1773)

Little work seems to have been done for a year after Labelye's report; no doubt the Commissioners were gathering their resources. But in the autumn of 1749 they obtained a harrow and in January 1750 they ordered a dredger to be built in accordance with designs made by Jacob Spencely, a Sunderland chandler.[28] It cost £213 and came into operation in June 1750, with an overseer and workmen employed under Spencely.[29]

For the next five months, and probably again in the summer of 1751, the dredger was engaged in deepening the Stell or North Channel. Clearly this was an attempt to improve the existing navigation but, as Spencely himself later admitted, the benefits were of short duration.[30]

William Vincent (*d.*1754), appointed harbour engineer June 1752 and died in office April 1754.

For about a year after Wiliam Lellam's death, work on the extension of the Old Pier at Scarborough was continued by Robert Wilkins, but the Minutes

Plate 2 Ayton Bridge, built by Robert Shout and William Ellis 1771-74. Photo by author.

for 3 December 1734 report:[31]

> That Mr. Vincent is thought a proper person for Carrying on the Works at the Pier and it is agreed by this House that Mr. Vincent shall succeed Mr. Wilkins and shall be appointed Ingineer . . . in the Harbour of Scarborough and shall be allowed the Sume of Seventy pounds a year

For the next ten years work went steadily ahead, with thirty or forty men regularly employed under Vincent's direction. However, in 1744, owing to financial difficulties, the labour force had to be reduced, and Vincent was ordered not to advance the work further than the foundations already laid. This meant closing the head of the pier in 1746, by which time the extension (known to the present day as 'Vincent's Pier') had reached a length of 320 ft,[32] built entirely beyond low water mark.

Meanwhile, Vincent had produced a plan for building a great East Pier, to form an outer harbour and to act as a breakwater.[33] In 1746 he was authorized to start construction,[34] and by the time he left Scarborough in 1752 he had successfully completed 100 ft length of this massive structure. It was continued under the Commissioners appointed in that year and by 1800 had reached almost exactly the position shown on Vincent's plan, a length of 850 ft having been built since 1746. During the period 1801-1826 it was extended and completed, as we see it today, by Chapman.[35]

While at Scarborough, Vincent produced a fine map of the harbour and adjoining parts of the town, surveyed in 1745. It was engraved and published by him in 1747.[36] An interesting note on Vincent's intellectual attainments or aspirations is provided by the fact that he was one of the very few engineers whose names appear among the subscribers to Desaguliers' *Course of Experimental Philosophy* (2 vols., published 1734 and 1744) to which, as is well known, Labelye contributed several important articles.

The presence at Scarborough of so able an engineer as Vincent naturally did not escape the attention of the authorities at Sunderland. He visited there by invitation in 1747,[37] and in June 1749 the Commissioners wrote again to 'Mr. Vincent the Engineer at Scarborough to desire he will be pleased to come here . . . to give his further Opinion in regard to the harbour'. And on 30 August 1749 it is recorded in the Sunderland Minute Book that:

The Commissioners from the Good Opinion they have of Mr. Vincent & his judgement, agree to give him £120 p. annum . . . so long as there will be any work to do for the preservation and improvement of the port and haven.

Probably as he was now engaged in building the East Pier at Scarborough, Vincent did not accept this tempting offer. But the Wear Commissioners in December 1750 asked both Labelye and Vincent for their opinions on a plan for a jetty and a year later the Commissioners wrote to the Scarborough authorities[38] requesting 'that they will be pleased to give Mr. Vincent leave to come to give his opinion & advice as to the Situation of the proposed pier'. This presumably refers to the proposal for a North Pier at Sunderland.

In July 1751 William Vincent of Scarborough and John Turner of Gosport were consulted on the construction of a masonry pier at Ramsgate harbour; and when in January 1752 the Ramsgate Trustees advertised the post of harbour engineer, at £200 per annum, Vincent was one of the applicants. He produced ample testimonials, but the job went to William Etheridge, not, as events were to prove, a judicious choice.[39]

Vincent's readiness to leave Scarborough in 1752, by contrast with his earlier determination to stay, can be attributed not only to the higher salary offered at Ramsgate but also to the rapidly deteriorating state of affairs at Scarborough where the mismanagement, and even the corruption, of the Corporation had in recent years reached such a pitch that the shipowners and inhabitants of Scarborough petitioned Parliament in January 1752 for a Bill to remove authority for new works at the Harbour from the Corporation and to set up an independent body of Commissioners.[40] Vincent, who had in November 1751 suffered the indignity of an attempt to reduce his salary,[41] gave evidence on the progress of the works to the Parliamentary committee.[42] The Bill was duly passed, by a large majority, and received the Royal Assent on 26 March 1752.

The Wear Commissioners, doubtless aware of these events, wrote again to Vincent on 8 April repeating their previous offer of £120 per annum, now to be guaranteed for seven years. In May he attended the Commissioners at Sunderland and promised 'at midsummer next to enter upon the place of Engineer and Superintendent of Works'.[43]

Immediately after settling in Sunderland Vincent began his new duties with vigour: putting the dredger into repair, opening a quarry, and building a workshop (probably near point A in Fig. 2). Moreover, when dredging started under his direction in the autumn of 1752, its object was to enlarge and deepen the South Channel, and he continued this operation without abatement throughout the following year, as shown by his returns of work done.[44] Indeed the opening up of this channel for navigation was Vincent's important contribution to the improvement of Sunderland harbour. But it was his only contribution as, before he had the opportunity of extending the South Pier, or building a north pier, he died in April 1754 after a short illness.[45] He left all his estate to his wife Mary; this included a house near West Sand Gate Street in Scarborough, where the family presumably lived before moving to Sunderland.[46]

Joseph Robson (*c.*1704-1780), appointed harbour engineer June 1755 and retired September 1779

Vincent's death came as a shock to the Commissioners. As a temporary measure Joseph Taylor, his foreman, was put in charge of the South Channel improvements, at an increased salary of £1.05 per week,[47] and advertisements for the post of engineer were published in June 1754. In October it was decided to write to Labelye for his opinion on the persons who had offered their services, but whether he ever received the letter, or replied from France, I do not know. However, by December the Commissioners were making enquiries in London regarding Joseph Robson, and a month later they invited him to carry out a survey at Sunderland.[48] This he did, and received £31.50 for his fees and expenses.[49] He also decided to accept the job if offered to him. His appointment was confirmed on 24 June 1755 at a salary of £120 per annum and he remained as harbour engineer for twenty-four years. A three-storey building on the quay was converted into a dwelling house for the engineer in the autumn of 1755.

Robson's principal works at Sunderland can be summarized as follows:

(i) He continued dredging the South Channel and cut through an outcrop of rock 'which lay in the way of Shipping'. Already by 1759 he could report a substantial improvement in the harbour as a result of the work then completed.[50] In a printed paper, *A Retrospective View of the State of Sunderland Harbour from the year 1755 to 1774*, it is stated that the cutting, 4½ ft deep through the rock outcrop, extended 500 ft in length and 180 ft in breadth; and that 'ships from 12 to 14 keels, or about 270 tons burthen, that formerly could but take in from four to six keels, or about 160 tons, within the harbour, now go out to sea full laden'.[51]

Fig. 3

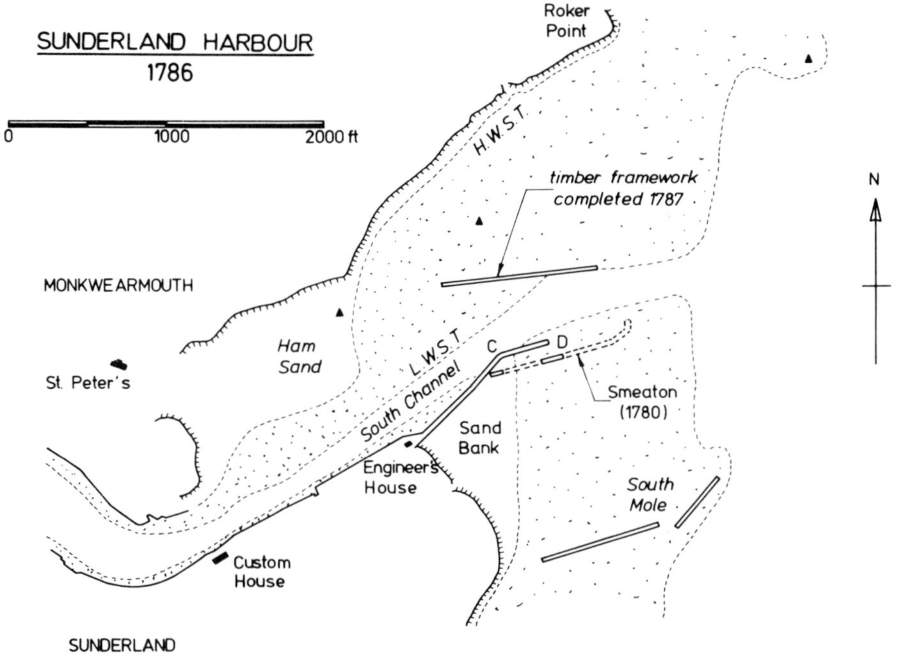

(ii) Part of this improvement was attributed to an alteration in the South Pier. As early as July 1755 Robson advised rebuilding the end of the pier. After further consideration he proposed that a length of 200 ft should be taken down and 'new built in an east-north-east direction' which would direct in a better manner the set of the ebb tide along the channel. This proposal received strong support from the shipping interest in Sunderland. The Commissioners gave their approval for work to begin in April 1757. By 1759 much progress had been made,[52] and this new portion of the South Pier, about 320 ft in length (CD in Fig. 3), was probably completed in 1762. The state of the harbour at that time is summarized as follows:

> a large Stone Pier and Key have been erected near the Mouth of the River, the Channel of the Port hath been much deepened, several considerable and necessary Alterations and Additions made in the said Pier, and divers other Works done for the Improvement and Security of the Port, Haven and River . . . at an Expence amounting, in the Whole, on 21st December 1762, to £48,000 and upwards.

It may be noted that the extension, as built by Robson, originally joined the old South Pier at an angle of about 150 degrees. At some time between 1786 and 1792 the junction was rebuilt by Robert Shout in the form of a more gradual transition, as shown in Fig. 4.

(iii) In his evidence to the Parliamentary committee in 1759, Robson referred to a scheme for building a pier or mole on the South Rocks, with the object of reducing littoral drift of sand into the harbour entrance. The plans lay dormant for several years, but in May 1767 he reported that, despite constant dredging, the depth of water over the bar had scarcely increased, and he again recommended construction of a mole on the South Rocks.[54] Before embarking on so large an undertaking the Commissioners sought advice from the eminent engineer John Wooler.[55] The content of his report[56] and the subsequent history of this project are best told in the words of John Murray:

> Mr. John Wooler agreed nearly with Mr. Robson in many of the principal points relating to the pier, and the effects to be produced by it; but differed with him on the quality and quantity of the limestone, with which the pier was proposed to be built. In consequence, Mr. Wooler's estimate far exceeded Mr. Robson's; but, notwithstanding this, the Commissioners appear to have had confidence in the opinion of their own engineer, who forthwith proceeded to build the pier in the manner projected by him, with stone chiefly quarried from the adjacent rocks.

Considerable progress was made in this work; for some portions of it remain at the present day, on the outer rocks, although many of the stones were subsequently removed to build the extension of the south pier. The reasons for abandoning the work do not now appear.

Murray goes on to criticize the scheme as being over-ambitious in relation to the funds available at the time, and concludes 'there can be no doubt that the enthusiasm of the engineer carried him far beyond the bounds of prudence'. It seems only fair to add, however, that in Robson's plan for the south mole, and for a similar structure to the north which he contemplated but never began, we can see the first hints of the great outer piers (completed in 1903 and 1914) which form such important features of the modern harbour.

Early in his term of office, in 1758-59, Robson was also concerned, together with John Smith of Sheffield, the well-known specialist in river navigation works, in plans for extending the navigation of the River Wear from Biddick Ford to Newbridge, a distance of about two-and-a-half miles.[57] But the estimated cost (around £7,000), for building a lock and dam one-and-a-quarter miles below Newbridge and deepening the river bed downstream, was considered to be too great in relation to the benefits likely to be gained, and in the 1759 Act the upper limit of the Commissioners' jurisdiction was moved down to Biddick Ford. At the same time Durham Corporation obtained an Act for making the river navigable from Biddick Ford nearly up to the city itself, a distance of seven miles. Smeaton produced the plans for this scheme, but it was never carried out.[58]

At a meeting of the Wear Commissioners in September 1779 Robson signified that 'from his Age and present State of Health it is necessary for him to have some Assistance in the Execution of his Office, and recommended Mr. James Shout of Gateshead as a Proper Person to be concerned with him for that purpose'. The Commissioners agreed to appoint James Shout as Assistant Engineer, but it is clear from the Minutes and Accounts that he immediately took over all the responsibilities and that Robson retired.

Robson's most important work outside Sunderland was Stockton Bridge. In August 1760 he obtained leave of absence 'to go to Stockton to give his opinion on building a bridge there over the River Tees'. Smeaton also visited the site in that

Fig. 4

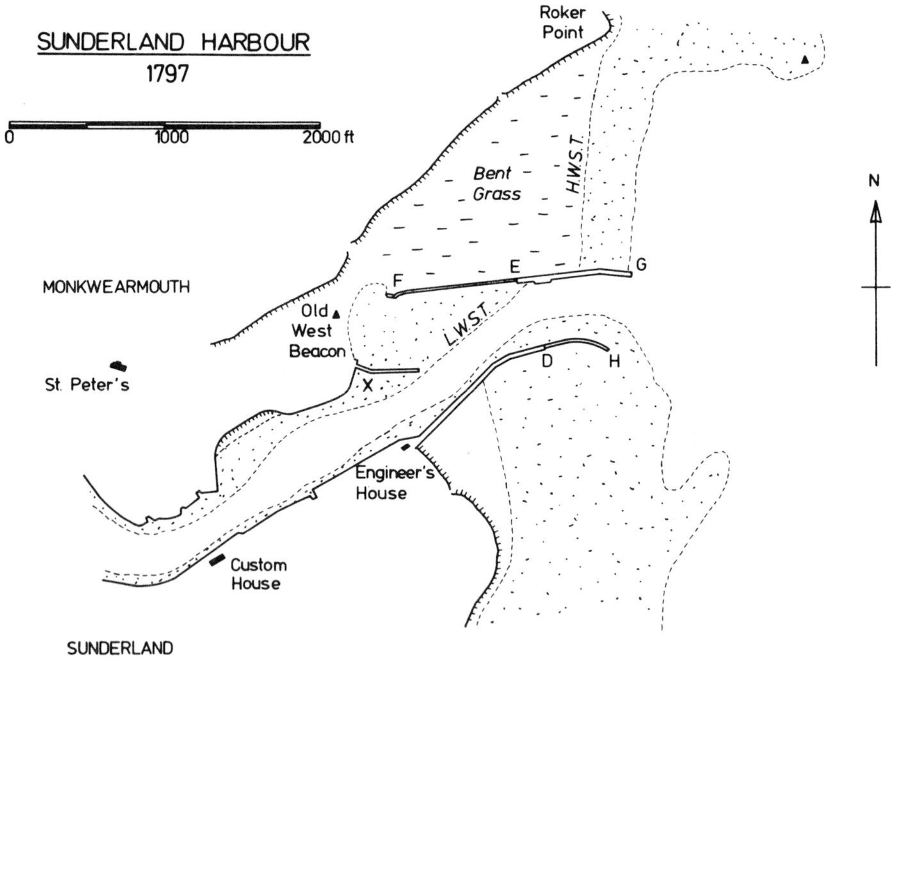

year, producing a plan and estimate, and it was he who attended Parliament two years later on the Bill.[59] The Act received the Royal Assent in March 1762. However, the Trustees then turned to Robson. His design, which differed very materially from Smeaton's, was accepted in March 1763. Before construction commenced, Robert Mylne was asked to 'examine & approve the place when the bridge is intended to be built'. In his report, dated 26 October 1763, Mylne says

> I have in investigating these circumstances been greatly obliged to the good Nature and Candour of Mr. Robson. He and I had a conference also upon the method of putting it into Execution, and after finding his Opinion coincide with mine as to the necessity of going down to the Clay with the foundations of the Piers, we then agreed upon a manner which is Specified . . .

Then follows a detailed report, signed jointly by Mylne and Robson, and also dated 26 October 1763, entitled *A method of laying the foundations of the Piers of Stockton Bridge*. The contract, signed by Robert Corney, Robert Shout and Robert Nellson, bears the same date.

Work began in 1764, the last arch was completed in September 1768 and the opening ceremony took place on 29 April 1769. The bridge, built at a cost of about £8,000, had five quasi-elliptical arches with a central span of 72 ft and side spans of 60 ft and 44 ft.[60] It was replaced by a more modern structure in 1887.

Robson's fees and expenses in connection with the design amounted to £38.92 (Smeaton had charged £26.25 plus £42 for attending Parliament), but a few years after completion of the bridge Robson received a further payment of £104.57. This suggests that he had made several visits to the site during construction (Plate 1).

Robson and Mylne again worked together when they were asked to advise on the best position for the proposed Legal Quays at Hull, and their report on this subject is dated 23 April 1767.[61] Another of Robson's jobs outside Sunderland is recorded in October 1765 when 'a new machine of a particular construction, built under the direction of Mr. Robson, a very able and experienced engineer, for cleansing the River Tyne, was launched off the Quay at Newcastle'.[62] He also gave advice on the construction of a dry dock at Tynemouth in 1758.

After retiring in 1779 Robson moved to Bishopwearmouth. He died in November 1780 and administration of his estate was granted to his eldest son William Robson, mariner.[63]

James Shout, a temporary appointment as harbour engineer from September 1779 to December 1780

In 1774 Robert Shout the elder of Helmsley and his three sons, Robert and John (masons) and James[64] (carpenter), contracted to build the south part of Tyne Bridge over the river between Gateshead and Newcastle.[65] The north part was to be built by other contractors, the joint engineers for the whole bridge being Robert Mylne and John Wooler acting respectively for the Bishop of Durham and the Corporation of Newcastle. By 1779 the main structure was completed, and James Shout must have felt free to take on the Sunderland job (at Robson's recommendation), though his appointment was not intended to last after December 1780.[66]

The seaward end of the South Pier had suffered some damage in the great flood of November 1771—the same occasion on which the old Tyne Bridge was destroyed. No doubt Robson carried out repairs but James Shout, when he came to examine it eight years later, decided that rather than making further repairs this part of the pier should be entirely rebuilt, and on a line parallel to Robson's but lying further south. Before agreeing to this somewhat radical proposal the Commissioners called in John Smeaton.

Smeaton duly arrived and, 'having carefully viewed the harbour and pier of Sunderland, and received the information of several eminent masters of ships belonging to that port, and having also considered the plan produced to me by Mr. Shout', he delivered

The Report of John Smeaton, Engineer, upon Mr. James Shout's Plan for Rebuilding and Extending the Old Pier of the Harbour of Sunderland.

The report is dated Gateshead, 16 Jan 1780; this suggests that Smeaton, while writing his report, stayed with Shout, who had not moved house. His fee and expenses amounted to £24.50.

Smeaton agreed with the plan, except that he advised bringing the end of the pier round to the north as shown in Fig. 3. In explanation he says:

> as the direction of the last stretch of the old pier [i.e. Robson's] has been now proved by many years' experience to afford a sufficiency of shelter to the shipping from east winds and seas, and also to be in a commodious direction for going in and out; I am of opinion, that what has been shewn by experience to be right, ought not to be essentially altered; that is, that the head of the advanced pier should be in the line of direction as the present pier.

Some regret may be felt that this scheme was not adopted; Jessop in 1807 certainly expressed the opinion that Smeaton's plan might have proved

Plate 3 View from North Pier, Sunderland Harbour. Engraving by George Hollis after E. Swinburne, published 1823.

better than the one eventually followed. However, although the Commissioners accepted Smeaton's report, and work actually began in accordance with it, Robson's pier was in fact retained, as we shall see, and later extended along a line even further to the north.

John Shout (1738-1781), appointed harbour engineer December 1780 and died in office June 1781

John Shout worked as a mason on Stockton Bridge under his father,[67] and seems to have settled in that town. In 1768 he designed and erected the elegant Doric column in the market place there.[68] Then follows a gap in my knowledge of his activities, for five or six years, until the Tyne Bridge contract.

When James Shout left Sunderland harbour in December 1780, his brother John immediately took over, as if this had been pre-arranged.[69] It is perhaps significant that Tyne Bridge was now very near completion—its official opening took place in April 1781.

There is little to report on his short career as harbour engineer, as just under six months after his appointment he died. The *Newcastle Courant* for 23 June 1781 gives the following obituary notice:

> Died last week at Stockton upon Tees, Mr. John Shout, Architect and Principal Engineer at Sunderland, under the Commissioners for improving the harbour there; he was much respected for his integrity and great skill.

Apparently he had not yet taken up permanent residence in the engineer's house at Sunderland.

Robert Shout (1734-1797), appointed harbour engineer June 1781 and retired August 1795

Born in 1734[71] the eldest surviving son of Robert Shout of Helmsley (1703-1774), the future engineer of Sunderland harbour was trained as a mason, and undoubtedly worked on the construction of several of the fine bridges in the North Riding designed and built by his father. These include Newsham Bridge (1752-53) near Malton and Scawton Bridge (1755-57) near Rievaulx Abbey, both of which exist today practically unaltered after more than two centuries, and Egton Bridge (1760-61), which was destroyed by a flood in 1930.[72]

The younger Robert Shout's first responsible job was as the contractor, jointly with William Ellis, for building Sleights Bridge (1762-63), a handsome 3-arch structure over the River Esk, planned by Robert Shout senior who at that time was Surveyor of the Bridges for the eastern part of the Riding.[73] He worked for a time, again under his father, on Stockton Bridge, and then with Ellis and Robert Corney took on the contract for Whitby Bridge (1765-68). This comprised five masonry piers with a timber deck. Next he designed and built two single-arch bridges on the road from Pickering to Helmsley. The one at Welburn, known as Tile-house Bridge, is a beautiful structure still in excellent condition; the other, at Sinnington, is a simpler 'country' bridge of 40 ft span, the only decoration being a well proportioned string course. Both these bridges were built 1768-69.

John Carr of York became Surveyor of the North Riding bridges in July 1772, and one of his earliest designs was for Rutherford Bridge, south of Barnard Castle. It had a single span of 82 ft. Robert Shout junior signed the contract on 6 October 1772 and completed the bridge just over twelve months later. Meanwhile in October 1771 he and William Ellis had started construction of Ayton Bridge, over the River Derwent. Carr in 1772 introduced some modifications and a new contract was let. By 1774 Shout and Ellis had virtually completed the structure but could not quite succeed in finishing the job at the contract price; it was taken over by the Riding using direct labour. Ayton Bridge, like all the surviving bridges built by Robert Shout and by his father, displays a high standard of workmanship[74] and careful attention to design details (Plate 2).

Shout's last and his biggest bridge job was the construction of the southern part of Tyne Bridge, under the direction of Robert Mylne. The contract, as previously mentioned, was taken on by Robert Shout senior and his three sons. But the father died in June 1774, only three months after the contract had been signed, and Robert, as the eldest and most experienced of the brothers, presumably undertook the main responsibility for building this important structure. It was completed at a cost (for the south part) of about £9,000.

Through their work on Tyne Bridge the Shout family probably became well known in that part of the world, and when John Shout died, it is not surprising that the choice for his successor at Sunderland harbour fell on so able an engineer as his brother Robert. His appointment dates from 25 June 1781, less than a fortnight after John's death.[75]

In his first working season Shout continued building the new South Pier, but in 1782 the Commissioners ordered construction to cease. Clearly they were worried by shortage of funds, and in 1783 preparations began for another petition to

Parliament. Work was resumed in May 1784, although a few months later instructions were issued, for the sake of economy, to take stone from the south mole. When Shout gave evidence to the Parliamentary committee the following year, he said:[76]

> That a new Pier is actually begun, and some progress made in it, and he has not the least Doubt but the new intended Pier will be of great Utility to the Harbour, by standing more to the Southward, and making the Harbour easier of Access than it is at present; and that a greater Depth of Water will be obtained upon the Bar.

But he added the ominous words 'the Expence of performing the new Works will amount to £60,000 and upwards; to answer which the present Duties are very inadequate'.

The Act having received the Royal Assent in April 1785, work on the pier continued. Nature then took a hand in the affairs of Sunderland harbour, with rather dramatic consequences. Matthew Shout provides the following brief account of what happened, in a letter written on 28 November 1816 and quoted by Surtees:[77]

> In the winter of 1785 the entrance of the Harbour was warped up by a large sand-bed, which extended quite across the Haven mouth, leaving scarcely depth of water for the entrance of a light vessel. It was suggested, in consequence, by Mr. Shout, the resident Engineer . . . that a temporary wooden Pier should be erected, in order to contract the channel, and enable the ebb tide to scour itself a deeper bed. The plan was immediately adopted; and in a few months there was a deep and spacious channel. This beneficial result induced the Commissioners to commence the building of a permanent Pier of stone; but, owing to the nature of the ground, loose shifting sand and gravel, the work proved extremely tedious and expensive. A length, however, of near seven hundred feet of solid pier, built upon piles, was completed by Mr. Shout, and ninety feet more were added by his successor, Mr. Pickernel.

No doubt Matthew, as a schoolboy, watched these operations with keen interest and pride in his father's achievement.

Certainly Robert Shout acted with great expedition. By January 1786 the Minutes record the success of the first stage of the temporary works, and a decision to place a row of stone-filled timber caissons, or 'carcases'. A length of about 1,000 ft had been built, in the position shown in Fig. 3, by the winter of 1787. Work then began on the masonry structure and the 700 ft length, men-

Fig. 5

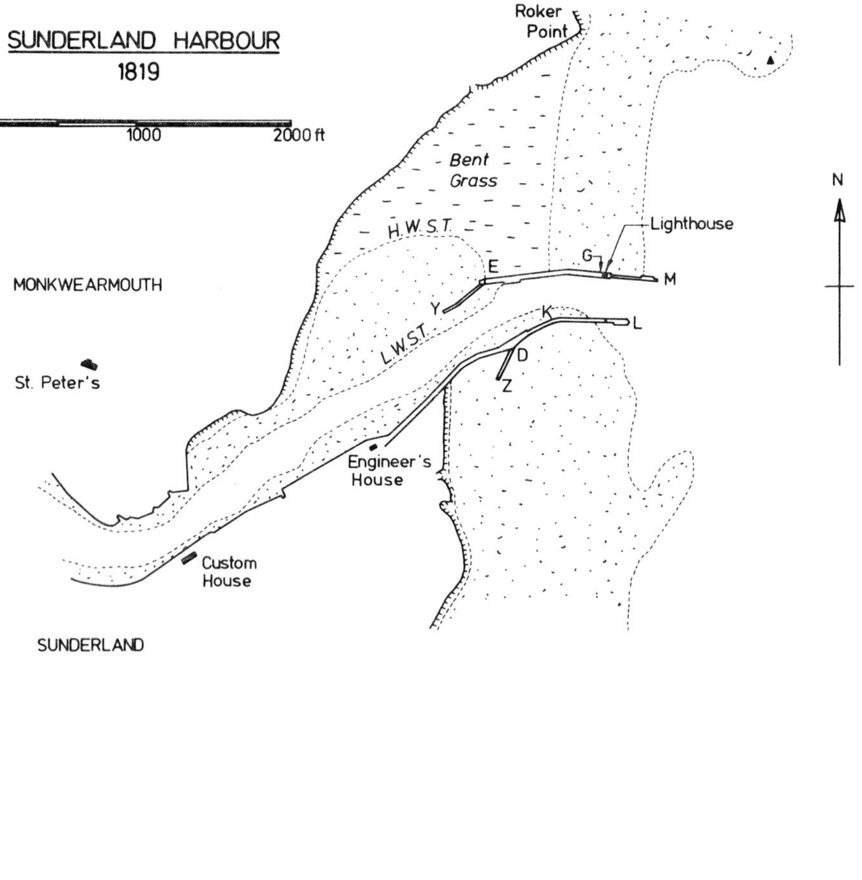

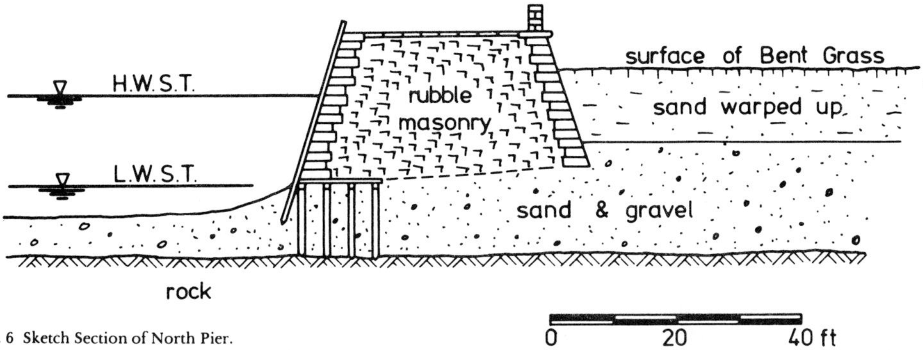

Fig. 6 Sketch Section of North Pier.

tioned by Matthew Shout, was completed in 1793 (EG in Fig. 4). A section of the pier is shown in Fig. 6.

This first (and major) part of the North Pier was Robert Shout's great contribution to Sunderland harbour. The need for a pier on the north side had been recognized by Fawcett and Labelye, and indeed was one of the main points in the 1747 Act. Nevertheless, attention had subsequently been directed almost entirely towards lengthening the South Pier despite the fact that one effect of this was to check the flood tide from passing southward, and consequently to send it up the river loaded with sand. Under normal conditions the ebb current was just about strong enough to force this deposit back to the sea, but the balance could be disturbed in a strong northerly storm; and this happened in 1785. It is, of course, easy to be wise in retrospect. What has to be recognized is that, faced with a catastrophic situation, Shout took the right course of action and followed this by creating an extremely important improvement in the harbour.

While concentrating his efforts and the resources of the Commissioners in building the North Pier there is naturally little to report as regards other works. A new dredger came into operation in 1787, worked under contract at 5p per ton of material raised.[78] The timber frames on the west side of the North Pier were extended further towards the shore (to point F in Fig. 4), to prevent sand being washed into the channel,[79] and repairs were carried out at the corner of the old South Pier (at point C in Fig. 3) between 1786 and 1792.

These repairs having been effected, and with the North Pier under construction, there was no justification for resuming work on the new South Pier. Instead, the fragments built before 1786 were left for the time being and in 1793 the decision was made to build a temporary extension of Robson's pier by placing stone-filled timber frames, some of which were to be those already used in the earlier North Pier works.[80] This extension DH is seen in Fig. 4.

Also in 1793, a part of the masonry of the North Pier being in danger of collapse as a result of undermining of the foundations by erosion, work began on a strong buttress near the west end, as shown in Fig. 4.[81] Finally, in 1795, Shout recommended the construction of two jetties.[82] One, at point X in Fig. 4, was built soon afterwards to improve the set of the ebb tide. The other, at the inner end of the North Pier, from E to Y in Fig. 5, had as its main purpose the prevention of damage to ships being driven ashore on the sands to the west. It was approved by the Commissioners but not constructed until about 1806.

Already in December 1794 Robert Shout's health gave cause for concern, and by May 1795 advertisements for a successor were published. He retired three months later, and died on 25 March 1797. His burial in Sunderland churchyard took place on 28 March near the south-west corner of the church.[83]

From the Helmsley parish registers and records we learn that Robert Shout, like his father before him, was a churchwarden (in 1769) and his signature appears among those of the Jurymen approving the church accounts from 1773 to 1779. In April 1762 he married Mary Moorhouse (1739-1806), a Helmsley girl. They had four daughters and a son, all born at Helmsley. The marriages of all four daughters took place at Sunderland between 1786 and 1793. Ann, the eldest, married Jurdison Davie, sailmaker and shipowner. Mary, the second daughter, married Joshua Ashton, another shipowner. Margaret's husband was John Harcastle, boatbuilder. Jane, the youngest girl, married Thomas Mordey, ship's captain; her son Dr. William Mordey became a well-known surgeon in Sunderland and a pioneer in public health.[84]

In his will, Robert Shout left to Matthew 'my mahogany desk, book-case and books therein'.[85] The rest of the furniture and all the plate, pictures, linen and china went to his wife. She and Matthew were the residuary legatees, and inherited a third share in the ship *Affinity*. The daughters each received a small legacy of £20.

While at Sunderland, Robert Shout was employed as a consultant on several occasions. In 1784 he and William Fordon were asked for their opinion on the line to be followed in building the East Pier at Scarborough.[86] In 1788 Shout submitted a notable design for a single arch 170 ft span bridge over the Wear.[87] He was given ten days leave of absence in 1790 to advise about building a pier at Bridlington.[88] In the following year he similarly received permission from the Commissioners to visit Stockton to advise on improving the navigation of the Tees, and lastly, in 1792, he was consulted on the best means of removing a sand bank in the Tyne, the growth of which threatened the port of Shields.[89] He reported that such a large body of sand in such a position could never be removed effectively by sluicing, and therefore recommended the use of dredge barges as employed at Sunderland. These have a crew of five men, three of whom work the winch (exactly as described by Labelye in 1748), and they raise 40 or 50 tons in one tide. The work is done by contract, at 5p per ton raised and carried away. Here we have one of the rare references to the performance of bag and spoon dredgers.

I know of only two occasions during Shout's term of office when reports were requested from independent consultants. On 15 July 1793 it was agreed 'that Mr. Pickernell, Engineer at the Port of Whitby be wrote to and requested to come to Sunderland to give his opinion on the works'. This was Jonathan Pickernell senior, who had been Smeaton's resident engineer on Hexham Bridge and from 1781 until his death in 1812 was harbour engineer at Whitby.[90] He reported on 30 August and received £21. No copy of his report appears to have survived, but from a brief entry in the Sunderland minutes Pickernell seems to have been advising on measures to strengthen the North Pier, probably in connection with the buttress. The second report, by Ralph Dodd, dated 28 July 1794, is discursive and of little interest except for the suggestion of building a dock on the north side of the channel.[91] Dodd was given £31.50 for his report.

Jonathan Pickernell (*c.*1765-1814), appointed harbour engineer August 1795 and dismissed May 1804

The Wear Commissioners advertised for a successor to Robert Shout in May 1795 and on 25 August appointed Jonathan Pickernell the younger, son of the Whitby harbour engineer,[92] at a salary of £120 per annum 'with House and Fire' provided. Ordered to increase the dredging facilities at Sunderland, Pickernell got in touch with Boulton & Watt in June 1796 to supply a steam engine for this purpose. The 4 hp engine arrived in July 1797 and was then mounted on a barge or 'keel' to operate (probably two) bag and spoon dredges. Pickernell also designed a sailing hopper vessel to carry the dredgings out to sea. The dredger began operations in the summer of 1798 and was still working in 1801; but the engine and gearing and the hopper vessel were sold piecemeal in 1804. It therefore appears not to have been much of a success, but its place in engineering history is assured as the first steam dredger engaged in river and harbour works.[93]

On 16 April 1798 Pickernell presented *A report on the Present State of the Harbour of Sunderland . . . with Remarks on the Measures to be adopted for its Improvement*. This was printed at Sunderland later that year. He pointed out that with an annual revenue of £5,000, of which £1,500 was required for maintenance, only £3,500 was available for new works. But a dock would cost at least £50,000. Thus it was more realistic to increase accommodation for the larger ships then using the port by deepening the water at the quays by dredging, at a charge of say £1,500 per annum, and to use the balance for essential improvements of the piers. The new works in this category ought to include an extension of the North Pier by 250 ft; and the construction of a wall up to high tide level, from E to X in Fig. 4, to concentrate tidal flow along the channel and at the same time to prevent vessels from being driven ashore on the flat ground behind. The pier extensions were to be advanced alternatively, working first for one or two seasons on the North and then on the South, leaving a distance between the pier heads nearly equal to the narrowest part at present, i.e. about 400 ft.

Work began on lengthening the North Pier in 1799, in accordance with detailed plans submitted by Pickernell on 5 June 1798. By 1801 about 100 ft had been constructed.[94] As the next step he designed an elegant octagonal lighthouse (his drawings were produced on 19 January 1802), which he built in freestone near the end of his North Pier extension (Plate 3). This had a height of 60 ft from base to cornice and a total height of 76 ft. It was completed in 1803 at a cost of £1,400.[95] Meanwhile, in December 1802, his scheme for extending the South Pier had been approved and he began preparing the necessary materials.

During the next season work proceeded, presumably involving the removal of some of the timber frames, before laying the foundations for the new masonry; and in June 1803 Pickernell received a gratuity of £50 for his 'extra attention as Engineer'. In April 1804, however, he was accused of misapplying wood and other property of the Commissioners for the use of himself and others, and on 4 May 1804 a majority voted for his removal.

What happened to Pickernell during the next few years I do not know, but in 1808, while at Falmouth, as an unsuccessful candidate for the post of engineer there, he was promised a post on board the naval gun brig of which his brother Peter Giles Pickernell had command. However, on the voyage from Falmouth to Spithead to take up this appointment the ship on which he sailed was captured by the French, and after nearly six years as a prisoner of war he died in France on 20 January 1814.[96]

His brother, who served in H.M.S. *Revenge* at Trafalgar, rose to the rank of Commander.[97] In a letter of 13 May 1814 Peter Giles referred to his nephew Francis Pickernell of Whitby. This is probably the Francis Pickernell (*c*.1796-1871) who was harbour engineer at Whitby from 1822 to 1861 and may well have been the son of Jonathan.[98]

Matthew Shout (1774-1817), appointed harbour engineer June 1804 and died in office March 1817

Matthew Shout, son of Robert the Sunderland harbour engineer[99] and grandson of Robert Shout of Helmsley, was living at Blyth in Northumberland before he became resident engineer at Scarborough harbour in October 1801.[100] There he worked for three years to the great satisfaction of the Commissioners and William Chapman their chief engineer.

Following Pickernell's dismissal from Sunderland, the Wear Commissioners advertised for a new engineer on 8 May 1804. Ten days later they decided to appoint Matthew Shout. He took up his duties on 30 June, at a salary of £120 with free rent of the Engineer's house. A year later he received £80 as a reward 'for his diligence and attention', and by 1811 his salary had been increased to £300 per annum.

Although the Sunderland Minutes are singularly uninformative about the progress of the works at this period, it is evident from a plan accompanying Jessop's report in 1807, and from the report itself, that Matthew Shout attacked the problems of improving the harbour with remarkable vigour. From the start he must have been convinced that the entrance was too wide, exposed as it was to all easterly gales. He therefore completed the removal of the South Pier extension and commenced building from the end of Robson's pier in a new, more northerly direction. By the end of 1807 he had reached point K (Fig. 5), extended the North Pier about 120 ft, using timber framing filled with stone work up to high tide level, and in addition he built the jetty EY (Fig. 5) as proposed by his father.

Where possible Shout used existing materials from the old South Pier extension, but the costs were obviously mounting rapidly, and in May 1807 the Wear Commissioners decided to make preparations for obtaining a new Act greatly to increase their revenue. In a report dated 30 June 1807[101] Shout recommended three necessary improvements, in the following order (i) further extension of the North Pier by about 300 ft, (ii) the construction of a dock or basin to hold 200 or 300 ships: 'an object of the greatest importance to the Shipping Interest', and (iii) dredging and deepening the harbour 'which cannot now take place on account of its present crowded state of Ships and Craft which has been the Chief reason of its failing when before attempted'.

In October the Commissioners agreed to authorize an extension of the North Pier by 150 ft (probably involving two seasons' work); but they then decided to have the whole planning of the harbour looked into by a leading consultant. On 25 November 1807 it was therefore ordered 'That the Engineer do request Mr. Jessop to attend to Survey the River, port and Haven and the works carried on therein, for the preservation and improvement of the Navigation', and a list of five questions was drawn up.

Jessop met the Commissioners on 2 December. Three weeks later he submitted *A Report on the Present State of the Piers of Sunderland Harbour and the Means recommended for their Improvement*. This was dated at Butterley, 19 Dec 1807, with a postscript of 12 Jan 1808. The report was printed at Sunderland. Jessop begins by giving a brief description of the harbour when Matthew Shout entered the Commissioners' service, and of the works carried out during the past three-and-a-half years. He says: 'So far what Mr. Shout has done, has been manifestly productive of advantage to the Harbour, and under the circumstances in which he found it, I do not know that he could have done better'. But clearly a critical decision now had to be made concerning the direction of the South Pier.

Jessop's advice was to swing it round from K in a line almost parallel to the North Pier, and in the first place to build as far as a point due south (by the compass) of the existing end of this pier. Both piers should then be extended by a further 400 ft, following the same lines and leaving a width of 250 ft between the pier heads.

The second question related to improvements within the harbour. Jessop expressed the opinion that dredging should be delayed until the piers were completed, but advised building a wall virtually identical to Pickernell's proposal with the addition of a pair of gates with sluices near point X in Fig. 4. In this way a basin would be formed, the water from which could be discharged at low water to aid in scouring the channel and bar.

On the third question, concerning a dock, Jessop felt unable to say anything encouraging. A dock with an entrance basin could certainly be built, but at a high cost, and there was still room above the bridge for a considerable extension of moorings. He gave some straightforward advice on the fourth question, for improvements of the upper part of the river, and under the fifth question, where he was asked to suggest any other measures of advantage, Jessop made the interesting proposal to reclaim a large area south of the South Pier by building a spur or groyne. Here we see the origins of a scheme which was later to be carried out with brilliant success in forming the ground on which the South Docks were to be constructed.[102] Finally, he gave careful estimates of the cost of the various proposals and strongly urged the Commissioners to include in the Act powers to open their own quarries for stone.

For this masterly report and for his expenses Jessop received £105. The Minutes of 31 December 1807, recording this payment, add that the Treasurer is also 'to pay to Mr. Shout ten Guineas [£10.50] for the Entertainment of Mr. Jessop at his House'.

Work began on the South Pier on Jessop's line in May 1808. By November the Bill had been drafted, and in February 1809 arrangements were made for three Commissioners, the Clerk and the Engineer to go to London to give evidence. The new Act

Plate 4 View of South Pier, Sunderland Harbour. Engraving by and after William Daniell, published 1822.

received the Royal Assent in May. By October 1810 the South Pier had been extended about 250 ft from K and a tide light erected near its eastern end (Plate 4).

Meanwhile, in September 1810, the Minutes record that 'The Engineer is to order of Mr. Jessop a Machine for Dredging the Harbour and send him such information as may be necessary towards erecting the same'. Evidently no time was wasted, as the Butterley Company completed the engine and machinery (including a rock-breaking chisel) for an 8 hp bucket-ladder steam dredger by July 1811, at a cost of about £1,500;[103] the equipment was then assembled in a vessel at Sunderland and the dredger had its first trial in October.[104]

The exact sequence of construction after 1810 is not easy to follow from the Minutes, but probably the masonry end of the South Pier, about 120 ft in length, was finished in 1813; thus completing the first stage of the extension from K as proposed by Jessop. Certainly in January 1814 Shout received orders to carry out the North Pier 'to an extent according to a plan by him', and the work seems to have reached point M by the end of the 1816 season, a total length of about 270 ft having been added on to the end of the pier built by Pickernell. Most, if not all, of this extension was built in timber framing, stone-filled to high-tide level, of the type already used by Shout before 1807 and approved by Jessop, with the modification that the facing stones should be 4 ft on bed, rather than 3 ft.

Considerable progress had also been made by 1816 with the groyne or jetty (DZ in Fig. 5); two cottages for the lighthouse keepers had recently been built on the North Pier, and the original lights were replaced in 1814 by nine Argand lamps placed in 18-inch diameter reflectors.

In his letter of 28 November 1816, quoted by Surtees,[105] from which some of the foregoing details are taken, Shout says that 'the Commissioners, aware, from the effect already produced, of the advantage likely to result, are proceeding, as fast as circumstances will permit, to carry into effect Mr. Jessop's plan to its fullest extent . . .' and indeed on 15 January 1817 the Minutes record that 'the Engineer's report on extending the North Pier having been read by him, Ordered that his plan for extending the same be carried into execution'.

Clearly, then, the Commissioners at the beginning of 1817 intended to proceed with further works on the piers. But Shout died on 14 March 1817, at the tragically early age of 42, and the impetus seems to have been lost—at least for several years.

By that time, however, a great deal had been achieved, and Sunderland now ranked as one of the major ports in Britain. In 1817, for example, nearly seven thousand vessels sailed from the harbour; about one million tons of coal, apart from other goods, were shipped;[106] the minimum depth of water in the channel and over the bar at ordinary spring tides was 17 ft;[107] and the sand bar at the entrance had been so much reduced in width that ships often preferred to use Sunderland, rather than the Tyne, as a haven.

As an independent consultant Matthew Shout advised on the reconstruction of Hartlepool pier. He reported in 1809, 1810 and 1812. His masonry specification is of interest as reflecting very probably the practice at Sunderland: facing stones not less than 4 ft on bed and 2 ft between joints, laid in regular courses backed with rubble stonework, the upper 6 ft of which is to be mortared.[108] Work began in 1811-12, supported by local subscriptions, and was completed under an Act obtained in April 1813.[109]

Of Matthew Shout's personal life, we know that he was born at Helmsley in 1774, as previously mentioned, and obviously came with his sisters and parents to Sunderland in 1781. Probably he received some training under his father and may well have been engaged on port or harbour works at Blyth before taking the Scarborough job in 1801. That his ability was exceptional is revealed by the following record in the Scarborough Harbour Commissioners' Minutes for 9 December 1802, written only just over a year after his appointment:

> Mr. Chapman having strongly recommended an Increase in Mr. Shout's Salary and having Minutely examined the mode of his carrying on the Works and believing the same to have been managed by him with great Economy and Industry, We do readily and with pleasure comply with the recommendation . . .

Moreover, he was scarcely thirty years of age when he took on the very responsible and demanding job at Sunderland. There he evidently acquitted himself to everyone's satisfaction; and it is worthy of note that the Wear Commissioners, in advertising for a successor,[110] go out of their way to refer to 'Mr. Shout, their much respected Engineer, lately deceased'. The chairman of the Hartlepool harbour Commissioners, writing in 1815,[111] also refers to 'Mr. Shout of Sunderland, a gentleman of great practical experience, and acknowledged judgment'.

He married Jane Dougal, daughter of a London merchant. They had one child, Robert, who was born in 1807 and died at the age of 17 as a midshipman in the East India service.[112] Matthew was buried beside his father in the churchyard of Holy Trinity, Sunderland.[113] He died intestate, and administration was granted to his widow.[114] She returned to London to live with her brother Commander George Dougal, R.N.

Acknowledgements

I am grateful to Mr. White of Sunderland Library for a great deal of help. My wife accompanied me on all the field work and contributed much in sketching, copying maps, etc. Mr. Ruddock of Edinburgh University kindly drew my attention to the documents concerning Stockton Bridge and Tyne Bridge. Finally, I wish to thank the Association for Industrial Archaeology for the honour of being asked to give the first L. T. C. Rolt Memorial lecture.

Note on Sources

The main primary sources, with their abbreviations, are as follows:
Sunderland Minutes, Wear Commissioners Minute Books (No. 1, 1747-1774; No. 2, 1775-1795; No. 3, 1795-1820).

Sunderland Accounts. Wear Commissioners Account Book 1747-1803.

Sunderland Parish Registers. The registers of Holy Trinity, Sunderland. These, together with a collection of engineers' reports and plans, were in Sunderland Reference Library but are now in the Tyne-Wear County Record Office [C.R.O.].

Originally there were Accounts of the Wear Commissioners for the period 1717-38 but these have been lost; indeed they were not available to John Murray in 1847.

Helmsley Parish Registers. The registers of All Saints, Helmsley. These are kept at the church and I was able to examine them by courtesy of the vicar, the Revd. D. G. C. Senior.

Durham Wills. Original wills and administration bonds proved in the Consistory Court of Durham. Department of Palaeography, University of Durham.

Scarborough Records. Scarborough Corporation Records: Elections & Orders, Books A.4 and A.11; Pier Papers, Box 32, F.2; Pier Accounts, Box 32, F.3. These cover the period 1731-52, and are kept in Scarborough Town Hall.

Scarborough Minutes. Harbour Commissioners Minute Book, No. 1, 1732-1833, Scarborough Reference Library.
J.H.C. Journal of the House of Commons.

All other references are given in full, including the location of manuscript sources.

Plans. A list of maps or plans of Sunderland harbour, covering the period of study, is given in Table 1. Original drawings or engravings include three large maps of the harbour and river by professional surveyors: Thompson 1736, Mackintosh 1792 and Giles 1819-22 (the harbour survey dating from 1819). There is an unsigned sketch plan dated June 1786, perhaps by Robert Shout, and two engraved plans of an explanatory nature, rather than accurate surveys, accompanying reports by Pickernell and Jessop.

In addition there are seven drawings, to a uniform scale of 200 ft to an inch, bound in a folio belonging to the Wear Commissioners, in the Tyne-Wear C.R.O. These show the harbour at different stages of development from 1718 to 1846; though unsigned, they were certainly prepared by John Murray in the latter year and used for the 600 ft to an inch engravings (some of which are not very accurate) in his 1847 paper to the Institution of Civil Engineers. Murray had access to several plans no longer available, the most important being Fawcett's survey of 1718 and Labelye's proposals of 1748. Figures 1 to 5 of the present paper are based essentially on Murray's 1846 drawings, with modifications introduced from the plans of 1786 and 1792 which he seems not to have known.

Views. Sunderland Museum has a fine collection of Victorian paintings of the harbour, but so far as I am aware there are only two views showing the piers as completed in 1817, and no earlier representations at all. (i) An aquatint drawn and engraved by William Daniell, published 1822 in his *Voyage round Great Britain*, vol. 6. The view (Plate 4) shows practically the whole of that part of the South Pier built by Matthew Shout between 1804 and 1813. Some alteration or repair work is in progress about half way along. A short length of the North Pier masonry can be seen, and Pickernell's lighthouse, but the outer end of the pier is mostly hidden behind the ships' rigging. (ii) A line engraving by George Hollis after a drawing by E. Swinburne, published 1823 and included as an illustration in Robert Surtees' *History of the County of Durham*. The view (Plate 3) is taken from a point near the end of the North Pier looking into the harbour at high tide.

Table 1. *Maps of Sunderland Harbour*

	Original		Engraving (e) or redrawn (r)		Location	Plate in Murray (1847)
	Surveyed	Scale ft/inch	Date	Scale ft/inch		
James Fawcett	1718		(r) 1846	200	RWC	24
			(e) 1858	600	Summers (1858)	
Isaac Thompson	1736		(e) 1737	600	Sunderland Library	
Charles Labelye	1748		(r) 1846	200	RWC	25
Robert Shout	1786	176	—	—	Sunderland Museum	
James Mackintosh	1792	50	—	—	Tyne-Wear C.R.O.	
Jonathan Pickernell	1797		(e) 1798	300	Pickernell (1798)	
			(r) 1846	200	RWC	27
Matthew Shout	1807		(e) 1808	200	Jessop (1808)	
			(r) 1846	200	RWC	28
	1817		(e) 1817	750	Garbutt (1819)	
Francis Giles	1819		(e) 1826	500	Sunderland Museum	
			(r) 1846	200	RWC	29

References

1 Biddick Ford is about eight miles upstream of Sunderland bridge and just over nine miles from the harbour entrance.

2 John Murray, 'An account of the progressive improvement of Sunderland Harbour and the River Wear', *Min. Proc. Institution of Civil Engineers*, vol. 6 (1847), 256-77. Murray was harbour engineer at Sunderland from 1832 to 1845.

3 Murray appears to have been unaware that this map was by Fawcett. However, the correct attribution is made by J. W. Summers, *The History and Antiquities of Sunderland* (Sunderland, 1858).

4 *J.H.C.*, 20 Feb. 1746/7, Burleigh's evidence.

5 *Durham Wills*, the will of James Fawcett of Sunderland near the Sea, dated 20 March 1720, proved 1721. He left all his estate to his wife Elizabeth.

6 *Sunderland Parish Registers* record the burial of 'James Fawcett Ingeneer' on 23 March 1720/21.

7 This date is given by Murray, loc. cit. p. 260.

8 J. S. Purvis, *Bridlington Charters, Court Rolls and Papers* London, 1926), pp. 194-95.

9 The Act (5 Geo. I) for rebuilding Bridlington South Pier received the Royal Assent in March 1718/19.

10 *J.H.C.*, 3 March 1726/7, Lellam's evidence.

11 *J.H.C.*, 20 Feb. 1746/7.

12 Murray, loc. cit. p. 260.

13 *Scarborough Records*, Book A.4, 10 Dec. 1731, reporting the appointment of 'Mr. Lelam for an Engineer'.

14 *J.H.C.*, 16 Feb. 1731/2, Lellam's evidence.

15 *Scarborough Records*, Box 32, F.2. 'William Lellam Engineer, Book of Accompts for Building the New Pier at Scarborough beginning at Christmas 1732'.

16 Parish Registers of St. Mary, Scarborough record on 3 June 1733 the burial of 'Mr. William Lelam Engineer'. I am indebted to the Revd. J. K. Fraser, Vicar of Scarborough, for this reference.

17 Tyne-Wear C.R.O., Wear Commissioners *Old Papers*, vol. 1.

18 *J.H.C.*, 29 Jan. 1732/3 and 6 Feb. 1732/3.

19 *J.H.C.*, 14 March 1736/7, Thomas's evidence.

20 *J.H.C.*, 20 Feb. 1746/7.

21 *J.H.C.*, 7 Dec. 1721, Thomas's evidence.

22 *J.H.C.*, 23 Jan. 1746/7.

23 *J.H.C.*, 20 Feb. 1746/7, Burleigh's and Thompson's evidence.

24 M. A. Richardson, *The Local Historian's Table Book*, vol. 2 (1843), 242. Thompson, a Quaker, died 6 Jan. 1776, aged 72.

25 *J.H.C.*, 14 March 1759, Thompson's evidence.

26 *Sunderland Minutes*, 23 Sept. 1747.

27 From 1738 to 1750 Labelye was employed on the design and construction of Westminster Bridge, but on occasions he undertook outside consulting work, notably for the Admiralty on Sandwich harbour (1744), for the Duke of Bedford in the Fens (1745), and on the harbours of Yarmouth (1747) and Sunderland (1748). He went to the south of France, for the sake of his health, in 1752 and died in Paris ten years later.

28 Sunderland Ref. Library, J. W. Corder, *Sunderland Family Records* (MS. *c.*1947).

29 *Sunderland Accounts*, 8 June 1750 'for building a Dredging Lighter £213.11.9'. Spencely's receipts 'for mens wages in taking up ballast' are entered from 25 June 1750.

30 *J.H.C.*, 14 March 1759, Spencely's evidence.

31 *Scarborough Records*, Book A.4, 3 Dec. 1734.

32 *J.H.C.*, 26 Feb. 1752, Vincent's evidence. He says that the pier extension 'was intended, by the Plan of the First Engineer, to be carried out Forty Yards further'. This of course refers to Lellam's plan of 1732.

33 The proposed East Pier is shown on a plan of the harbour among Smeaton's collection of drawings in the Royal Society library, vol. 5, f. 111v. From the stage of the works on the Old Pier extension the plan can be dated with certainty to *c.*1740. Smeaton was consulted in 1765; his own plan is in vol. 5, f. 109v.

34 *Scarborough Records*, Book A.11, 10 Sept. 1746.

35 William Chapman, *Report on the Antient, the Intermediate, and Present State of the Harbour of Scarborough* (Newcastle, 1829).

36 *A Plan of Scarborough (by) William Vincent Engineer for Building the New Pier, 1745*. Scale 132 ft to an inch.

II

37 *Sunderland Accounts*, 17 July 1747, 'Mr. Vincent Engineer for his Journey from Scarborough £29.8.0'.

38 *Sunderland Minutes*, 21 Dec. 1751.

39 John Smeaton, *An Historical Report on Ramsgate Harbour* (1791).

40 *J.H.C.*, 16 Jan. 1752.

41 *Scarborough Records*, Book A.21, 19 Nov. 1751.

42 *J.H.C.*, 26 Feb. 1752, Vincent's evidence.

43 *Sunderland Minutes*, 13 May 1752.

44 *Sunderland Accounts* for 1753 include numerous entries relating to payments by 'Mr. Vincent the Engineer, for deepening the South Channel'.

45 *Sunderland Parish Registers* record the burial of 'William Vincent, Engineer' on 16 April 1754. From the Wear Commissioners Minute Book it is certain that Vincent was working at least to the end of January 1754.

46 *Durham Wills*. The will of William Vincent, Engineer, in the Port of Sunderland, dated 12 March 1754, proved 1754. The will also mentions Vincent's eldest son, who was not yet twenty years old.

47 *Sunderland Minutes*, 22 May and 24 June 1754; and *Accounts*, 10 July 1754, 'Joseph Taylor for workmen deepening South Channel £33'.

48 *Sunderland Minutes*, 22 Jan. 1755.

49 *Sunderland Accounts*, 19 March 1755, 'Joseph Robson, gratuity in coming from London £31.10.0'.

50 *J.H.C.*, 14 March 1759, Robson's evidence.

51 Quoted by William Fordyce, *History and Antiquities of the County of Durham* (Newcastle, 1857), vol. 2, p. 500.

52 *J.H.C.*, 14 March 1759. Robson stated in evidence that 'the direction of the Pier has been altered, and an Addition made to it', and that he would further extend the pier.

53 North Yorks C.R.O., Cholmley Papers 2CG VIII/3/6, *The Case of the Petitioners the Coal Owners and Fitters of Sunderland* [1763].

54 Robson's report of 6 May 1767 is summarized by Murray, op. cit., pp. 264-65. The portions of the south mole actually built are shown in Fig. 3.

55 John Wooler, at this period harbour engineer at Whitby, had previously been consulted by the Wear Commissioners in 1763 on the probable effects on the regime of the river, resulting from the proposed construction of some new quay walls. He received £21 for his report (*Sunderland Minutes*, 1 Dec. 1763).

56 *Sunderland Minutes*, 29 Sept. 1767, 'Mr John Wooler to be paid £50 for his Trouble and Expence in coming to Sunderland to give his opinion concerning the Pier proposed to be Erected on the South Rocks. And his Report to be printed'.

57 *J.H.C.*, 14 March 1759, evidence by Robson and Smith, including estimates. John Smith senior, who built most of the locks on the Don Navigation, and had recently completed Naburn lock and dam on the Ouse, received a fee of £21 from the Wear Commissioners for his report (*Sunderland Minutes*, 8 Nov. 1758).

58 Charles Hadfield, *The Canals of Yorkshire and North East England* (Newton Abbot, 1972), vol. 1, p. 116.

59 The following information is taken from the Bridge Order Book, Treasurer's Accounts and List of Workmen employed on Stockton Bridge (Durham C.R.O.). Smeaton's charge for his work in 1760 and 1762 and a covering letter dated 7 Jan. 1764 are enclosed as loose sheets in the Order Book. He was paid in 1767.

60 Thomas Richmond, *Local Records of Stockton* (Stockton, 1868), pp. 72 and 75. The bridge is figured in a pleasant but not very accurate lithograph in John Brewster's *History and Antiquities of Stockton-upon-Tees* (Stockton, 1829). (See plate 1.)

61 M. W. Baldwin, 'The engineering history of Hull's earliest docks', *Trans. Newcomen Soc.*, vol. 46 (1973), pp. 1-12.

62 John Sykes, *Local Records*, vol. 1, p. 254 (Newcastle, 1833).

63 *Durham Wills*. Admon (1781, A49) of the estate of Joseph Robson heretofore of the Parish of Sunderland but last of the Parish of Bishopwearmouth. Dated 2 July 1781. The burial of Joseph Robson, age 76, is recorded in *Sunderland Parish Registers*, 29 Nov. 1780.

64 *Helmsley Parish Registers*, 'James son of Robert Shout' was baptized on 5 Feb. 1750/1. I have not found the date of his death.

65 The contract documents, dated 15 March 1774, are signed by Robert Shout the elder, Robert Shout the younger of Helmsley, masons, John Shout of Stockton, mason, and James Shout of Helmsley, joiner. (University of Durham, Dept. of Palaeography. ee Box 134, 220998.)

66 *Sunderland Minutes*, 1 Sept. 1779, recording the appointment of James Shout of Gateshead; the appointment was to terminate in December 1780.

67 *Helmsley Parish Registers*. The baptism of 'John son of Robert Shout' is entered on 23 Sept. 1738.

68 John Brewster, op. cit., reproduces a line engraving of the Market Place at Stockton with the column prominently displayed. Its date is given by Richmond, op. cit., p. 74.

69 *Sunderland Minutes*, 21 Dec. 1780, simply say that Mr. John Shout is appointed engineer at a salary of £100 p.a., with no mention of previous enquiries or advertisements.

70 Wyatt Papworth, *Dictionary of Architecture* (1887), vol. 7, gives the additional information that John Shout died on 13 June 1781, age 42.

71 *Helmsley Parish Registers* record the baptism of 'Robert son of Robert Shout' on 30 Dec. 1734.

72 Dates and contracts etc. for these and other North Riding bridges by Robert Shout and his son Robert are given in the *Quarter Sessions Order Books*. Drawings of all except Whitby Bridge are in John Carr's *Bridge Book* (*c.*1800). See also J. P. Utley, *North Riding Bridges* (MS notes). All these are in North Yorks. C.R.O.

73 *N. R. Quarter Sessions Order Book*, 19 April 1757, 'Ordered that Mr. Robert Shout be the Surveyor of all the Bridges, within this Riding on the East Side of Hambleton'. He held this position until 1765.

74 John Carr, writing about Newsham Bridge, *c.*1800, nearly fifty years after its completion, says it is 'well built and with very good stone and probably from the workmanship thereof by Robert Shout of Helmsley'. He was quite correct, as the *Quarter Sessions Order Books* prove.

75 *Sunderland Minutes*, 25 June 1781. The Commissioners appoint 'Robert Shout, in the room of the late John Shout deceased, to be Engineer and Superintendent of the Works at a salary of £100 p. annum'.

76 *J.H.C.*, 4 March 1785, Shout's evidence.

77 Robert Surtees, *The History and Antiquities of the County Palatine of Durham* (1820), vol. 2, pp. 1-2.

78 *Sunderland Minutes*, 12 Sept. 1787.

79 The order for this extension was minuted on 24 Sept. 1788.

80 *Sunderland Minutes*, 1 July 1793.

81 *Sunderland Minutes*, 6 Sept. 1793.

82 *Sunderland Minutes*, 21 Jan. 1795.

83 *Newcastle Courant*, 1 April 1797: 'Died, on Saturday last, Mr. Shout, late Engineer in Sunderland'. *Sunderland Parish Registers*, 28 March 1797, record the burial of Robert Shout. All headstones have now been removed, but their original positions and summary of inscriptions are given in a manuscript *List of Names recorded in Sunderland Parish Churchyard* (Sunderland Reference Library).

84 J. W. Corder, *Sunderland Family Records*. Corder suggests that there was an elder son, Robert, a grocer. This is probably correct, but there is no mention of him or his family in the will of Robert Shout the engineer.

85 *Durham Wills*. The will of Robert Shout, of Sunderland near the Sea, Engineer, dated 3 March 1797, proved 13 April 1798.

86 *Scarborough Minutes*, the joint report by Shout and Fordon is dated 30 April 1784.

87 Soane Museum.

88 *Sunderland Minutes*, 8 Oct. 1790.

89 A copy of his report to the Mayor and Aldermen of Newcastle, dated Sunderland 29 Feb. 1792, is among the Wear Commissioners' archives.

90 North Yorks. C.R.O., Whitby Harbour Trustees' Minutes.

91 This is the first of Ralph Dodd's numerous reports on civil engineering projects. It was printed at Sunderland in 1794.

92 Borthwick Institute, York. The will of Jonathan Pickernell, senior, dated 1 June 1812, mentions his two sons, Jonathan and Peter Giles and a married daughter, Mrs. Dickinson.

93 A. W. Skempton, 'A history of the steam dredger 1798-1830', *Trans. Newcomen Soc.*, vol. 47 (1975), in press.

94 Matthew Shout, in the letter of 1816 previously quoted, says that Pickernell added 90 ft to the North Pier; Murray gives 115 ft. But the *Sunderland Minutes*, 6 Nov. 1798, record the decision 'to extend the North Pier to the east by 100 feet', and this is probably the correct figure.

95 John Murray, 'An account of the removal of the light-house at Sunderland', *Min. Proc. Institution of Civil Engineers*, vol. 3 (1844), 342-55. This paper describes the remarkable feat of transporting the lighthouse bodily to the head of the rebuilt outer part of the North Pier in 1841.

96 W. Fordyce, *History and Antiquities of the County of Durham* (Newcastle, 1857), vol. 2, pp. 500-1.

97 The career of Peter Giles Pickernell is summarized in O'Byrne's *Naval Biographical Dictionary* (London, 1849). He died at Blackheath in 1859, age 87.

98 This is stated to be the case by H. B. Browne, *Chapters of Whitby History 1823-1946* (Hull, 1946), pp. 91-2; but Francis is not mentioned in his grandfather's will, and I have not found a record of his baptism or birth.

99 *Helmsley Parish Registers:* 'Matthew son of Robert Shout baptized 6 Oct. 1774'.

100 *Scarborough Minutes*, 21 Oct. 1801 record the appointment of Matthew Shout at a salary of £84. This was increased from 1 Jan. 1803, on Chapman's recommendation, to £105 p.a.

101 This handwritten report is among the Wear Commissioners' archives (Tyne-Wear C.R.O.).

102 John Murray, 'On the progressive construction of the Sunderland Docks', *Min. Proc. Institution of Civil Engineers*, vol. 15 (1856), 418-44.

103 Derbyshire C.R.O., Butterley Co. Ledger Book B.

104 John Sykes, *Local Records* (Newcastle, 1833), vol. 2, p. 64.

105 Robert Surtees, op. cit., pp. 1-2.

106 G. Garbutt, *A Historical and Descriptive View of the Parishes of Monkwearmouth & Bishopwearmouth and the Port and Borough of Sunderland* (Sunderland, 1819), p. 396.

107 From soundings taken in 1819 by Francis Giles.

108 The specification is included in his 1809 report, a copy of which is among the Wear Commissioners' archives.

109 Cuthbert Sharp, *A History of Hartlepool* (Durham, 1816), p. 138. See also Richmond, op. cit., pp. 116 and 120.

110 *Newcastle Courant*, 5 April 1817.

111 Cuthbert Sharp, op. cit., p. 137.

112 J. W. Corder, *Sunderland Family Records*.

113 *Sunderland Parish Registers* and *List of Names Recorded on Monuments in Sunderland Parish Churchyard*.

114 *Durham Wills*. Admon (1817, A39) of the estate of Matthew Shout of Sunderland, Engineer, dated 7 April 1817 and 4 April 1818.

III

Early Members of The Smeatonian Society of Civil Engineers

BY

A. W. SKEMPTON, F.R.S. (Vice-President) and
ESTHER CLARK WRIGHT, Ph.D.

Introduction

Notes are given on all members of the Society elected before 1800 who have been identified with certainty and practised as engineers or engineering surveyors. There are thirty-eight of them and their works span a period extending roughly from 1750 to 1820. In addition, four other members who had particularly close connections with engineering are included.

The profession of civil engineering, as distinct from the much older calling of the military engineer, may be said to have been formally recognised for the first time in this country by the foundation of the Society in 1771. Under the leadership of John Smeaton, the greatest engineer of his day, the membership included a highly representative cross-section; resident engineers, surveyors and consultants with a small practice as well as such famous men as Rennie, Watt, Jessop, Mylne and Smeaton himself. Among the members there were also several engineers, obviously important, but now almost forgotten despite the fact that they contributed much to the profession in its early years. This interesting group includes Thomas Yeoman, the first President of the Society; John Grundy, a friend and colleague of Smeaton; and Robert Whitworth, a very able canal engineer who, as a young man, was Brindley's principal assistant. In the next generation William Chapman is another engineer strangely neglected by historians. He displayed considerable talents and versatility, and had an extensive practice in the north of England. There is ample scope for further research on these, and indeed on practically all the members of the Society mentioned in this Paper.

In the notes the name of each member is prefixed by the year of his election and is followed, where relevant, by a statement of any special connection with the Society. Well established dates of birth and death are given without qualification. A *circa* date indicates that the year of birth is deduced from clear evidence, such as age at a known date of death. Less certain dates are denoted by the addition of a question mark.

Part I. Original Members

At the first meeting of the Society, held on 15 March 1771 at the King's Head Tavern, Holborn, John Smeaton and six others were present. Thomas Yeoman took the chair and it was

> "Agreed that the Civil Engineers of this Kingdom do form themselves into a Society consisting of a President, Vice-President, Treasurer and Secretary and other Members who shall meet once a fortnight . . . at seven o'clock from Christmas . . . to the end of the sitting of Parliament."

The Minutes of this meeting were written up by Yeoman, certainly before 12 April 1771, in a book which he purchased and which still exists. On the cover of this book appear the words

EARLY MEMBERS OF THE SMEATONIAN SOCIETY

An°. 1771 *Society of Civil Engineers*

At the next meeting on 29 March 1771 Robert Mylne was chosen Vice-President and three more members were admitted. A fortnight later another member joined, bringing the number of original members to eleven. No further elections took place for nearly a year.

1771. THOMAS YEOMAN, F.R.S. (1708(?)–1781)

President 1771–80 and Treasurer 1771–77. He also wrote up the early Minutes and compiled the first list of members.

Yeoman erected the machinery in the world's first water-powered cotton mill. The mill, at Northampton, on the river Nene, came into operation in 1743, using machines of the type invented by Lewis Paul. Yeoman remained in Northampton until 1757 surveying enclosures and turnpike roads, making agricultural machinery and installing a weighbridge (*c.* 1748) in accordance with plans by his friend John Wyatt (1700–66). He also established something of a national reputation for building and installing the very successful ventilators invented by Dr. Stephen Hales, in hospitals, prisons and ships, and some of his own design in the Houses of Parliament.

He joined the Northampton Philosophical Society shortly after its formation in 1743 and was President of that Society several times before he moved to London. In 1760 he became a member of the Society for Encouragement of Arts, Manufactures and Commerce, and in 1763 was chosen to be joint chairman of the Committee on Mechanics. This position he held until 1778. He was elected F.R.S. in 1764.

While still at Northampton, Yeoman became involved with at least three civil engineering projects. In 1753 he surveyed and made an estimate for extending the navigation of the River Nene from Thrapston to Northampton, and in 1756 he attended the Parliamentary proceedings for the Act, under which the work was carried out. In 1754 he surveyed the proposed Stroudwater Canal and three years later he surveyed the River Ivel. From about 1762 he was engaged extensively on work connected with rivers and canals. In that year and again in 1765 he submitted plans for the River Chelmer navigation. Also in 1762 he and Joseph Nickalls made a survey of the River Lea, in the neighbourhood of Waltham Abbey. Yeoman's first report on the River Stour in Kent, for improvements in land drainage of the adjacent low grounds, is dated 1765; and he was consulted during the execution of the works there (1776–78). In 1766 he reported on the Ancholme drainage scheme. In conjunction with other engineers (Mylne, Smeaton, Nickalls, Wooler, Whitworth, Erskine) he was consulted by the City of London on problems connected with the Thames. We have noted such reports and plans dated 1767, 1770, 1774 and 1777.

Meanwhile, in July 1766, Yeoman was appointed to assist Smeaton in surveying, planning and estimating for major improvements in the River Lea navigation from the Thames up to Hertford. Later that year he proposed several alterations which were approved, including the Limehouse Cut. The Act was obtained in June 1767 and Yeoman then carried out the works as engineer until 1771. Although the records are scanty, it is certain that Yeoman was also engineer for the River Stort navigation (1766–69). While these works were in progress he found time to report on the Forth and Clyde Canal, in September 1768, and to make a survey of the Nene outfall below Wisbech in 1769. Towards the end of that year he acted as consulting engineer on the Coventry Canal, and again in 1771. He prepared estimates for the proposed Leeds and Selby Canal in 1773 and finally, apart from the River Stour improvements already noted, he drew up the plans and saw the Act through Parliament for the Stroudwater Canal, which was built in accordance with his designs 1776–79.

The Society's Minutes on 28 April 1780 record that "Unfortunately our worthy President, being ill and not capable of attending—in hopes of his future attendance, we postpone this meeting to this day fortnight being the 12 of May". In fact Yeoman was not seen again at the meetings of the

Society, the formation of which may have been due to his suggestion. He died in January 1781. It is interesting to note that in Mortimer's *Universal Director* (1763) Yeoman is described as "Surveyor and Civil Engineer": an extremely early use of the latter term.

1771. JOHN SMEATON, F.R.S. (1724–92)

The Society was formed in 1771 with Smeaton's approval and active support. He attended the meetings very frequently until the end of the 1792 session, a few months before he died. Sir Joseph Banks, a member of the Society, then purchased all of Smeaton's manuscripts, reports and drawings. The drawings are in the Royal Society Library, and the Society of Civil Engineers, under Mylne and Rennie, published the entire set of reports in three large quarto volumes (1797 and 1812). No other engineer has been honoured in such a way. In 1830 the Society added the prefix "Smeatonian" to its title. The fine portrait of Smeaton, painted by Romney, was purchased by subscriptions from the members in 1926, and very recently, to mark the duo-centenary of the first meeting, a senior member presented to the Society an exquisite model in silver and perspex of the Eddystone lighthouse.

Smeaton's first engineering essay, on a method for re-building the sunken pier of Westminster Bridge, dates from 1748. In 1752 he began his classic research on the power of water and wind. The results were published in *Phil. Trans.* 1759, in a paper which initiated the experimental study of fluid mechanics and which gained for its author the high distinction of a Copley Medal. During the summer of 1755 Smeaton spent two months in the Low Countries chiefly examining civil engineering works. In January 1756 he was recommended by the President of the Royal Society as the most suitable person to build a new lighthouse on the Eddystone rock. This work he completed in a masterly manner in 1759, creating the protype of all subsequent masonry lighthouses built in the open sea.

From 1759 until 1783 Smeaton was engaged as a consulting engineer in England, Scotland and Ireland, reporting with remarkable clarity and thoroughness on a wide range of subjects and carrying out a distinguished series of engineering works. Among these we may mention the Calder and Hebble navigation (1759–64 and 1769–70) and the Forth and Clyde Canal (1768–78, completed 1785–90 by Robert Whitworth); three fine masonry bridges, at Coldstream on the River Tweed (1763–66), at Perth on the River Tay (1766–71) and at Banff (1772–80); the lighthouse, built 1773–76, on the sands at Spurn Point; and the following harbours constructed during the period 1766–84: St. Ives, Eyemouth, Aberdeen, Portpatrick and Cromarty, in that chronological order. As a mechanical engineer he designed numerous mills, starting in 1753 with a watermill at Halton, near Lancaster, and including the great waterwheel for London Bridge waterworks (1767–68); he also designed machinery for the Carron Ironworks (waterwheels, blowing engines and a boring mill) chiefly in the years 1768–71; and his improvements in the steam engine, before Watt came on the scene, found practical expression notably in two engines, of about 40 h.p. and 70 h.p. respectively, at Long Benton Colliery (1773) and at Chasewater Mine (1775).

Smeaton's largest bridge, a noble 9-arch structure at Hexham over the River Tyne, built 1777–80, was destroyed by a monstrous flood in 1782. Next year his health deteriorated. The combination of these events led to a decision to enter upon no new undertakings, and from 1784 he devoted much time to writing a book on the Eddystone lighthouse (published 1791) including an account of his research on hydraulic limes, which led to the first basic discovery in cement chemistry. Nevertheless his advice in this period was sought, and given, on various schemes. For instance, in 1787 he was persuaded by his friend Alexander Aubert to accept the position of engineer for the Ramsgate harbour works. Also, in 1790 he carried out surveys for several new lighthouses, at the request of Trinity House.

EARLY MEMBERS OF THE SMEATONIAN SOCIETY

Smeaton lived and worked for most of the time after 1759, at Austhorpe Lodge, his house in Whitkirk near Leeds, where he had been born. His two daughters, Ann and Mary, frequently helped their father by finishing the drawings in indian ink and wash. He was associated professionally on several occasions with John Wooler and Thomas Yeoman, and had particularly close contacts with John Grundy. His pupil and assistant, William Jessop, became a member of the Society in 1773. Another assistant, Henry Eastburn, was elected in 1789. Joseph Nickalls, one of Smeaton's early resident engineers, was a founder member of the Society, and President 1783–92. Robert Mackell (d. 1779), his resident on the Forth and Clyde Canal, took over from Smeaton as engineer of that work in 1773. John Gwyn deserves special mention as resident engineer on six of Smeaton's major works, from 1766 when he made the site investigation borings for Perth Bridge, to his death in 1789 while still engaged on the Ramsgate harbour works.

Smeaton was beyond question the most eminent civil engineer in the eighteenth century. He combined, to a degree which has rarely been equalled, great intellectual powers and outstanding practical ability. The engineers of this country are proud to acknowledge him as the founder of their profession.

1771. Robert Mylne, F.R.S. (1733–1811)

Vice-President 1771–77, 1790–92. Treasurer 1791–1810. Principal member of the committee of four, which reorganised the Society in 1793 and of the special Committee for the publication of Smeaton's *Reports.*

Mylne came of a long line of Scottish master masons and architects. After his training in Edinburgh he spent four years on the Continent, chiefly in Italy. He returned to London in July 1759 just in time to enter the open competition for designing a new bridge over the Thames at Blackfriars. This he won, against such strong competitors as John Gwynn,* George Dance senior, Smeaton, and Sir William Chambers. The bridge was built 1760-69 under his personal supervision, and ranks as one of the masterpieces of civil engineering.

After Blackfriars, his most notable bridge was the handsome structure over the River Tyne at Newcastle, designed 1772–74 and built 1774–81, with John Wooler as joint engineer. Mylne advised on the rebuilding of Hexham Bridge (which was completed in 1793, faithfully following Smeaton's original architectural scheme). Smaller bridges designed by him include the graceful Aray Bridge at Inverary (1773–75), Romsey Bridge over the River Test (1782–84) and Ridley Hall Bridge on the South Tyne (1788–93).

Mylne achieved distinction as an architect, as well as a bridge engineer. His buildings displayed perfect proportions and faultless detail. Some of the best known are Tusmore House, Oxfordshire (1766–69), the City of London Lying-in Hospital (1770–73), The Wick, Richmond Hill (a beautiful house built in 1775 for Lady St. Aubyn) and the fine east front of Stationers' Hall, London (1800–2). For many years he held the post of Surveyor to St. Paul's Cathedral, where he is buried near Wren.

Though possessed of a talent amounting to genius as a structural and architectural designer, Mylne devoted a great deal of his time to hydraulic engineering. As early as 1771 he was appointed engineer to the New River Company; in which post he was active for the rest of his life. He advised on at least four harbour schemes and, jointly with Rennie and Capt. Huddart, he planned the London Docks in 1801. He wrote excellent reports on the Thames navigation, between Lechlade and Abingdon (1791 and again in 1802) and was consulted on repairs to the Sapperton tunnel on the Thames and Severn Canal in 1790.

Two major works were, for various reasons, not completed in his life-time: the Gloucester and Berkeley ship canal, for which he was engineer from 1794 to 1797, and the Eau Brink Cut on the

* John Gwynn, R. A. (d. 1786), not to be confused with Smeaton's resident engineer John Gwyn (*c.* 1733–89).

River Ouse above Kings Lynn. For the latter he got as far as laying out the work, but, in the event, construction did not begin until 1818 under John Rennie. Part, but only a part, of the trouble on these (and on some other) jobs arose from Mylne's irascible temperament. One of his workmen said that "Mr. Mylne was a rare jintleman, but as hot as pepper and as proud as Lucifer". Nevertheless, though at times very critical of his contemporaries, he enjoyed their company and he did a great deal for the Society.

His son, William Chadwell Mylne, F.R.S. (1781–1863), succeeded him as engineer to the New River Company and succeeded Rennie as Treasurer of the Society; a post which, in turn, was held by his grandson, Robert W. Mylne, F.R.S. (1817–90).

1771. JOHN GRUNDY (1719–83)

"John Grundy of Spalding, Engineer", as he usually described himself, was the son of John Grundy senior (*c*. 1696–1748), land surveyor, teacher of mathematics and one of the early civil engineers in England.

Our first note on Grundy refers to his survey of the Trent near Newark in 1741. Next year, under his father's direction, he carried out repairs of breaches in sea walls (embankments) on the Norfolk coast south of Happisburgh. In 1749 he was employed by the Earl of Lincoln to report on the drainage of Borough Fen near Crowland. Two years later he built a sea-sluice on the south side of the Humber Estuary at Stallingborough. These were typical of his smaller works, which also included water supply schemes, notably at Grimsthorpe for the Duke of Ancaster.

In 1743 he and his father made a detailed survey of the River Witham from Boston to Lincoln and in the following year they published a report on means for improving navigation on the river and draining the adjacent lowlands. After long discussions, so typical of Fen drainage in the eighteenth century, further action was taken in 1753 when Grundy put forward another set of proposals. Langley Edwards in 1760 recommended a different situation for the main sluice. Smeaton was then called in, and the three engineers wrote a joint report in 1761 to which was attached a map by Grundy made early in 1762; the Bill followed in May. The principal works were completed about 1770, with Edwards as engineer, the Grand Sluice at Boston being in the position indicated in the joint report and a long cut on the river following essentially the lines proposed by Grundy in 1753. This shortened the course of the river between Boston and Chapel Hill from 13½ miles to 11 miles. The sluice at Boston, with three pairs of pointing sea doors and a navigation lock, was probably the largest structure of its kind so far built in England.

Another important scheme involved the drainage of low grounds in Holderness, east of the River Hull. Smeaton and Grundy visited the area in 1763 and Grundy drew up plans to which Smeaton suggested some alternative modifications. Grundy attended the Parliamentary proceedings early in 1764, when the Act was obtained. The first stage of the works, completed in 1767, under Grundy's direction, included a new main drain with an outfall sluice just upstream of the bridge at Hull, and raising the eastern bank of the river over a length of many miles. Joseph Page* was resident engineer and Charles Tate did the surveying work.

Grundy was also engineer for the Driffield Navigation (1767–70) and for the first part of the Market Weighton Canal (1772–77). Samuel Allam of Spalding worked as resident engineer on both those jobs. In 1770 Grundy presented a decisive opinion on the proposed cut to improve the Nene outfall below Wisbech. He was planning engineer for the first dock at Hull, the largest commercial dock of the eighteenth century, (built 1775–78) but for its construction Henry Berry (*c*. 1720–1812) was consulting engineer with Luke Holt, on Smeaton's recommendation, as resident. Grundy's

* Probably the Joseph Page who joined the Society in 1774; but no address of this member is given in the Minute Book.

last major report which we have been able to trace proposes a scheme for draining East Fen and adjoining lands near Wainfleet (November 1774).

John Grundy was perhaps the first civil engineer in this country to have been trained for the profession and to work as a consultant, purely in civil engineering, throughout the greater part of his career. He was a personal friend of Smeaton, who had a high regard for his abilities.

1771. Joseph Nickalls (1725(?)–1793)
President 1783–92.

Nickalls began as a millwright, and in 1758 was sufficiently well known to be asked, along with Thomas Yeoman, to give an expert opinion on mill-work to the Society of Arts. In 1759 Smeaton brought him up from London to work as resident engineer on the Calder and Hebble navigation; a post he held until 1761. Nickalls built the watermill pumping engine at Stratford, on the River Lea, in 1763, to Smeaton's designs. As Smeaton's "Instructions for Mr. Nicholls to proceed with the Waterwheel and Spurwheels for the great Engine . . . at London Bridge" are on record (1767) it is safe to assume that he built this too. We may remark that his name was often spelt, or rather misspelt, in this way by his contemporaries; even by those who knew him well.

In 1762 he was surveying on the River Lea, and again in 1766 and 1771, on behalf of the mill owners. For the Thames Commissioners in 1770 he surveyed the river and drew up plans for improving the navigation between Maidenhead and Reading. In February 1771 he successfully presented the case in Parliament against opposing canal schemes. The works, including eight locks, were duly carried out (1771–73) under his direction, as General Surveyor to the Commissioners. His report on Dover Harbour (1777) contains an early use of the term "civil engineer". He also reported on Wells Harbour (1782) and Yarmouth Haven (1783); on the River Severn (1786) and, jointly with Robert Mylne, Capt. Huddart and Sir Thomas Hyde Page, on Margate Harbour (1788). In 1787 he was consulted by the Bristol Society of Merchants and as a result put forward a proposal for turning the River Avon into a floating harbour with a dam and entrance locks at Black Rock. This was followed by reports from Jessop (1788) and Smeaton (1789), both of whom, while attracted to Nickalls' scheme, favoured a dam at Rownham Mead, about 1½ miles further upstream. In September 1789 all three engineers visited Bristol together, and eighteen borings made later that year at the Rownham site were presumably carried out at their request. In 1790 Jessop reported on the dam and entrance locks, and Nickalls considered in some detail the problems of land floods and silting. Nothing directly came from all this, however, and it was not until 1804 that work actually began on the Floating Harbour, under Jessop's direction, in accordance with a different set of plans.

1771. John Golborne (*c.* 1724–83)

Golborne lived at Chester, and for many years was engineer to the River Dee Company. In this capacity he continued the works of river training and land reclamation begun in the 1730s by Nathaniel Kinderley (1673–1742). He was consultant for improvements on the River Weaver Navigation (1759–60) and in 1768 was recommended to the Forth and Clyde Canal proprietors, by Brindley, as "the properest person for considering the navigation of the rivers Carron and Clyde". Immediately after reporting on these matters he was asked by Glasgow to prepare a scheme for deepening the river along a length of twelve miles below the city. This proved to be an outstanding success and on its completion in 1775 he was presented with a silver cup.

Meanwhile in 1769 Golborne and Yeoman both reported on the drainage of the North Level and on the Nene outfall, followed by Grundy in 1770. As a result a new cut below Wisbech was made in 1773–74, with Golborne as consulting engineer, the work being carried out by Richard Dunthorne and William Elstobb. It was named "Kinderley's Cut" in honour of the same Nathaniel Kinderley who first proposed a scheme of this kind in 1721.

Golborne again appears in the Fens, with an able report on the Middle and South Levels and the Ouse outfall (1777), based on surveys made at his request by Robert Whitworth. In the previous year his nephew and former pupil, James Golborne, who had been his assistant on the Clyde works, was appointed Superintendent General to the Bedford Level Corporation. James Golborne became a member of the Society in 1793.

1771. Robert Whitworth (*c.* 1733–99)

A member of the committee of four who reorganised the Society in 1793.

From 1767 to 1772 Whitworth was Brindley's principal assistant. He drew the maps and probably made the surveys for the Coventry Canal (1767), the Birmingham Canal (1767) and the Oxford Canal (1768). He surveyed the line of the Leeds and Liverpool Canal in 1768 when Brindley was checking John Longbotham's plans, and again in 1769. His skill can be seen in the plan, profile and soundings of the Thames from Maidenhead to Mortlake (1770) and his draughtsmanship is displayed in the well known map of Brindley's canals, published in 1769. Already in 1768, with Brindley's approval, Whitworth submitted his first independent report, on the Lagan Navigation in Ireland.

After 1772, when Brindley died, Whitworth developed a substantial practice. For the city of London he surveyed and reported on projected canals from Waltham Abbey to Marylebone (1773) and from Bishop's Stortford to Cambridge (1779–80). He made the first plans for the Herefordshire and Gloucestershire Canal (1777) and for the canal from Ashby-de-la-Zouch to Griff (1781). On the occasion of his survey of the Ouse from Denver to the sea in 1777, John Golborne referred to "Mr. Whitworth, who levelled for that great Genius Mr. Brindley, and who has had more Experience in that Branch than any Man of his Profession". The Thames and Severn Canal was built in accordance with his design, made in 1782, and he personally set out the summit-level with the 2 mile long Sapperton Tunnel. Josiah Clowes (d. 1795) was construction engineer, and the canal was opened in 1789.

In 1785 Whitworth entered the third phase of his career when he became chief engineer of the Forth and Clyde Canal. Work on that important undertaking had come to a halt in 1778, due to lack of funds, but by 1784 sufficient capital had been raised for its completion. Whitworth re-planned the western part, on which the Kelvin aqueduct bridge was built (1786-89) to his design and under his supervision; 400 ft. in length and rising 70 ft. above the river, this was the greatest work of its kind in Britain or Europe at that time. He was elected an honorary burgess of Glasgow, and the city of Edinburgh took advantage of his presence in the locality to obtain from him a report on Leith harbour (1786).

His next big construction job was on the Leeds and Liverpool Canal. This, too, had come to a halt but in 1789 preparations were under way for obtaining a new Act for its continuation. Whitworth came down from Scotland and reported in October. The Bill passed in 1790. He was appointed chief engineer; work began early in 1791, and he planned a new route to the west of the summit past Burnley. By 1796 Foulridge tunnel had been completed. When he died in 1799 the canal was practically at Enfield, an extension of about 30 miles then having been completed by him, which included, as well as the tunnel, the first of the summit-level reservoirs and a massive embankment at Burnley.

Whilst engaged on these works Whitworth reported again on the Ashby Canal, and in 1793 he helped his son William on the surveys and design of the Wilts and Berks Canal. By 1795 he was able to leave part of the Leeds and Liverpool work to Samuel Fletcher, the resident engineer, accepting a lower salary in consequence, and agreed to take over direction of the Herefordshire and Gloucestershire Canal after the death of Josiah Clowes, who had been in charge since 1793. The canal was completed from Gloucester as far as Ledbury in 1798.

His brief obituary notice in the *European Magazine* (April 1799) refers simply but very properly to "Mr. Robert Whitworth, one of the most able engineers in England".

His son Robert Whitworth junior, had worked with him on the Forth and Clyde, and, with his father, was engineer of the Ashby Canal from 1794 to 1797. William Whitworth was assistant on the Leeds and Liverpool, and after his father's death seems to have been retained as a consultant. He built the Wilts and Berks Canal (1795–1810).

1771. HUGH HENSHALL (*c.* 1734–1816)

In 1765 James Brindley married Henshall's young sister Anne, and from that year, if not earlier, Henshall was closely associated with the celebrated canal engineer. Probably trained by his father, a land-surveyor, Henshall took part in the surveys for the Staffs and Worcestershire Canal and he drew the parliamentary plan for the Trent and Mersey Canal. On the latter he was resident engineer, under Brindley, for the first six years of construction (1766–72). After Brindley's death he succeeded to the post of chief engineer on the Trent and Mersey and also became consulting engineer to the Chesterfield Canal. Both works were successfully completed under his direction in 1777. Later, Henshall carried out several surveys for canals and river navigations, but probably devoted much of his time to the business affairs of Hugh Henshall and Co., the carrying subsidiary of the Trent and Mersey Canal Company. He also had an interest in a pottery at Longport.

PART II. MEMBERS ELECTED 1772–92

The Society soon increased to an effective membership of about twenty, at which strength it remained for many years. Among the new members, as well as engineers and surveyors, noted below, we find instrument-makers and other craftsmen. In the Minutes for 3 May 1782 we read that

> "This evening was spent with the usual harmony and very politely presented with a Treatise called an Experimental Enquiry concerning the Natural Powers of Water and Wind by our worthy Member Mr. Smeaton, and Mr. Walford as politely presented this Society with Englands Improvement by Sea and Land by Yarranton."

This was the beginning of what became a substantial collection of books, drawings and reports. The greater part of this collection was eventually given to the Institution of Civil Engineers in 1845. Useful lists of members were drawn up from time to time, notably in 1786. Visitors first attended the dinners by invitation in 1781.

1772. JOHN SMITH

John Smith, elected to the Smeatonians in 1772, was John Smith jnr. of Attercliffe in Yorkshire, the engineer-contractor of the Western Division of the River Nene Navigation (1758–61). This extended from Thrapston up to Northampton, and included twenty locks. Thomas Yeoman, who had given evidence on the Bill before Parliament in 1756, was called in by the Navigation Commissioners to approve Smith's plans and also to examine the works in progress. Smith had been trained by his father, John Smith senior, engineer of the River Don Navigation (1731–51) and they were both engaged in building Naburn Lock on the Yorkshire Ouse *c.* 1755. John Smith jnr. later carried out works on the Ouse above York, including Linton Lock and Dam, on which he sought Smeaton's advice, and on the Swale; though the latter navigation was not completed. There can be little doubt that he was the John Smith who reported jointly with James Creassy on the River Witham in 1776. According to the Society's Minute Book he died before 1784.

III

EARLY MEMBERS OF THE SMEATONIAN SOCIETY

1772. JOHN HOLMES (*c.* 1727–1797)

A kinsman and life-long friend of Smeaton, and his first biographer, Holmes had been trained by Henry Hindley of York, and became a well-known clockmaker in London. Though not an engineer, he is included here as he was associated with Smeaton in several of his water-works undertakings (e.g. at Deptford and York) and was an able "mechanician". In 1780 he received a gold medal from the Society of Arts for his invention of a valve water-cock.

1773. LANGLEY EDWARDS (*d.* 1774)

In 1757 Langley Edwards gave evidence to a House of Commons Committee on the River Ivel. He was consulted on the River Witham improvements and reported in 1760. With Grundy and Smeaton he was joint author of a report on the Witham made in the following year, and it was Edwards who carried out the work. He had the honour of formally opening the Grand Sluice at Boston in 1766.

In the winter of 1763 the worst floods ever remembered covered the fens west of Boston, known as the Black Sluice District. Langley Edwards put forward a scheme (1764) for improving the drainage, which was accepted, and work began under his direction, probably soon after the Act had been obtained in 1765. Four years later he seems to have been replaced by Edward Hare. A report by him on the Middle and North Levels of the Fens is dated 1771. He attended his last meeting of the Society in April 1774 and died in the summer of that year.

1773. JOHN GOTT (1720–93)

For about 30 years, until his retirement in 1792, Gott was resident engineer on the Aire and Calder navigation. He also built several bridges, including one at Hexham (1767–70) which suffered badly in a disastrous flood of the River Tyne in November 1771.

Gott lived at Calverley, near Leeds. He rebuilt the school there. His son William (1745–1810) followed him as a bridge engineer and on the Aire and Calder. His younger son, Benjamin Gott (1762–1840), became one of the leading woollen manufacturers in Leeds.

1773. JOHN LONGBOTHAM (*d.* 1801)

Longbotham is important as the planner and first engineer of the Leeds and Liverpool Canal. His surveys were made in the years 1767–68. Brindley, with Whitworth's able assistance, then checked the route and estimates. After further investigations of an alternative route on the Lancashire side, the Act was obtained in 1770, essentially in accordance with Longbotham's plans. During the next five years he was responsible for all the engineering works, designing the locks and aqueducts, preparing the contracts and supervising construction. By 1774 over fifty miles of canal had been built, from Liverpool to Gathurst and from Gargrave to Shipley, with a branch canal to Bradford for which he was consultant. Funds were now beginning to run out. Longbotham left in 1775, apparently after a dispute with the proprietors, and the comparatively short lengths still left to reach Leeds and Wigan were completed in 1777 and 1781 respectively. Work on the summit-level did not begin until 1791, under Robert Whitworth.

Our knowledge of Longbotham's career after 1775 is fragmentary. He died in poverty in 1801, when the canal company paid his funeral expenses.

1773. JOSEPH PRIESTLEY (1741–1817)

Born near Halifax, Priestly joined the staff of the Leeds and Liverpool Canal in 1770. Two years later he moved to Bradford, as superintendent of the Canal, and remained in this post until his death at the age of 76. He attended the Society's meetings up to 1809.

EARLY MEMBERS OF THE SMEATONIAN SOCIETY

1773. William Jessop (1745–1814)

Secretary 1774–92. One of the committee for reorganising the Society in 1793 and a member of the special Committee for the publication of Smeaton's *Reports*.

Josias Jessop, William's father, who had been a foreman shipwright in Plymouth Dockyard, took practical charge of building the Eddystone lighthouse (1756–59) under Smeaton's direction. He died in 1761. His son became Smeaton's pupil in 1759 and assistant in 1768. In that capacity we hear of him surveying on the River Ure Navigation in 1770–71 and again in 1773 when he accompanied Smeaton to Ireland in connection with the Grand Canal, but he must have been associated with the great engineer on many jobs in the preceding six years.

Jessop's first works on his own were carried out under the Act of 1774 for the Selby Canal (completed 1778) and for improvements on the River Aire navigation (completed 1785) with John Gott as resident engineer. At the same time Jessop had responsibility for works on the Calder and Hebble (1776–83). Between 1783 and 1787 he was engineer for improvements on 67 miles of the Trent, from Wilden Ferry to Gainsborough, and thereafter held the position of permanent engineer to the navigation company. Meanwhile, as a consultant, he had been engaged on the Loughborough and on the Thames navigations, and had reported on harbours at Workington and Rye, among other projects.

With his reputation already established, Jessop now entered a period of intense activity in canal engineering. In chronological order, he was engineer for the Cromford Canal (1789–94, Benjamin Outram as assistant), the Leicester Navigation (1791–94, Christopher Staveley resident engineer); the Barnsley Canal (1793–1802, Samuel Hartley resident engineer); the Grand Junction Canal (construction proceeded from 1793–97 with Jessop as engineer and James Barnes as resident; it was completed 1797–1805 with Barnes as engineer and Jessop as consultant); the Ellesmere Canal (Jessop engineer and Telford resident engineer 1793–95, Telford then took over as engineer 1795–1805 with Jessop as consultant until 1801); and the Rochdale Canal (the line planned by Rennie; Jessop laid out the works, which were built under his direction 1794–1804 with William Crosley as resident engineer, followed by his son William Crosley jnr.). In Ireland he was consultant to the Grand Canal from *c.* 1789, and for the canal docks at Dublin (*c.* 1792–1796). Even this formidable list is not complete, and three of his most important works were still to come.

In 1799 the West India Dock company was formed. Jessop and Ralph Walker were appointed engineer and resident engineer, respectively. Under their joint efforts the great Import Dock, of 30 acres extent and a depth of 29 feet, reached completion by September 1802. The 25-acre Export Dock alongside followed in 1805. Two large basins with the necessary tide-locks provided entrances at Blackwall and Limehouse, and warehouses (designed by Joseph Gwilt) ranged along the dock sides. Nothing on this scale had been seen before.

In June 1801, when works at the West India Docks were well under way, Jessop was appointed engineer for the Surrey Iron Railway: the first public railway in the world, sanctioned by Parliament, and quite independent of a canal. His contributions to early railway engineering began in 1790 when he advised that railways should be built on the steep gradient portions of the proposed Forest Line of the Leicester Navigation between Loughborough and Coleorton. These were constructed under his direction 1793–94, using cast iron edge-rails, the forerunner of modern rails. His advice was later sought, and taken, in connection with railways on the Derby Canal extension from Little Eaton to Denby (opened 1795) and at Blisworth on the Grand Junction Canal (1800). The two latter railways were constructed by Benjamin Outram (1764–1805), who engineered the Derby Canal and also the Peak Forest Canal, which included a notable railway (1795–96). On all these lines Outram used cast iron flanged plate-rails. Similar rails, in an improved form, were employed by Jessop on the Surrey Iron Railway. This ran for ten miles from Croydon to a basin on the Thames at Wandsworth, and was double track throughout. The line to Croydon was opened in July 1803, with George Leather as resident engineer. Work started soon afterwards on an eight mile extension to Merstham, completed 1805.

Jessop's last major work was at Bristol. During the period 1765–1800 numerous proposals had been made for docks there. Smeaton, Nickalls and Jessop himself were involved; but the most imaginative idea came from a local vicar, William Milton, in 1791. Eventually Jessop broke through all the arguments and produced in August 1802 a detailed scheme, based on the Rev. Mr. Milton's suggestion, for diverting the River Avon through a 2 mile cut and transforming the old course of the river into a floating harbour covering 70 acres. This he insisted must be accepted, and it was. The Bill passed in August 1803; Jessop was appointed engineer, with his son Josias as deputy and resident engineer. Work began in 1804 and the first ships entered the harbour through the Cumberland Basin in 1809.

Finally, at Telford's special request, Jessop was appointed consultant to the Caledonian Canal at the beginning of this vast undertaking, in 1803. Jessop went over the whole line and subsequently visited the canal works every year until 1812.

In addition to these professional engineering activities Jessop had several business interests, one of which was a partnership with Benjamin Outram in the Butterley ironworks, Derbyshire. The firm began work in 1792 under Outram's management. After Jessop's death in 1814 his son William Jessop jnr. (*c.* 1783–1852) became a partner and technical director, and the Butterley Company (as it was called from 1806) established a leading reputation for their foundry and mechanical engineering work. William Jessop jnr. joined the Society in 1825. His elder brother, Josias Jessop (1781-1826), after working at Bristol, developed a substantial practice as a consulting civil engineer. He was elected to the Smeatonians in 1813.

While in Smeaton's employment Jessop presumably lived at Whitkirk. He then moved to Fairburn near Pontefract and from *c.* 1782 to 1805 his home was at Newark. During most of this time he also had a London address in Southampton Street. From 1805 he lived at Butterley Hall, which had been Outram's residence. A modest and fair minded man, Jessop was highly respected by his brother engineers.

1774. Major Henry Watson, F.R.S. (1737–1786)

From time to time the Society elects to its membership distinguished military or naval engineers, and first of these was Major Henry Watson. He trained at Woolwich Academy. After a few years as a young engineer officer he went to India and in 1765 was appointed by Lord Clive chief engineer of Bengal. There he planned and began construction of the first docks at Calcutta, work on which was finally brought to a halt by lack of funds. An able mathematician, Watson published a translation, with his own notes, of Euler's "Compleat Theory of the Construction and Properties of Vessels", and he designed two ships which are said to have proved the fastest vessels then built. Further details of his career can be found in D.N.B.

1776. John Cooper (*c.* 1742–1792)

John Cooper of Portpool Lane, Holborn, and his brother James Cooper of Poplar were the leading London millwrights in the last quarter of the eighteenth century. They undertook a considerable range of industrial mill-work, but specialised in the mechanical equipment of breweries. From 1784 onwards they played a significant part in introducing Boulton & Watt rotative engines in place of the traditional horse-wheels.

1776. Thomas Fruin

Thomas Fruin was engineer for the sluice near Ferriby on the River Ancholme (1768–69), and is very probably the 'Mr. Fruin' who, as contractor, carried out works on the Stour in Kent under Yeoman's direction (1776–78).

EARLY MEMBERS OF THE SMEATONIAN SOCIETY

1776. William Faden (1749–1836)

At the meeting on 14 March 1777 "Mr. President, with the hearty concurrence of the Members present, gave the thanks of the Society, to Mr. Faden, for a Copperplate of Summons's and fifty impressions thereof, for the use of this Society; which he requested, might be accepted as a testimony of his esteem, and regard for its honour and prosperity." In 1794 he engraved a new plate for the Summons or invitation card, which is still in use. He published the first volume of Smeaton's *Reports* in 1797.

As successor to the famous Thomas Jefferys, and in his own right, Faden became at an early age one of the best known London cartographers. He was appointed Geographer to the King in 1783. Many of of the Smeatonians' maps and plans were engraved and printed by him: Hodskinson's map of Suffolk (1783), Whitworth's Thames and Severn Canal (1784), Rennie's Lancaster Canal (1792), Jessop's Bristol Harbour (1803), to mention only a few examples. He organized, financed and published surveys of Hampshire (1791) and Norfolk (1797) both on the scale of one inch to a mile.

1776. Samuel Phillips

Samuel Phillips, "Engine Maker to his Majesty's Board of Works", was elected to the Society of Arts in 1773. In the membership lists of that Society his address changes from Goodge Street to St. George's Road, Blackfriars, in 1775. The latter address also appears in the Smeatonian Minute Book in 1784 and 1786, but changes to "Surrey Road, Blackfriars Bridge" in the 1789 list. He remained an active member of the Smeatonian Society until 1804. Rennie stayed with Phillips when he first came to London, to work at the Albion Mill, in 1784.

1777. Thomas Morris

In 1763 Morris was working under Brindley on the Duke of Bridgewater's canal, and in 1767 he appears to have been in charge of construction of the locks at Runcorn, which were completed towards the end of 1772. During most of 1774 and 1775 he was engineer for the Chester canal but, like others before and after him, he ran into trouble with the proprietors. When elected to the Society in 1777 he was still living at Chester.

Thomas Morris, who built a dock at Glasson on the Lune estuary, (1783–87) and was Liverpool's dock engineer from 1789 to 1799, in succession to Henry Berry, may have been his son.

1777. Joseph Hodskinson (1735(?)–1812)

Treasurer 1778–81, Vice President 1781–89

By 1767 Hodskinson had become one of the most expert land-surveyors in England. Jointly with John Ainslie and Thomas Donald he was in that year chosen by Thomas Jefferys, Geographer to the King, to survey the county of Yorkshire on the scale of one inch to a mile. This famous map was published in 1772. For his own map of Suffolk, on the same scale, published in 1783, Hodskinson was awarded the rare honour of a gold medal by the Society of Arts.

An observant and thoughtful man, Hodskinson graduated, as it were, from surveying to civil engineering and his first report, on Wells Harbour (1782), received the approbation of John Grundy. Later he was consulted on Shoreham harbour (1789) and Scarborough harbour (1791). His sensible proposals for extending the River Stour navigation to Canterbury (1792) were unfortunately not acted upon. He was engaged on coast protection works at Cherry Cob Sands on the Humber estuary (1793–99) and advised on the land drainage outfall of the neighbouring Keyingham Level (1795). In the same year, he gave evidence in the Parliamentary hearings on the Eau Brink Cut, on behalf of the port of Kings Lynn, and in 1796 he represented the interests of the Legal Quays in the Port of London enquiry.

He was assisted by his nephew Joseph Hodskinson junior, after whose death in 1800 he took as partner George Bassett.

1777. John Pinkerton (d. 1813)
Pinkerton worked at first in partnership with his brother James, and after *c.* 1780 on his own, as one of the early civil engineering contractors. He and some others, such as Thomas Sheasby, represent a stage in the development from the usual 18th century gang-piecework system to the general contractor of the 19th century, of which the firm of Jolliffe & Banks is the first outstanding example.

In a period when unexpected difficulties so often occurred on site and estimating was notoriously inaccurate, Pinkerton often ran into financial difficulties and disputes, and on more than one occasion he was guilty of skimping the workmanship. Nevertheless, he carried out a considerable amount of canal construction, in the north and the Midlands and as far south as the Basingstoke Canal. He also undertook a large part of the work on Rennie's great scheme for draining the East, West and Wildmore Fens in Lincolnshire.

1780. James Cooper (d. 1801)
James Cooper of Poplar and his brother John had a high reputation for their work as industrial millwrights. Their election to the Smeatonians provides interesting evidence of the status of this important but neglected group of engineers to whom Dr. Tann has recently drawn attention.

1780. Matthew Boulton, F.R.S. (1728–1809)
A man of undoubted genius, and the greatest industrialist of his day. As the active and imaginative partner of James Watt he contributed almost as much as Watt himself to the development of the steam engine. His life and business career have been set out in the admirable biography by Dr. H. W. Dickinson.

1781. Samuel Wyatt (1737–1807)
Wyatt designed and built the celebrated Albion Mill at Blackfriars, erected 1783–86. It was the first mill to be planned from the start to incorporate Watt's double-acting rotative engines, and two of these, each of 50 h.p., were installed. If not the first, it was certainly one of the earliest large buildings in the world to be founded on a structural raft. The young John Rennie installed the mill work.

Wyatt had an engaging personality and a wide range of talents. As an architect he is remembered for several excellent neo-classical buildings: Doddington Hall, Cheshire, the Commissioner's House in Portsmouth Dockyard and Trinity House, London, are perhaps the best known. He also designed James Watt's house at Heathfield, Birmingham, and was a personal friend of Matthew Boulton.

He designed and built Dungeness lighthouse in 1791. The following year he was appointed Surveyor to Trinity House Corporation, and he succeeded Smeaton as Director of Works at Ramsgate Harbour. Both positions he held until his death in 1807. He submitted plans for docks on the Thames at the Isle of Dogs (1796), not dissimilar to those later built as the West India Docks, and he was consulted on the first (but uncompleted) Thames tunnel at Gravesend in 1799. His last work was the lighthouse at Flamborough Head (1806).

1783. John Snape (*c.* 1737–1816)
His survey for the Birmingham Canal extensions met with Smeaton's approval in 1782. Other surveys by him include those for the Dudley Canal (1785 and 1792), for the Worcester and Birmingham Canal (1789), and for the Stratford Canal (1792), and the Grand Junction (1802).

EARLY MEMBERS OF THE SMEATONIAN SOCIETY

1783. Thomas Dadford

Dadford was construction engineer on the Staffs and Worcestershire Canal (1768–71) under James Brindley. In 1776 he became chief engineer for the Dudley Canal, the first section of which was built under his direction, and later he acted as consulting engineer on the extension up to the Birmingham Canal. As engineer of the Trent and Mersey Canal he supervised construction of the northern part of the Coventry Canal from Fradley to Whittington (1785–87).

In 1789 Dadford and Thomas Sheasby joined forces as contractors for the Cromford Canal, but they gave this up after a year or so in order to take on the work, as engineer-contractors, of building the Glamorganshire Canal (1790–94). In fact Dadford completed this on his own, as Sheasby left in 1792 to construct the Neath Canal.

Dadford's sons, Thomas, James and John, were capable engineers and had been trained by him.

1785. John Rennie, F.R.S. (1761–1821)

Treasurer 1811–21. Member of the committee for reorganising the Society in 1793 and of the Special Committee for the publication of Smeaton's *Reports*. On behalf of the Society, he edited Smeaton's *Miscellaneous Papers* (1814)

Rennie was born in East Lothian, the son of a well-to-do farmer. He attended Dunbar High School and Edinburgh University, and gained practical experience under Andrew Meikle (*c*.1719–1811). In 1784 he had the good fortune to be chosen by James Watt to design and erect the millwork in the Albion Mill. This and other mechanical engineering jobs, including some work for Boulton & Watt, occupied most of his time until 1789. During this period, Peter Ewart (1767–1842) was Rennie's assistant; Ewart became a member of the Society in 1822.

In 1790 a rather dramatic change took place in Rennie's career. Before the end of that year he had carried out three canal surveys, one of them being an early version of the Kennet and Avon. In 1791 he started planning his first major work, the Lancaster Canal. Construction began on the Lancaster at the end of 1792. Meanwhile he made surveys for the Rochdale and Crinan canals and on the river Chelmer. 1793 saw further surveys on the Kennet and Avon and the Rochdale. In 1794 he settled the line of the Rochdale Canal; he made an important report on the Thames navigation; jointly with Jessop he reported on the Somerset Coal Canal (William Smith, later to become the great geologist, being the surveyor); and work began under his direction, and to his plans, on the Kennet and Avon and the Crinan Canals. On the Lancaster Canal the impressive aqueduct bridge over the River Lune was erected 1794–96 to his design. In 1797 he started work on the first dock in Grimsby, where the foundation conditions were particularly difficult. Two years later he designed the beautifully proportioned five arch bridge at Kelso, and in 1800 he prepared a masterly scheme for the drainage of a very large area of fenland in Lincolnshire, embracing the East, West and Wildmore Fens.

All these works were completed under Rennie's direction, except the Rochdale Canal, which was constructed by Jessop. From 1800 until his death in 1821 Rennie designed and carried out an unprecedented number of highly distinguished works. Here we can only give a selected list of the more important ones (with their construction dates), including for ease of reference those already mentioned: the Lancaster Canal (1792–1803); the Crinan Canal (1794–1809); the Kennet and Avon Canal (1794–1810); Grimsby Dock (1798–1801); Kelso Bridge (1800–4); East, West and Wildmore Fens (1802–10); Leith Docks (1801–17); London Docks (planned 1801 by Rennie, Mylne and Huddart, built 1802–5 under Rennie); the Humber Dock at Hull (1803–9, with Chapman as joint planning engineer); East India Docks (1803–6, jointly with Ralph Walker); Boston Bridge (1803–7, an early iron structure); Greenock Harbour (1807-9); Howth Harbour (1809-19); Holyhead Harbour (1810–24); Waterloo Bridge(1811–17, Rennie's masterpiece and perhaps the finest large masonry bridge ever built in this or any other country); Plymouth Breakwater (planned jointly by Rennie and

Whidbey, who superintended construction from the start in 1812 until 1830); Sheerness Dockyard (1813–23); Southwark Bridge (1814–19, a fine iron arch bridge over the Thames); the Eau Brink Cut (1818–21); Ken Bridge,New Galloway (1821–23); and his last design, for London Bridge (1821).

The logic and clarity of his reports, the scale and the number of his works, and above all the beauty of his structures, place Rennie in a class of his own. His sons, George Rennie, F.R.S. (1791–1866) and Sir John Rennie, F.R.S. (1794–1874), both became well known engineers and both joined the Society in 1822. Rennie's grandson, George Banks Rennie, was Treasurer of the Society 1883–1902 and his great-grandson, John A. Rennie, held this position from 1903 to 1948.

Rennie's assistants and resident engineers include, as well as his two sons, many able men who deserve more attention than they have yet received from historians. We may mention James Hollinsworth who worked on Grimsby Dock, London Docks, the Crinan Canal and was resident engineer on Waterloo Bridge; Anthony Bower, surveyor and resident engineer on the Lincolnshire fen drainage scheme; Joseph Whidbey, F.R.S. (*c.* 1755–1833) who carried out the coastal surveys during Vancouver's voyage round the world (1791–95), became a personal friend of Rennie, joined the Society in 1809, was Master-Attendant at Woolwich Dockyard and superintended the construction of Plymouth Breakwater for eighteen years; John Aird, resident engineer on the harbour works at Greenock, Howth and Dunleary; John Thomas, of whom Rennie had a high opinion for his work at Sheerness; and Francis Giles (*c.* 1787–1847), for many years Rennie's principal surveyor, later a consulting engineer, and a member of the Society from 1828.

1786. George Young (1750–1820)

A land surveyor, Young produced a map of Worcester in 1779. He surveyed the River Severn in 1786 and again in 1790 when he was associated with Robert Mylne on works for improving the navigation. In 1793 he made the survey for the Shrewsbury Canal.

1789. Henry Eastburn (1753–1821)

Eastburn was Smeaton's pupil from 1768 and his assistant from 1776. He worked 1792–93 as resident engineer on the Basingstoke Canal, under Jessop, and in 1794 he was surveying the rivers Derwent and Rye in East Yorkshire for Earl Fitzwilliam. Later that year he became resident engineer on the southern part of the Lancaster Canal, where he was in charge until 1798. Rennie approved of his work there and in 1801 recommended him as resident for the construction of the London Docks.

1789. James Watt, F.R.S. (1736–1819)

On 27 March 1789 the Minutes record that "Mr. Watt Engineer was proposed and admitted a Member of the Society". With Nickalls in the chair, this meeting was attended, among others, by Smeaton, Mylne, Rennie and the new member James Watt. Thus the four most distinguished early members were present, whose names are still honoured at every meeting of the Society by the toast to "The Memory of our late worthy Brothers Mr. Smeaton, Mr. Mylne, Watt and Mr. Rennie".

The career of James Watt is so well known, and so fully documented, that no further details need be given here. His son, James Watt junior, F.R.S. (1769–1848) was elected to the Society in 1822.

Part III. Members Elected 1793–1800

A reorganisation of the Society in "a better and more respectable form" took place in 1793. This was carried out by a committee consisting of Mylne, Jessop, Whitworth and Rennie: the four leading civil engineers in the country at that time. Mylne became the sole officer of the Society, as Treasurer, and he held this post until 1810. The engineers constituted the First Class of members who "shall on proper occasions, assume the Style and Title of Members of the Society of Civil Engineers in London". In addition "there shall be a class of Gentlemen, forming part of the Society, under the denomination of honorary Members", and also as honorary members "there shall be admitted, various Artists, whose professions and employments, are necessary or useful thereto as well as connected with Civil Engineering".

In its new form the Society had twenty-four members, of whom twelve were elected in 1793. The President of the Royal Society, Sir Joseph Banks, appears among the "Gentlemen", and he attended the meetings with regularity. The new elected "Artists" included two instrument makers: the great Jesse Ramsden, F.R.S. (1735–1800) and John Troughton (*c.* 1739–1807) who in 1782 founded the famous firm in Fleet Street, later known as Troughton & Simms.

A Special Committee was set up in 1795, consisting of Mylne, Banks, Jessop, Huddart and Rennie, to publish Smeaton's *Reports*, with Dr. Charles Hutton, F.R.S. (1737–1823), an honorary member of the Society, as editorial assistant under Mylne.

1793. JAMES GOLBORNE (1746–1819)

Golborne worked on the R. Clyde improvements under his uncle, John Golborne, with whom he had served a seven year 'apprenticeship'. In 1776 he succeeded Richard Dunthorne (1711–75) as Superintendent General to the Bedford Level Corporation. In this post, which he held for over thirty years, Golborne was responsible for making and repairing river banks, drainage cuts, bridges, land surveying, etc. He reported on the navigation from Peterborough to Wisbech (1783) and on the outfall of the River Ouse (1791). He made the borings for the Eau Brink Cut investigations (1794) and was otherwise much engaged on this project, for which he and Robert Mylne prepared the estimates in 1801. Occasionally, Golborne reported on works outside the Fens, e.g. on Keyingham Level (1799).

1793. SIR THOMAS HYDE PAGE, F.R.S. (1746–1821)

A military engineer, trained at Woolwich, Page was involved in various civil engineering works, notably at Dublin Harbour and in the Fens. His career is summarised in D.N.B.

1793. JOHN DUNCOMBE (d. 1810)

Duncombe assisted Jessop in planning and surveying the Ellesmere Canal and was assistant engineer during construction until 1803. He was then appointed by Telford as general superintendent of work on the Highland roads in Scotland. Here he seems not to have been a success, at least towards the end of his career. In 1810 he ran into some sort of trouble and died in Inverness gaol.

1793. SIR SAMUEL BENTHAM (1757–1831)

Elected in the class of "Gentlemen honorary members". Bentham's brilliant but somewhat eccentric career as a naval architect and engineer is well known, and we have nothing to add to the long account given in D.N.B.

1793. JOHN FOULDS

Elected as a "millwright" in the "Artist" class of honorary members. Foulds was engineer to the London Bridge Waterworks from *c.* 1765 at least until 1800.

1794. Captain JOSEPH HUDDART, F.R.S. (1741–1816)

A member of the Society's Special Committee for the publication of Smeaton's *Reports*.

Huddart was captain of an East India ship for ten years and also produced some excellent hydrographic surveys, notably of St. George's Channel (1779) and Hasborough Gatt off the Norfolk Coast

(1790). In 1791 he was elected an Elder Brother of Trinity House Corporation, for whom he acted as adviser on lighthouses, beacons and floating lights. He made surveys and charts and wrote sailing directions for the Longships (1795), for Sunk Sand in the Thames estuary (1802), for Flamborough Head (1806) and South Stack at Holyhead (1809).

His experience as a sailor and skill as a hydrographer led to consulting work in the planning of harbours. Reports on Boston (1793), Leith (1800), Swansea (1804), Whitehaven (1804) and Howth (1808) are perhaps the most important. With Rennie and Mylne he planned the London Docks in 1801.

Huddart's talents extended into mechanical engineering; he is remembered as the inventor and designer of the very successful rope-making machinery, which came into production at the Limehouse manufactory in 1800. Four years later he published an account of the scientific principles involved in the design. He attended the meetings of the Society with regularity until a year before his death, and was, as Smiles says, "a singularly estimable character".

1795. JOHN WATTÉ

Elected as a Land Surveyor in the "Artist" class of honorary members.

Watté was highly regarded by the Fens engineers for his accurate surveying. He laid out and built the sea banks on the east side of the Nene estuary. He was much involved with Mylne and James Golborne in preparatory work for the Eau Brink Cut.

1795. WILLIAM CHAPMAN, M.R.I.A. (1749–1832)

Chapman, who was born at Whitby, and as a young man spent several years at sea, worked from about 1777 to 1783 in the Newcastle district, where he and his father and his brother John owned leases on collieries at Byker and Wallsend. By the summer of 1783, however, with the encouragement of John Wooler, a friend of his father, Chapman had practically decided to enter the field of civil engineering and try his fortune in Ireland. Before taking such a momentous decision he wrote to Smeaton for advice. A copy has survived of the interesting and kindly letter written by the great engineer in reply; Chapman left for Ireland later that year, and after a period of little employment his earliest construction job was as engineer for the Kildare Canal, built 1787–89.

While on the Kildare Canal Chapman worked out, and applied in practice, the correct geometrical construction for skew or oblique arch bridges. In 1790 work began to his plans on the River Barrow Navigation, and in 1791 he took charge of building the difficult section of the Grand Canal across the Bog of Allen. Some simple experiments carried out by him at about this time are perhaps the first ever made in an attempt to predict the settlement of peat under load. The Circular Line of the Grand Canal around Dublin was constructed largely in accordance with Chapman's plans; indeed his earliest report (1785) is on this subject. Moreover, he was engaged in the building of the Dublin docks, probably as engineer, with Jessop as consultant; and his brother Edward Walton Chapman (1762–1847) was certainly resident engineer from 1793 to 1796, when the work reached completion.

Towards the end of 1794 Chapman returned to Newcastle, where he lived for the rest of his life, except for a period around 1812 in County Durham. During 1795 and 1796 much surveying, planning and discussion took place on projects for linking the Tyne and Solway Firth by a canal. Chapman, Jessop, Whitworth and John Sutcliffe were all involved, but eventually the idea was abandoned. His first important undertaking in England was the Beverley and Barmston scheme for draining about 13,000 acres of low ground west of the River Hull. His report is dated October 1796 and the work was executed according to his plans and under his close supervision during the years 1799–1810, with William Settle as resident engineer and Samuel Dickinson as the principal land surveyor. Another carefully designed scheme by Chapman of a rather similar nature and only slightly smaller extent, the Muston and Yedingham Drainage in the Vale of Pickering, was carried out from 1800 to 1808. This included the Sea Cut from the River Derwent to Scalby. In 1801 he became chief engineer for Scarborough Harbour. There he completed the great East Pier (in 1826), extended the middle pier (built by William Vincent in the mid 18th century,) and constructed the West Pier. Apart from an enlargement of the latter, in 1880, the harbour was left by Chapman essentially in its modern form. The resident engineers at Scarborough were Matthew Shout (1801–4), Roger Nixon (1804–22) and William Barry (1822–38).

Chapman's other civil engineering works include the Sheffield Canal (1816–19), improvements at Shoreham Harbour (1817–21), and at Leith (planned 1824, carried out 1827–31), the Carlisle Canal (1820–23), and the inner and outer harbours at Seaham (1828–32). In 1824–5 he also played an active part in planning the Newcastle and Carlisle Railway.

About 1806, when the works in East Yorkshire were well under way, Chapman became part owner of East Kenton Colliery, near Newcastle. In 1807 he invented a method of loading coal into ships from the staiths to avoid the breakage caused in the traditional coal shutes. At least two of these ingeneous self-acting machines were built, but in 1812 the design was improved by Benjamin Thompson. Meanwhile, after returning from Ireland, Edward Chapman took over the Willington Ropery in partnership with William, and the two brothers paid much attention to the development of rope-making machinery. In 1812 they took out a patent for a chain-haulage locomotive, the most interesting point of which was the first known application of the bogie. An eight wheel locomotive, on two bogies, was tried out in 1814 on the railway from Lambton Colliery to the River Wear, and in the following year William Hedley modified his pioneer locomotives on the Wylam Colliery line in accordance with this design.

Chapman attended the Society's meetings for over thirty years. He was the author of several books and pamphlets, e.g. on canals (1797), on cordage (1808), on the importance of keeping proper records of colliery workings (1815), and on preservation of timber (1817). We have seen many of his printed reports. If somewhat lengthy they nevertheless reveal an honest and usually successful attempt to arrive at the truth, and also a willingness to modify his plans as new evidence becomes available.

1795. James Cockshutt, F.R.S. (*c.* 1742–1819)

"James Cockshutt, Esqr. of Wortley Iron Works near Sheffield, Civil Engineer, a Gentleman well versed in many theoretical as well as practical branches of natural knowledge, having been under the Instruction & frequently in the Employment of the late Mr. John Smeaton": so runs the citation in his certificate for election to the Royal Society (1804).

In 1787, in partnership with Richard Crawshay, Cockshutt designed the first bar rolling mill built at the Cyfartha works, used in the manufacture of wrought iron by the new method of puddling and rolling invented by Henry Cort. Later he introduced these processes at Wortley, where, under his management, the works attained their maximum development. Cockshutt supported Cort's son in his efforts (1812) to gain recognition for his father's invention, and refuted the disgraceful evidence given by Samuel Homfray against this petition.

1799. Colonel William Mudge, F.R.S. (1762–1820)

We conclude this study by noting a member of the Society who, though not an engineer, contributed much to engineering, as Director of the Ordnance Survey and as the first Examiner at Addiscombe College, and with whose family Smeaton had been closely associated.

After his training at Woolwich, William Mudge received a commission in the Royal Artillery in 1779 and then served abroad for several years. On returning to England he made a special study of mathematics under Dr. Charles Hutton, by whom, in 1791, he was recommended for a post in the newly formed Ordnance Survey. He became Director of the Survey in 1798, and until 1809 was extremely active both in the field and at headquarters. By then the primary and secondary triangulation of the whole of England and Wales had been completed, except for three eastern counties, and publication of the O.S. one inch to a mile maps was under way, the first (of Kent) having appeared in January 1801.

In 1809 he became Lieutenant-Governor of Woolwich Academy, and in the following year he was asked to undertake the additional duty of Examiner at Addiscombe. This college had just been established for training the cadets for the Artillery and Engineers of the East India Company's service. Mudge retained the directorship of the Survey, but to an increasing extent during the ten years before he died the work was in charge of Thomas Colby, F.R.S. (1784–1852). Colby joined the Survey in 1802 and succeeded Mudge in 1820 as Director. He became a member of the Society in 1822.

Mudge's father, Dr. John Mudge, F.R.S., a physician in Plymouth and a distinguished amateur astronomer, was a contemporary and intimate friend of Smeaton, who became acquainted with the family while working on the Eddystone lighthouse. His grandfather, the Rev. Zachariah Mudge, celebrated as a divine, was vicar of St. Andrew's Plymouth. In 1759 when the lighthouse had been completed, Smeaton and the old gentleman ascended to the lantern "and joined together in singing the grand Old Hundredth Psalm, as a thanksgiving for the successful conclusion of this arduous undertaking".

Acknowledgments

We are grateful to Miss Anthea Beamish and Mr. W. N. Slatcher for clarifying various points on the history of the Aire and the Calder Navigations. Mr. Charles Hadfield kindly checked our notes on Jessop, and helped in many other ways. Mr. Gordon Biddle checked the notes on Whitworth. Mr. John Boyes generously gave much information on river navigations in eastern England. Mr. Trevor Turner has placed his expert knowledge of Smeaton freely at our disposal and in particular he supplied dates for Smeaton's bridges in Scotland. Confusion on the respective roles played by Grundy and Henry Berry at Hull dock has been resolved by Mr. Mark Baldwin, to whom we express our thanks for access to his research files. Finally we wish especially to acknowledge Miss Joyce Brown for her biographical work on several members of the Society.

Bibliography

Primary Sources (MSS.) Those marked * are in the Library of the Institution of Civil Engineers.

* Society of Civil Engineers. First Minute Book (1771–92).
* Society of Civil Engineers. Second Minute Book (1793–1821).
* Society of Civil Engineers. Third Minute Book (1822–82).
* Reports of John Grundy, vol. 2 (1741–66).

Scarborough Harbour Commissioners. Minutes (1752–1833). Scarborough Ref. Library.
River Nene Commissioners Minutes. 3 vol., (1756-1825). Welland & Nene River Authority.
Holderness Drainage Trustees. Minute Book, vol. 1 (1764–68). East Riding R.O.
Society of Arts. Minute Books, Committee Minutes, Membership Lists.
* "Machine" Letter Books of John Smeaton, 4 vols. (1781–92).
Boulton & Watt Collection. Birmingham Ref. Library and Assay Office.
* Reports of John Rennie, 17 vols. (1790–1821).
Rennie Papers. National Library of Scotland.
Trinity House. Court Minutes and Accounts.
Mining records at Neville Hall (North of England Institute of Engineers) and at Northumberland C.R.O.
Beverley and Barmston Drainage Commissioners. Minutes (1798–1861). East Riding R.O.

Primary Sources (Printed)

J. Smeaton, *A Narrative of the Building . . . of the Edystone Lighthouse*, London, 1791.
Reports of the Late John Smeaton F.R.S., 3 vol., London, 1797 and 1812.
A Catalogue of the Civil and Mechanical Engineering Designs of John Smeaton (ed. H. W. Dickinson and A. A. Gomme), London, 1950.
Robert Mylne's Diary (1762–1810), ed. A. E. Richardson, in *Robert Mylne, Architect and Engineer*, London, 1955.

R. Renwick (ed.), *Records of the Burgh of Glasgow*, **7** (1760–80), Glasgow, 1912.
Printed reports mentioned in the text are either in the British Museum Reading Room or in the library of the Institution of Civil Engineers.
Printed maps, charts and plans are in the British Museum Map Room.
House of Commons, *Journals, Accounts & Papers* and *Reports*.

SECONDARY SOURCES

C. T. G. Boucher, *James Brindley, Engineer, 1716–1772*, Norwich, 1968.
R. A. Buchanan, "The construction of the Floating Harbour in Bristol 1804–1809", *Trans. Bristol Glos. Archael. Soc.*, **88** (1969), 184–204.
C. Close, *The Early Years of the Ordnance Survey*, first published 1926, reprinted Newton Abbot, 1969.
V. T. H. & D.R. Delany, *The Canals of the South of Ireland*, Newton Abbot, 1966.
H. W. Dickinson, *Matthew Boulton*, Cambridge, 1936.
H. W. Dickinson & Rhys Jenkins, *James Watt and the Steam Engine*, Oxford, 1927.
B. F. Duckham, "The Fitzwilliams and the navigation of the Yorkshire Derwent", *Northern History*, **2** (1967), 45–61.
J. Farey, *General View of the Agriculture of Derbyshire*, 3 vol, London, 1811–17.
A. Gibb, *The Story of Telford*, London, 1936.
R. Gough, *British Topography*, 2 vol., London, 1780.
C. Hadfield, *The Canals of South Wales and the Border*, Newton Abbot, 1967.
C. Hadfield, *The Canals of South and South East England*, Newton Abbot, 1969.
C. Hadfield, *The Canals of the West Midlands*, Newton Abbot, 1969.
C. Hadfield, *The Canals of the East Midlands*, Newton Abbot, 1970.
C. Hadfield & G. Biddle, *The Canals of North West England*, 2 vol., Newton Abbot, 1970.
J. B. Harley, "The Society of Arts and the survey of English Counties", *Journ. Roy. Soc. Arts*, **112** (1964), 269–75.
H. Household, *The Thames & Severn Canal*, Newton Abbot, 1969.
C. E. Lee, "Early railways in Surrey", *Trans. Newcomen Soc.*, **21** (1940), 49–79.
J. Lindsay, *The Canals of Scotland*, Newton Abbot, 1968.
C. F. D. Marshall, *A History of Railway Locomotives down to the end of the year 1831*, London, 1953.
R. S. Mylne, *The Master Masons to the Crown of Scotland*, Edinburgh, 1893.
J. Percy, *Metallurgy: Iron and Steel*, London, 1864.
J. Rennie, *An Historical, Practical and Theoretical Account of the Breakwater in Plymouth Sound*, London, 1848.
A. H. W. Robinson, *Marine Cartography in Britain*, Leicester, 1962.
E. Robinson, "The profession of civil engineer in the eighteenth century: a portrait of Thomas Yeoman", *Annals of Science*, **18** (1962), 195–215.
A. W. Skempton, "The publication of Smeaton's Reports", *Notes & Records of the Royal Society*, **26** (1971), 135–55.
A. W. Skempton, "Samuel Wyatt and the Albion Mill", *Architectural History*, **14** (1971), 53–73.
W. N. Slatcher, "The Barnsley Canal: its first twenty years", *Transport History*, **1** (1968), 48–66.
D. Swann, "The engineers of English port improvements 1660–1830", *Transport History*, vol. 1 (1968), 153–68, 260–76.
J. Tann, *The Development of the Factory*, London, 1970.
J. H. Thornton "The Northampton Cotton Industry, *Northants Nat Hist. & Field Club*, **33** (1959) 241-259.
T. Turner, "John Gwyn and the building of the North Bridge, Kingston upon Hull 1784–7", *Transport History*, **3** (1970), 154–63.
N. Walker & T. Craddock, *The History of Wisbech and the Fens*, Wisbech, 1849.
W. H. Wheeler, *A History of the Fens of South Lincolnshire*, 2nd edn, Boston, 1897.
A. F. Williams, "Bristol port plans and improvement schemes of the 18th century", *Trans. Bristol Glos. Archael. Soc.*, **81** (1962), 138–88.
E. C. Wright, "The early Smeatonians", *Trans. Newcomen Soc.* **18** (1938), 101–10.

SUPPLEMENTARY BIBLIOGRAPHY

C. R. Andrews, *The Story of Wortley Ironworks*, Mexborough, 1950.
C. T. G. Boucher, *John Rennie 1761–1821*, Manchester, 1963.
H. M. Colvin, *A Biographical Dictionary of English Architects 1660–1840*, London, 1954.
Dictionary of National Biography (D. N. B.).
J. Huddart, *Memoir of the Late Captain Joseph Huddart*, London, 1821.
S. Hughes, "Memoir of William Jessop" in *Weale's Quarterly Papers on Engineering*, **1** (2) 1843.
H. F. Killick, *Notes on the Early History of the Leeds and Liverpool Canal*, Bradford, 1900.
J. Rennie, *The Theory, Formation and Construction of British and Foreign Harbours*, 3 vol., London, 1854.
L. T. C. Rolt, "William Jessop, builder of canals and railways" in *Great Engineers*, London, 1962.
A. W. Skempton, *The Smeatonians. Duo-Centenary Notes on the Society of Civil Engineers*, London, 1971.
J. Simmons, "William Jessop, civil engineer", in *Parish and Empire*, London, 1952.
S. Smiles, *Lives of the Engineers*, vol. 2, London, 1861 [Smeaton and Rennie].
J. Weale (ed.), "Memoir of William Chapman", in *Weale's Quarterly Papers on Engineering*, **1** (1) 1843.

EARLY MEMBERS OF THE SMEATONIAN SOCIETY

POSTSCRIPT

At the time of writing this paper very little was known about earth dams built for canal resevoirs. From G.M. Binnie *Early Dam Builders in Britain* (1987) and A.W. Skempton, Historical development of British embankment dams (*Clay Barriers for Embankment Dams*, 1990, 15–52), information is now available on Jessop's Butterley dam (1792–95 for the Cromford Canal) and Rennies' Rudyard dam (1798–1801 for a branch of the Trent & Mersey Canal), both with a puddle-clay core, and on two of Jessop's homogenous clay-fill dams at Knipton and Hollingworth, built between 1795 and 1801 for the Grantham and Rochdale canals respectively. All these have a height of about 10m.

IV

THE PUBLICATION OF SMEATON'S REPORTS

THE collected Reports of John Smeaton (1724–1792, F.R.S.: 1753) were published by the Society of Civil Engineers in 1797 and 1812. The first edition consists of three large quarto volumes illustrated by seventy-two engraved plates and a portrait frontispiece. An unabridged, but less expensive second edition, in two volumes, appeared in 1837. For many years past, the Reports have been recognized as outstanding classics in the engineering literature, and quite recently the second edition has become available as a facsimile reprint (1).

The events leading to the publication of these Reports, comparatively few of which were printed in Smeaton's lifetime, constitute an interesting but practically unknown story, which it is the purpose of this paper to reconstruct. Fortunately there are ample primary sources, chiefly in the form of letters between Sir Joseph Banks and Robert Mylne (2) and the Second Minute Book (1793–1821) of the Society of Civil Engineers (3).

In March 1771 Smeaton and some of his friends met at the King's Head Tavern in Holborn, and 'agreed that the Civil Engineers of this Kingdom do form themselves into a Society. . . .' The Society, thus established, flourished and continues to flourish today; indeed, at the time of writing these words, it will soon be celebrating the duocentenary of that first meeting in Holborn (4). Robert Mylne (1733–1811, F.R.S.: 1767) was one of the founder members, and he succeeded Smeaton in 1793 as the leading figure in the Society. He, like so many of his contemporaries, held Smeaton in the highest esteem, and it was Mylne who played a principal part in publishing the Reports.

From its early days the Society included among its members men who were interested in, or connected in some way with engineering, and also several scientists who were friends of the engineers and enjoyed attending the dinners and taking part in the 'conversation, argument, and social communication of ideas and knowledge' which, as Mylne said, 'were, at the same time, the amusement and the business of the meetings' (5). Among this latter group a distinguished member of the Society, and one very regular in his attendance, was Sir Joseph Banks (1743–1820, President of the Royal Society from 1778 to 1820). He first became a member of the Engineers' Society (as it was then colloquially called) shortly after Smeaton died, but Banks had known Smeaton well (6) and, as events were to prove, had a great admiration for him.

This was soon to be demonstrated in a most practical manner; for Banks purchased from Smeaton's two daughters (Mrs Ann Brooke and Mrs Mary Dixon), his executrices, the seven books of his manuscript reports and the complete collection of his drawings. The collection of drawings is now one of the treasures of the Royal Society library (7). The earliest note we have relating to this transaction is in a letter from Cavendish to Banks (2), dated 23 September [1793] where, among other matters, he says:

> I have received authority from Mr. Brooks [*sic*] to deliver Mr. Smeaton's papers to you, which I shall do when you come to town, unless you chuse to have them sent to you.

Publication of the first volume, 1797

Nothing more is heard until 27 February 1795, when the following Minutes were recorded at a meeting of the Society of Civil Engineers:

> A conversation was held, on the Subject of all the Reports and papers, on Engineering business, left by the late Mr. Smeaton; now in the possession, & the property of Sir Joseph Banks. And it was proposed that a Committee of this Society should undertake the publication thereof at their own Risque and expence.
>
> Sir Joseph Banks, in a very handsome manner, offered the Copyright in the said papers, which he had purchased, at a considerable expence, without his receiving any consideration for the same; & retaining the original papers as his property after the publication thereof has been accomplished.
>
> It was understood at the same time, that the profits of such a publication (if any there should be) is to remain the property in expectancy, of the heirs &c, of the late Mr Smeaton; agreeable to the condition made for that purpose, by them with Sir Joseph Banks.
>
> Whereupon the following Members associated themselves into a Committee for this purpose: to Wit Sir Joseph Banks Bart., Willm. Jessop Esquire, Capt. Huddart & Mr Mylne.

John Rennie had not been present at this meeting, but on 13 March 1795, when he took the chair, we hear that:

> Mr Rennie offered himself as one of the Special Committee, for the publication of Mr. Smeatons papers; and it was thereby settled, the said Committee was compleat.

Apart from Banks and Mylne, then, the Committee comprised: William Jessop (1745–1814), Smeaton's former pupil and assistant and now one of the foremost civil engineers in England; Captain Joseph Huddart (1741–1816, F.R.S.: 1791), marine surveyor and consulting engineer; and the celebrated John Rennie (1761–1821, F.R.S.: 1798).

A few weeks later Banks sent the various papers to Mylne:

> New River Head, 14 May 1795. [Mylne to Banks.]
> I received, in due time, the parcel, Lists &c of Mr Smeaton's Reports. It would appear, little or nothing is to be done but to put these to the press . . . chronological order would be the best arrangement . . . [with] a copious Index, and arrangement of subjects . . . at the end of the Publication. All of which I humbly submit as the best way not to lose time.

> [Soho Square] 16 May 1795. [Banks to Mylne.]
> Being myself naturally of a hasty disposition, I have so often hurried myself into scrapes that I am possibly become too dilatory; of that my Friends of the Comm^ee will judge, & I will submit to their sentence.
>
> In the case of the papers of our late worthy friend, it appears to me that selection should be made of the best, some I apprehend will do him much greater credit than others. . . . Whether I am wrong or right in these sentiments, I shall, as I before said, leave to the judgement of the Comm^ee, of which you & I are the only persons who have as yet read the Reports; will you be so good therefore when you have fully satisfied yourself with the perusal of them, & make all such remarks &c as you may chuse to commit to paper, to forward them to some one of the other Members. As soon as we have all satisfied ourselves in private, we may meet as a body, & come to a final resolution.

The matter rested there for a while. Then, in the Engineers' Society Minutes for 13 November 1795, it is recorded that:

> Doctor Hutton being present and having undertaken to be the Editor, of Mr Smeaton's Reports in Civil Engineering, the Seven Volumes thereof were directed to be sent to him, for his perusal and report thereon.

This is Dr Charles Hutton (1737–1823, F.R.S.: 1774), professor of mathematics at Woolwich Academy, Foreign Secretary of the Royal Society 1779–1784 and, incidentally, author of a book on the *Principles of Bridges* (Newcastle 1772). He was elected to the Society of Civil Engineers in 1793.

Later, in November 1795, Hutton recommended that:

> Copies be taken of all Mr Smeatons Reports &c, before they are arranged by the Editor for Publication.

The manuscripts were duly returned to Mylne in December, and next month he wrote as follows:

> New River Head, 29 Jan. 1796. [Mylne to Banks.]
> I have lately done myself the pleasure to call on you twice; that I might have an opportunity of informing you, in particular, of having obtained a person to Copy Smeaton's papers, at the Rate of 18^{d} for a Score of pages, each of 72 words; Paper being found him. I thought *this* so reasonable, I have set him to work, on one Volume for a trial; under the hope of its being approved by the Committee at large.
>
> I shall make a Report at the next Meeting, where I hope most of the Com^{ee} will attend; and that you will do us that pleasure; that the whole may be put into a train.

> [Soho Square, 31 Jan. 1796.] [Banks to Mylne.]
> I am sorry I have been so unlucky as to be abroad when you have Called & I think myself more than commonly so, because you are always admitted when you do come & I have been of late seldom abroad.
>
> I conclude that your Copyist copys none but the Reports that are to be printed: it would be unnecessary expence to copy more; besides that I do not wish to run the Risk of those that we do not print falling into other hands.
>
> I always attend the Engineers' Club, if I am able & hope to be able next time: next Friday, however, being Lincolnshire Club, I cannot, so hope it will not fall on that day.

Banks did indeed attend meetings in January and also in February, March and April, but there is no record of any decisions about Smeaton's Reports; and it is clear from later evidence that Mylne, presumably with the rest of the Committee's support, had decided to go ahead with his original proposal of publishing all the Reports, arranged more or less chronologically and grouped in subjects, with an index. For this, we can be grateful to him.

Samuel Brooke, a well-known printer, and a member of the Engineers' Society in the special Class of 'Artists whose professions and employments are necessary & useful . . . [to] Civil Engineering', undertook to carry out the printing; and the work was to be published by William Faden, *Geographer to the King*, an 'Artist' member of the Society since 1776, and publisher of many of the engineers' plans of canals and other works, as well as the first (1 in./mile) Ordnance Survey map.

London, 2 Sept. 1796. [Mylne to Banks.]
I have sent you, inclosed, a Specimen, of the type, size of page, &c &c; which Dr Hutton and I have settled, for the Standard of our intended publication of Smeaton's Reports. I wish to have your opinion. The Doct^{r} computes it may turn out to be, about 2 thick Volumes. I think it will be 3 at least; with the printed Reports added, and interspersed according to the order of Dates.

I am looking out all the printed ones I have, to send to the Doctor, and I wish you would be so good, to do the same, or give directions for that purpose. Other friends and Engineers will do the same, and, as to Maps, or prints, or Plans, &c, that must be a subject of serious consideration at our first Meeting, next winter Session.

[Revesby Abbey] 5 Sept. 1796. [Banks to Mylne.]
. . . The Type & Form appear to me as good as any other. The Quantity of matter, however, startles me a little; it was my idea to have begun with a pretty close selection of unpublished reports, & to have compleated the Series by degrees as the Public might give encouragement: so that, in the end, the whole, published & unpublished, would, if the Public chose it, be put into their possession not in a chronological Series, but in one form'd from the comparative importance of the subject matter.

Regarding my Papers at London, no one can get them in my absence; & I apprehend that it will be time enough at my return in November to rummage among them; for I do not believe I have Printed Copies of many of the scarce Reports, as Smeaton was excessively careless about preserving them, & seemed to content himself with his MSS. when he had occasion to consult them.

However careless Smeaton may have been about the printed reports, he evidently kept the manuscripts in an orderly way in the seven volumes previously mentioned.

At the Society's meeting on 15 December 1796:

Doctor Hutton attended late—and brought with him, The first part of Smeatons Man^{spt} Reports, ready for the Printer; which were accordingly delivered to Mr Brook to be proceeded on, with dispatch.

Samuel Brooke worked efficiently, and by 5 May 1797 Mylne was able to report:

> that the *Printing* of Mr Smeaton's Manuscript Reports was so far proceeded with, that one Volume (a third part of the work) would be ready, in a short time, fit for publication. He therefore suggested the expedience, of paying by instalment, a Sum of money into the hands of Mr Faden the publisher, on behalf of the Committee, to pay the necessary expences.
>
> It was therefore ordered, that each member of the Committee shall pay £40 into the hands of Mr Faden, in trust for the purposes of the publication.

This they did, and so established a fund of £200 (8).

Robert Mylne composed a most interesting Preface to the first volume of the Reports (9). He gives a short history of the Society of Civil Engineers and an account of its present constitution, as reorganized after Smeaton's death; and makes the following remarks concerning the Reports:

> It is with much satisfaction, the COMMITTEE OF CIVIL-ENGINEERS have, at length, so far accomplished their long wished object, as now to present the public, with the first part of the works, of their late worthy and ingenious brother, Mr JOHN SMEATON: one of the greatest Engineers, that this, or perhaps any country, ever produced.
>
> As the members of *this Society* have interested themselves so greatly, in the publication of the present work; and as the author of it had so great a share in the first establishment, as well as, the subsequent management of the society, it may not be amiss in this place, to employ a few words, in giving a short account of it, before any thing is said more particularly of the publication itself.
>
> [Then follows the history of the Society.]
>
> It may be mentioned, as a mark of the Society's regard, that a tribute is always paid, after dinner, *to the memory of their late worthy brother*, JOHN SMEATON. The publication of this work shews their respect goes farther; by contributing a more substantial monument to his fame and character.
>
> The Society having learnt that Sir JOSEPH BANKS had, for a considerable sum, purchased all the manuscripts, designs, drawings of every sort, and all the papers of Mr SMEATON . . . it was proposed to him, and most handsomely acquiesced in, that *the Society* should undertake to perform the task of publishing the Reports only . . . and that the loss, if any, should be defrayed by themselves, as well as, that the profits, if any, should go to Mr. SMEATON'S representatives.

Plate 11

JOHN SMEATON, F.R.S. (1724–1792)
An engraving by W. Bromley from a painting by M. Brown

Facing page 140

IV

Plate 12

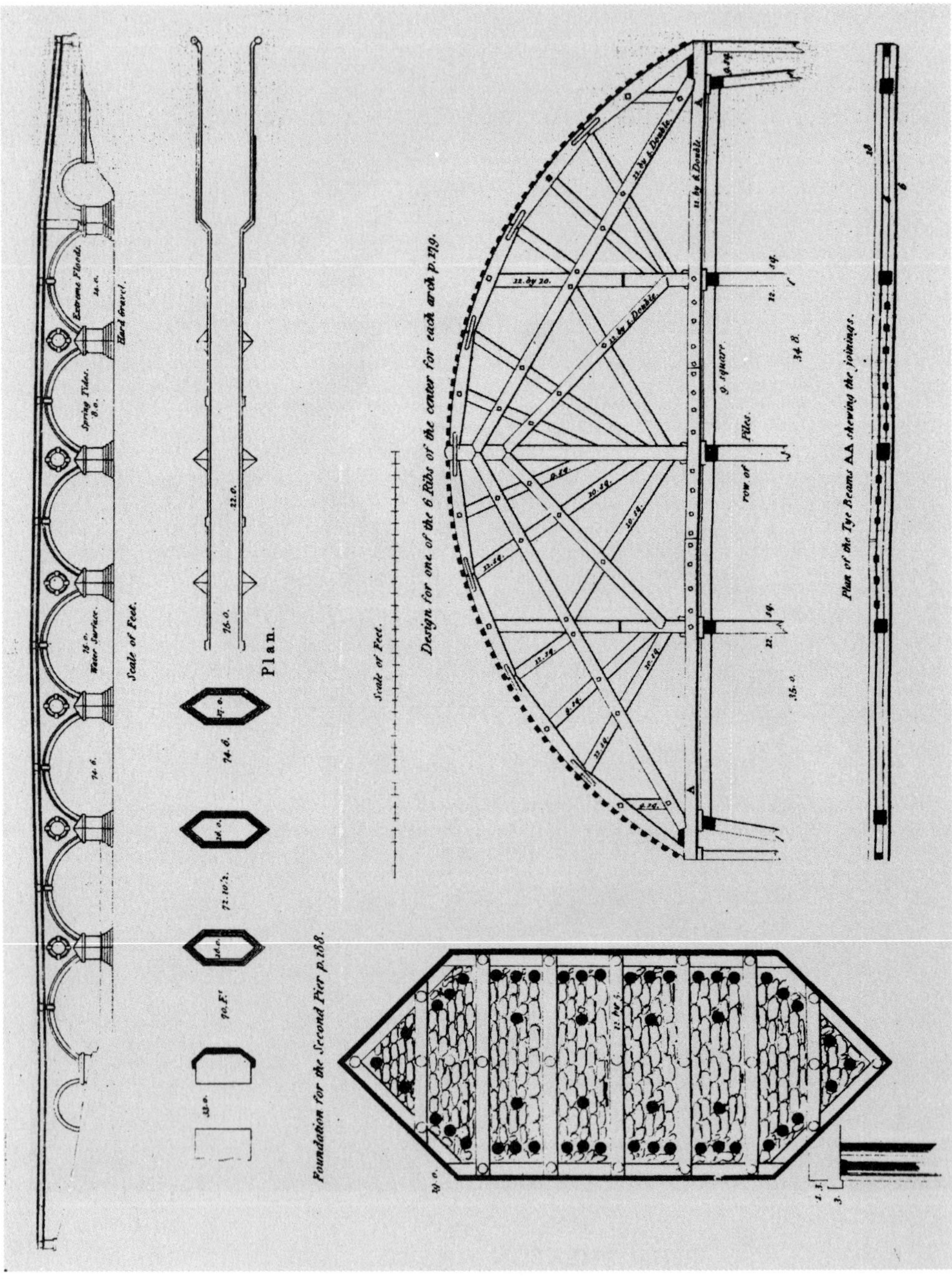

Smeaton's drawings for Perth Bridge engraved by Wilson Lowry

[The formation of the Special Committee is then described.]

The Reports, only, were the great object of this Society, and of their Special Committee. These, they thought, would be of the greatest use to the profession, to teach actual and practical knowledge; as well to conceive advice and opinion given, as to convey them, with perspicuity and energy to others.

Of these Reports, the present volume contains about one half; the remainder, being intended for a second volume, if the present shall be approved by public encouragement.

The manner in which the reports are here arranged, is in chronological order, or the time in which they occurred, for each subject; with this variation, that all the several reports on the same subject, when there are more than one, made at different times, are here brought together. . . .

As to Mr SMEATON's style and language, he had a particular, and in some degree a provincial way of expressing himself, and conveying his ideas, both in speaking and writing; a way which was very exact and impressive, though his diction was far from what may be called classical or elegant. . . . His language and words, therefore, and even the orthography, have been closely adhered to, without taking the liberty to make any alterations, unless perhaps sometimes in the change of a letter or a word, where a manifest deviation from grammar occurred, such as the author would of himself have altered, had he been the editor.

The Preface was sent to Banks, who wrote to Mylne on 3 October 1797 with criticism concerning the account of the Society's present organization and some minor matters.

The Minutes of 17 November 1797 record that the:

Business of the Publication of Mr Smeaton's Reports, was as follows. . . . A Portrait of Mr Smeaton was agreed to be prefixed to the first volume of the said Work. Mr Faden was directed to get the same done by Mr Bromley, the Artist, who engraved Mr Aubert's Plate; agreeable to his letter, of 26th September last to Mr Faden.

A Letter from Mrs Dixon by favour of Mr Rennie, was communicated to the Society, & thanks were directed to be returned, for her obliging communication. Ordered, that a letter be wrote to Mrs Dixon, to that effect—and that her letter be printed at the End of the Life, composed by Dr Hutton.

The portrait of Smeaton, referred to above, had been commissioned by his friend Alexander Aubert (1730–1805, F.R.S.: 1772) from the painter Mather Brown (1761–1831). It was presented by Aubert to the Royal Society in 1799. The engraving by William Bromley (1769–1842), after this portrait, appears as the frontispiece to Volume I of the *Reports* (10). It is reproduced in plate 11.

Following Mylne's Preface comes a section entitled 'Some Account of the Life, Character and Works of Mr. John Smeaton, F.R.S.' This consists of two parts, the first of which is a biographical memoir by Hutton based on the article in his *Mathematical and Philosophical Dictionary* (vol. 2, London, 1795, pp. 462–465) which, in turn, is largely derived from *A Short Narrative of the Genius, Life, and Works, of the late Mr. John Smeaton, civil engineer, F.R.S.* (London, 1793) written by Smeaton's life-long friend and kinsman, John Holmes (*c.* 1728–1797). The second part is a letter, dated 30 October 1797, written at the request of the Special Committee by Mrs Mary Dixon, in which she gives a lively sketch of her father's character (11).

With all this introductory material in hand, the production of Volume I could proceed.

On 19 January 1798 the Committee:

> Resolved, that the price of the 1st Volume of Smeaton's Reports in boards, be 18s. to the publick.

and on 23 March 1798:

> The publication of the 1st Volume of Mr Smeatons Reports was announced by Mr Faden, and that the same is favourable to the intentions of the Society.
>
> A copy, for each of Mr. Smeaton's daughters, was directed to be sent to them, in the name of the Society. A copy was also directed to be sent to Doctor Hutton, the Editor.

Later that year, in November, William Faden:

> Reported that 89 Copys, half bound, had been Sold, at 18 shill. each.

and on 7 December 1798:

> Mr Mylne laid before the Committee an Account of the Receipts and payments: for Printing and Publishing the First Volume of Smeaton's Reports.

	£	s.	d.
Received	£280	2	0
Paid or indebted to Mr Brook, Printer	395	15	3
Ballance, against it	£115	13	3

N.B. Allowance to Mr. Faden, not Stated therein

The selling price of 18s. coupled with Mr Brooke's bill for printing (and paper) indicate an edition of about 500 copies. The amount received agrees exactly with the £200 subscribed by the Special Committee plus the 89 copies sold.

The volume is quarto, with 412 pages of main text, as well as the Preface, etc. The paper is watermarked 1794 and 1795, and there are no plates other than the portrait frontispiece. The title page bears the date 1797. In this state Volume I is exceedingly rare, and few historians are aware of its existence (12). Most owners would have replaced it by Volume I of the uniform edition of 1812, which had 33 plates added. But the text remained unaltered, as we shall see, and the correct reference to the Preface and all other matter (except the plates) in Volume I of the Reports is therefore 1797, and not 1812 as invariably quoted.

Volumes II and III, and the plates

On 22 January 1802 William Faden reported the gift to the Society, by Mary Dixon's husband, of a collection of copper plates illustrating the Eddystone lighthouse and some other of Smeaton's works.

The letter of thanks, written by Robert Mylne, reads as follows:

London, 28 Feb. 1802.
We, of the Committee of the Engineers Society, for the publication of Mr Smeaton's Reports, beg leave to acknowledge, the Valuable present, of two boxes of engraved Copper Plates, 33 in Number, which you have transmitted for us to Mr Faden. And, we beg leave to offer to you (as representative of Mr Smeatons family) the grateful Sense which the Society feels on this Ocassion.

With these Materials, the Committee will be enabled, to pursue a more extensive Plan, after the publication of their Second *Volume of Reports*; by adding *such Plates*, as are necessary and can be procured, explainative of the Said Reports; by offering to the publick if it shall appear proper to do so a more useful and Convenient Edition *Quarto* of the Edystone Light-house; and also, by publishing Ramsgate-Harbour, uniform in Size, with these Several Works.

In performing these tasks (which they think of imposing on themselves) The Committee consider there cannot be erected, a more lasting Monument in honour of their departed friend; than giving to the World in a usefull Manner, and Circulating to Posterity, the labours of a man, who by instructing the age he lived in, transmits correct and practical knowledge on Subjects, peculiarly interesting to this Country.

> Your Contribution for these purposes, from *the family of Smeaton*, is a testimony of aprobation, which the object deserves, and the Committee prize in the highest degree.

Here, then, is a clear indication that the Society now had in mind the publication of the second volume and the re-publication of volume I, both to be illustrated by plates specially engraved; and also the publication, in quarto form, of Smeaton's well-known book on the Eddystone (13).

The first step towards achieving these objectives was taken on 6 May 1803, when:

> A Proposition was Settled, to the Extent of Engraving all the Drawings of Mr Smeaton (the property of Sir Joseph Banks) which may elucidate in a proper manner—The Subject matter of the 1st Volume of Reports Published in the name of this Society, and annexing the Prints thereof, to the Said Publication.

Meanwhile Mr Brooke, the printer, had died; and his son, Samuel Brooke junior, requested that the remaining copies of Volume I should be removed from his warehouse. In July 1803 Mylne offered to 'take care of them in safe custody and preservation'.

On 21 February 1806 Mylne reported on the expense of the publication of the remaining Reports and the engravings:

> whereupon it was proposed that a Subscription of 5 Guineas each should be entered into for the purpose above mentioned.

Those who at once put down their names for this subscription, include: Mylne, Jessop, Rennie, Huddart and Banks, of the Special Committee, together with William Vaughan, John Lloyd, William Faden, the Earl of Morton and William Chapman. A paper of Mylne's among the Society's archives shows that £115 had been subscribed before the end of the 1806 session.

Progress seems to have been slow, however, and it was not until 4 March 1808 that Mylne called for further subscriptions. At this meeting it was:

> Resolved that application be made to Sir Joseph Banks, that he would be pleased to permit the Society to have leave to take copies of such Drawings or Plans from his collection of Mr Smeatons papers as may be necessary to elucidate the Reports contained in the Vol. already published, and those papers remaining to be printed.

A year later, Mylne now having reached the age of 76, John Rennie took over the work.

145

17 March 1809
Mr Rennie reported his having had several conversations with Messrs Longman, Hurst & Co. respecting the publication of the two remaining volumes of Smeaton's Reports also for engraving the Copper plates for the first Vol. Likewise Sir Joseph Banks had obtained the consent of the Heirs of Mr Smeaton to give up their claim as to any advantage which might arise from the publication of these Reports. Mr Rennie further informed the meeting that Messrs Longman and Co. were willing to undertake the printing together with all the Disbursements attending the publication of the two volumes with their requisite Engravings and that they would give to the Society thirty copies of the reports complete on condition of their receiving five hundred copies of the first Vol.

It must have been shortly after this meeting that the drawings and other papers were handed over to John Farey (1766–1826); and the delicate work of copying and reducing the original drawings to quarto size was entrusted to his sons John Farey junior (1791–1851) and Joseph Farey (14). The copper plates, seventy-two in number, were all engraved during the years 1809, 1810 and 1812 by Wilson Lowry (1762–1824, F.R.S.: 1812), who brought the art of precise technical and scientific illustration in this medium to its highest peak. A characteristic example is reproduced in plate 12.

Mylne's final contribution to the enterprise is recorded in his diary (15):

7 Feb. 1810 Delivered all the remaining copies of Mr Smeaton's work to Messrs Longman.

and in the Minutes:

10 Feb. 1810 Mr Mylne informed the Society that he had delivered 120 Copies of Smeaton's Reports to Messrs Longman, Hurst, Rees & Orme.

Presumably some other unsold copies of Volume I were still with William Faden.

At the Engineers' Society on 31 January 1812 it was:

Resolved that it be recommended to Messrs Longman & Co. to print 750 Copies of Mr Smeaton's Reports.

The work was divided between Samuel Brooke (junior), who printed Volume II, and Strahan & Preston for Volume III. The paper in these volumes is watermarked 1809 and 1810, and the title pages bear the date 1812. At the same time Volume I was reissued with plates and a new title page uniform with Volumes II and III.

I have seen seven copies of the 1812 issue of Volume I and in all cases the paper is watermarked 1794 and 1795. The sheets therefore are those originally printed for the first (1797) edition; simply rebound, with the plates inserted, and a new title page (16).

The exact date of publication is not known, but on 21 May 1813 the Minutes recorded that:

> It was resolved that one Copy of the set of three volumes of Smeaton's Reports should be sent to each of the following
>
> | | |
> |---|---|
> | Mr Dixon | University of Oxford |
> | Mr Brooke | University of Cambridge |
> | The Royal Society | College of Edinburgh |
> | British Museum | College of Glasgow |
> | Advocates' Library | Royal Irish Academy |
>
> from the Society of Civil Engineers.

Finally, on 4 June 1813, it was:

> Ordered that a Copy of Smeaton's Works out of the 19 Vols belonging to the Society should be given to each of the Subscribers who have paid above Seven Guineas & that those who have only paid £5 5 0 should pay £2 2 0 additional before their volume was delivered to them (17).

The 'miscellaneous papers'

In addition to the Reports, the Society decided to publish, in a uniform quarto edition, a volume containing the eighteen papers which Smeaton had contributed to *Phil. Trans.* (18).

Copied into the Minutes for 17 January 1812 we find the following letter from Messrs Longman to John Rennie, dated 12 December 1811:

> Smeatons Papers in the Philosophical Transactions occupy 300 pages & 10 plates. If they are printed now in the same number of pages, the expence of 500 copies will be about £262; and if you can have the use of the copies of the Royal Society [plates] the expence will be about 80 guineas less.

Once the *Reports* had been published, this matter of the *Papers* was taken up actively:

> 4 June 1813. A letter from Messrs Longman & Co[y] was read proposing to publish the papers which Mr Smeaton sent to the Royal Society & pub-

147

lished in the Transactions in a volume to make a fourth to his works already published on the Condition that the Society would procure from the Royal Society the plates which were used at the publication of the papers and on the Society [of Civil Engineers] paying £94. The Society unanimously resolved to accede to this proposal.

The Council of the Royal Society granted the loan of the plates in November 1813.

On 24 June 1814 the Engineers' Society reported that:

It was moved and seconded that the Money in the hands of the Treasurer should be paid to Messrs Longman, Hurst & Co. on account of the publication of the fourth volume of Mr Smeaton's works & it was resolved unanimously and Mr Rennie was directed to write the Treasurer Mr Vaughan accordingly.

A fortnight later, on 8 July 1814:

Mr Rennie produced a Manuscript of a Preface intended to be prefixed to the Volume of Smeatons Papers printed in the Philosophical Transactions and the same being read by Mr Giddy was approved of and ordered to be the preface to the said volume.

This Preface, or *Advertisement*, dated 20 July 1814, reads as follows:

The papers which are here collected together include the whole of Mr SMEATON'S communications to the Royal Society, published in the Philosophical Transactions. Most of the articles have been reprinted verbatim from the originals in that work: a few of them have been printed from copies which had belonged to the author, and contained corrections and additions in his own hand writing. For the latter the editor has to make his acknowledgments to Mr SMEATON'S daughter, Mrs DIXON, who obligingly lent him the tracts.

The accompanying plates, with one or two exceptions, are the same as were originally given with the several articles in the Philosophical Transactions; the President and Council of the Royal Society having, in the handsomest manner, granted the loan of the coppers, in order to afford every facility in their power to the publication.

The present volume completes the series of Mr SMEATON's works. The account of the Edystone light-house was published during his lifetime, and has been lately reprinted. Since his death, the Committee of Civil

148

Engineers have published, with the liberal permission of Sir JOSEPH BANKS, who had purchased the whole of the manuscripts, three volumes of Mr SMEATON's Reports, which comprise every paper of any consequence composed by him, in the way of his profession, during the long period of his public life. To these volumes, the present may be considered as an important and indispensable appendage; and the whole together will be found to contain a most valuable fund of practical instruction in the several branches of Science and the Arts, which had occupied the attention of the ingenious author, and upon which he has raised for himself a high and lasting reputation.

Second edition of the reports

It remains only to mention the second edition of the *Reports*, published in 1837. This contains the unabridged text of the first edition, the portrait frontispiece and all seventy-two plates. The latter were printed from Lowry's original engravings, with new page references. Four pages of Introductory Remarks were written by the editor, W. M. Higgins, F.G.S., Architect and Engineer; formerly Professor of Natural Philosophy, Guy's Hospital. He admirably summarizes the importance of the *Reports* in the following words:

> Considering the science of construction . . . we know of no work that has a character so universal as SMEATON's Reports. . . . They are intrinsically valuable for the engineering information they contain . . . and as the records of great national works, we find in every page some evidence of the wealth, energy, and industry of our native country. . . . As the labours of one man they have an interest of no common order . . [and] as long as an accurate discernment, extensive observation, genius, modesty, and inflexible honesty, are valued, SMEATON's Reports will be read with advantage and followed as models.

Acknowledgments

I am grateful to Sir John Wrightson, Bart., Hon. Treasurer of the Society of Civil Engineers, for the loan of the Society's Second Minute Book. I have received much help from the staff of the Royal Society library, and from Mr Richardson, Librarian of the Institution of Civil Engineers. Miss Joyce Brown, of Imperial College, has been of great assistance in her work for me at the British Museum.

Appendix I

Contents of the Reports

The following analysis of the contents of Smeaton's *Reports* is not complete, but includes most of the more important items. It conveys in the briefest possible manner some idea of the extraordinary range of his work.

Of the reports printed in Smeaton's lifetime, those marked (*) are in the British Museum or the Institution of Civil Engineers. Other printed reports marked (†) I have not seen, but they are recorded in a list compiled by John Rennie for Sir Joseph Banks (*n.d.*, B.M.: B.276). Page references are given for the three volumes of the first edition (1797–1812). Dates refer to the year in which the reports were written; those placed in square brackets [ ] have been obtained from evidence such as dated drawings etc.

Inland navigation			
River Calder	1757	I	20– 21
River Calder	1767–69	II	125–145
Trent and Mersey Canal	1761	I	13– 17
*River Witham (with John Grundy and Langley Edwards)	1761	I	26– 35
Fossdyke (with John Grundy)	1762	I	55– 71
*Forth and Clyde Canal, 1st Report	1764	II	31– 65
†Forth and Clyde Canal, 2nd Report	1767	II	73– 97
*Forth and Clyde Canal (Smeaton's *Observations* on reports by Brindley, Yeoman and Golborne)	1768	II	98–120
River Devon	1767	I	154–169
River Lea	1766–67	II	155–163
River Ure	1770–72	II	172–184
Dublin Grand Canal	1773–75	II	247–277
Birmingham Canal	1782	II	189–212
Fen drainage			
Potterick Carr	1762–64	I	39– 54
Holderness Level	1764	I	88– 98
*Lewes Laughton Level	1768	I	340–355
†North Level of Fens	1768	II	280–298
Hatfield Chase	1776	II	299–315

Harbours			
Christchurch	1762	I	192–198
*Rye	1763	I	103–116
Whitby (with John Wooler)	1765	I	230–233
†Bristol	1765	I	218–222
St Ives	1766	I	295–300
Eyemouth	1767	I	302–305
†King's Lynn	1767	III	1– 11
*Dover	1769	III	134–146
Dover (sluice gates)	1771	III	146–150
Port Patrick	1770–74	III	60– 73
Aberdeen	1770	III	38– 46
Aberdeen	1788	III	47– 50
Workington	1777–78	III	170–176
*Bridlington	1778	III	187–193
*Sunderland	1780	III	194–195
Scarborough	1781	III	198–199
*Wells	1782	III	18– 37
St Hellier and St Aubin	1788	III	206–214
Sandwich Haven	1789	III	129–133
*Ramsgate	1791	III	74–128
Lighthouse			
Spurn Point	1767–78	I	252–278
Docks			
Hull	1773	III	155–162
Hull (with John Wooler)	1778–79	III	163–169
North Shields	1775	III	200–203
Plymouth Dockyard	*n.d.*	III	177–186
Bridges			
London Bridge	1763–67	II	1– 26
Perth (including John Gwyn's report 1766)	1763–67	I	174–190
Coldstream	1763–66	III	235–242
Coldstream (including Robert Reid's report 1784)	1781–90	III	244–251
River Trent, Flood Bridge	1768	I	327–332
Edinburgh (with John Adam and John Baxter)	1769	III	281–222
Newcastle	1769–71	III	252–259
*Newcastle (with John Wooler)	[1772]	III	260–266
Banff	1772	III	349–354
Hexham	1777–80	III	267–293

Hexham failure	1782–84	III	293–344
(including Robert Mylne's report 1783 and † Smeaton's Memorial concerning Hexham Bridge, III, 299–320)			
Walton	1778	III	371–377
Harraton	1783	III	378–382
Dams			
River Carron	[1773]	I	379–381
River Coquet	1776	II	324–332
Water supply			
Halifax	1761	I	7– 11
Edinburgh	1780	III	228–234
Mills (*waterwheel*)			
Wakefield, flour mill	1754	II	430
Kilnhurst Forge, hammer mill	[1765]	II	424–425
Carron, blowing engine No. 2	1769	I	364–366
Carron, boring mill	1770	I	376–379
Hull, oil mill	1776	II	398–401
Pumping engines			
(a) *Waterwheel*			
London Bridge Waterworks	1767	I	27– 28
Deptford Waterworks	1779	II	367–372
(b) *Horse-gin*			
H.M. Victualling Yard, Gosport	1778–80	II	381–391
(c) *Steam*			
Engine for Earl of Egmont	1765	I	223–229
Chasewater Mine	1775	II	347–359
Long Benton Colliery	1777	II	435–438
York Waterworks	1785	II	342–345

It is worth recording four early reports, not included in the collected edition:

A Scheme for taking down and rebuilding the damaged Pier of Westminster Bridge	1748
Report of J. Smeaton relative to the deepening of the River Clyde	Sept. 1755
Survey of the River Calder, from Wakefield to Brooksmouth . . . near Halifax	Nov. 1757
Mr. Smeaton's Answer to the Misrepresentation of his Plan for Black-Friars Bridge	Feb. 1760

Appendix II

Bibliography

Reports of the late Mr. John Smeaton, F.R.S., made on Various Occasions in the Course of his Employment of an Engineer.
Printed for a Select Committee of Civil-Engineers, and Sold by Mr [William] Faden, Geographer to his Majesty. Vol. I. London: Printed by S[amuel] Brooke, Warwick Lane, M.DCC.XCVII.
Preface [by Robert Mylne]; *Some Account of the Life, Character and Works of Mr John Smeaton F.R.S.* [by Charles Hutton and Mrs. Mary Dixon]; *Contents*; *Reports &c.*
Quarto. xxv+(viii)+412 pp. and portrait-frontisp.

Reports of the late John Smeaton, F.R.S. made on Various Occasions in the course of his employment as a Civil Engineer, In three Volumes.
London: Printed for Longman, Hurst, Rees, Orme and Brown, Paternoster-Row. 1812.
Vol. I reissue of 1797 edition, with new title page and 33 plates.
Vol. II *Contents; Reports.*
Quarto. xi+440 pp. and 23 plates. S[amuel] Brooke [junior], Printer, 35 Paternoster-Row, London.
Vol. III *Contents; Reports; Index.*
Quarto. vii+420 pp. and 16 plates. Strahan and Preston, Printers-Street, London.

The Miscellaneous Papers of John Smeaton, Civil Engineer, F.R.S. comprising his Communications to the Royal Society printed in the Philosophical Transactions, forming a Fourth Volume to his Reports.
Illustrated with Plates. London: Printed for Longman, Hurst, Rees, Orme, and Brown, Paternoster Row. 1814. S[amuel] Brooke [junior], Printer, 35 Paternoster-Row, London.
Advertisement [by John Rennie, editor]; *Contents; Treatises and Miscellaneous Papers.*
Quarto. viii+208 pp. and 12 plates.

Reports of the late John Smeaton, F.R.S. made on various occasions in the course of his employment as a Civil Engineer. Second Edition. In two Volumes.

153

London: Published by M. Taylor (nephew and successor to the late J. Taylor) Wellington Street, Strand. 1837. Printed by James Moyes, Castle Street, Leicester Square.

Vol. I *Contents; Introductory Remarks* [by W. M. Higgins, editor]; *Preface; Some Account of the Life . . .; Reports.*
Quarto. xxxv+423 pp. with portrait frontisp. and 38 plates.

Vol. II *Contents; Reports.*
Quarto. xii+423 pp. and 34 plates.

Notes

(1) The facsimile of the 2nd edition of Smeaton's *Reports* has been published by the Cornmarket Press, London (1969).

(2) *Sir Joseph Banks' Correspondence.* Dawson Turner Collection, Dept. of Botany, British Museum (Natural History). Vol. 8, p. 257; Vol. 9, pp. 214, 215; Vol. 10 (2), pp. 1, 2, 22, 23. Other letters are in B.M. Add. MSS. 33980.51 and 116. All are indexed in *The Banks Letters* (ed. W. R. Dawson), London, 1958.

(3) *Society of Civil Engineers.* [Second] *Treasurer's Minutes and Accounts, 1793–1821.* MS 258 pp. In possession of the Society and kept at the Institution of Civil Engineers.

(4) Since 1830 known as the 'Smeatonian Society of Civil Engineers'. For a short history of the Society, see A. W. Skempton, *The Smeatonians: Duo-Centenary Notes on the Society of Civil Engineers, 1771–1971* (London, 1971).

(5) Mylne's *Preface* to *The Reports of the late Mr John Smeaton F.R.S.* (London, 1797).

(6) To mention only one of the ways in which Banks and Smeaton were acquainted: they dined together several times a year at the Royal Society Club from 1770 (when Banks became a member) until a few months before Smeaton's death in 1792. Smeaton had been elected to the Club in 1765. (A. Geikie, *Annals of the Royal Society Club*, London, 1917.)

(7) H. W. Robinson, *Notes and Records Roy. Soc. Lond.*, **2**, 69–70 (1939).

(8) The Minutes for 22 January 1802 mention that the Special Committee had 'expended £200 of their own Money . . . in publishing the 1st Volume of Mr. Smeaton's Reports'. This would be the sum made up by contributions of £40 each from Banks, Jessop, Huddart, Rennie and Mylne.

(9) The *Preface* to Vol. I of *The Reports of the late Mr John Smeaton F.R.S.* (London, 1797) is unsigned, but there can be no doubt that it was written by Mylne. As the leading member of Society of Civil Engineers the task of composing this important *Preface* would naturally fall to him. By 1797 no other member of the Special Committee had personal knowledge of the formation of the Society in 1771, which is described in some detail; and it was to Mylne that Sir Joseph Banks sent various criticisms of the draft of the *Preface*.

(10) Alexander Aubert's letter to Sir Joseph Banks (R.S. Letters, 3.3), reads as follows:

Highbury House. 7 Nov. 1799.

I beg leave to present to the Royal Society a Picture of my late most esteemed Friend Mr. John Smeaton, Civil Engineer and Fellow of this Society, whose

> memory will ever be dear to all Lovers of Science & particularly to those who were acquainted with him. It was painted by Mr Mather Brown, a few years before his [Smeaton's] death & is allowed to be a striking resemblance of him. A print engraved from this picture by Mr Wm. Bromley, I had the pleasure of giving to many of his friends some years ago. The Society of Civil Engineers have also prefixed a print taken from it by the same artist to the first volume of his Reports from your kind communication to them of the Manuscripts, Designs Drawings &c which you so liberally purchased of his Representatives. . . . I present the Picture to the Society because the members of it will have more frequent opportunities of seeing it than if I continued to keep it in my possession & because it may contribute to perpetuate the memory of so valuable & so amiable a member.

It may be noted that the engraving by William Bromley was published in 1790. See F. O'Donoghue, *Catalogue of Engraved British Portraits . . . in the British Museum*, Vol. 4 (London, 1914). It scarcely does justice to the original.

(11) 'Mr Smeatons Daughters frequently assisted in the shadows and finishing, in indian ink' their father's drawings. This is recorded by John Farey junior in a draft note in Vol. I of Smeaton's *Designs* (Royal Society library). Mary Dixon also made the drawing for the frontispiece of the Eddystone folio (engraved 1789), and it was from her that John Rennie borrowed Smeaton's own copies of his *Phil. Trans.* papers, for the publication of Vol. IV (1814). She bequeathed to the Royal Society a portrait of Smeaton painted by J. Richardson (see list of Presents to the Society in *Phil. Trans.* for 1824).

(12) The only copy I have seen of Vol. I of the *Reports*, in the first state (1797), belonged to Sir Joseph Banks. It is in the British Museum (431.d.12). Vol. I of the 1812 set at the Institution of Civil Engineers has the 1797 title page bound in. This set was presented to the Institution by the Society of Civil Engineers.

(13) A quarto edition of the Eddystone book was not published. Instead, Messrs Longman & Co. reissued the folio in 1813 (see Note 17), presumably using the original copper plates presented to the Society by Mr Dixon. The Eddystone folio was first published in 1791, and a second edition in 1793. The report on Ramsgate Harbour was reprinted in Vol. III of the *Reports*.

(14) John Farey senior was well known to Banks, for whom he worked on several occasions as a practical geologist (*The Banks Letters*, *op cit.*; L. R. Cox, *Proc. Yorks. Geol. Soc.*, **25**, 39 (1942). In a letter dated 4 August 1810 (*Phil. Mag.*, **36**, 102–105 (1810)), Farey writes: 'My eldest son was a few days ago employed, in examining the miscellaneous bundles of papers, which belonged to the late ingenious Mr. John Smeaton, the civil engineer, now in Sir Joseph Banks's possession, with a view to the further publication by Messrs. Longman, Hurst, Rees & Co. of his Drawings and Reports on civil engineery, which so long and impatiently have been expected by those interested in this branch of the useful arts. . . .' Farey's sons, John and Joseph, had already established their reputations as superb engineering draughtsmen.

(15) Mylne's diary has been published by A. E. Richardson, *Robert Mylne, Architect and Engineer* (London, 1955).

(16) The seven sets of the 1812 edition of the *Reports*, which I have examined, are (i) Sir Joseph Banks' copies (B.M.: 431.d.13–15); the copies presented by the Society of Civil Engineers (ii) to the British Museum (B.M.: 62.c.16–18), (iii) to the Royal Society, and (iv) to the Institution of Civil Engineers; (v) the Society's own set; (vi) the set in

Mr R. Glossop's collection, previously belonging to the Glasgow Philosophical Society; (vii) my copies. The sheets of Vol. I in all these are identical with those in the first issue of 1797, and are watermarked 1794 and 1795. If this is generally true, the 1812 edition must have been limited in size to the number of 1797 copies still unsold.

(17) An advertisement by Messrs Longman, in 1814, reads: 'Lately Published. Mr Smeaton's Reports. . . . In 3 Vols. Quarto. Price £7 7s. Boards, illustrated with 74 [*sic*] Plates by LOWRY. Also, a New Edition of a Narrative of the Building . . . of the Edystone Lighthouse. . . . By John Smeaton, Civil Engineer, F.R.S. in Imperial Folio. Price £6 6s. half Bound.'

(18) Three of Smeaton's *Phil. Trans.* papers had been reprinted by I. & J. Taylor in 1794: *Experimental enquiry concerning the natural powers of water and wind* . . . (1759, Copley Medal paper); *An experimental examination of the . . . mechanical power necessary to be employed in giving different degrees of velocity to heavy bodies from a state of rest* (1776); *New fundamental experiments on the collision of bodies* (1782). The book, 'offered to the public at a moderate purchase', proved a success, and was reissued in 1796. Taylor approached the Society of Civil Engineers in January 1799 with a view to purchasing the remaining copies of Vol. I of the *Reports*, the manuscript copies of the remaining reports and a selection of Smeaton's drawings. The proposition was not accepted, but we may note that Taylor's nephew and successor published the second edition in 1837.

Postscript

A. P. Woolrich (*Notes & Records Roy. Soc. 35* (1980) 131–133) has discovered, from Longman's ledger books, that Mylne delivered 714 copies of volume 1 in 1810. Therefore it now appears that the puzzling figure of 120 recorded in the Minute Book (see p. 145) should have been 720, this probably being Mylne's estimate of the number handed over. The ledgers also reveal that 500 copies of the three-volume set were issued in 1812, and at no time did any further printing of volume 1 take place. Thus the suggestion (pp. 143 and 146) that the text of this volume in all copies dates from 1797 is confirmed.

V

The Engineering Works of John Grundy (1719–1783)

John Grundy died 200 years ago, on 15 June 1783, at about the time of his 64th birthday. He belonged to an early generation of English civil engineers which included John Smeaton, FRS and 'that great Genius Mr. Brindley'. In his own day Grundy gained a well-deserved reputation, and he contributed a good deal to the benefit of our country by improvements in the rivers and fenlands of Lincolnshire and Humberside. But by the mid-19th century he was almost forgotten, and remained so until 1938 when Dr Esther Wright drew attention to him and other little-known engineers of the period.[1] More recently, a considerable amount of research has been devoted to Grundy and to his father;[2] several important documents have come to light, and our present knowledge is well expressed in the booklet by Neil Wright.[3]

Here, then, it is necessary to touch only in the briefest way on Grundy's life and family history. My purpose is to examine his engineering works. These are listed in the Table at the end of the paper, together with a selection from the large number of his reports including some on schemes that were not carried into execution.

JOHN GRUNDY

He was baptised on 1 July 1719 at Congerstone in Leicestershire, the son of John Grundy and Elizabeth (neé Dalton) his wife. John Grundy senior (*c*. 1696–1748) 'of Congerstone, Land Surveyor and Teacher of the Mathematics', as he described himself in the 1730s, had a wide-spread practice as a surveyor, but became increasingly concerned with civil engineering matters involving fen drainage and rivers. In 1735–36 he was consulted on the drainage of Deeping Fen, and in April 1738 he and Humphrey Smith were jointly appointed 'Surveyors and Agents' (i.e. engineers) to the Adventurers, Owners and Proprietors of this large tract of fenland in Lincolnshire. At about that time he and his family moved to Spalding. In 1742 he was appointed sole Agent, a post he held, with freedom to undertake some outside work, until a year before his death.

Originally published in *Lincolnshire History and Archaeology*, vol. 19, 1984, pp. 65–82.

Fig. 1 John Grundy, 1719–83

There can be no doubt that 'our' John Grundy was trained as a land surveyor and civil engineer by his father, under whom he worked as assistant (with a few jobs of his own) in the early 1740s. His independent engineering career effectively begins in 1745 with a series of works at Grimsthorpe for the Duke of Ancaster. In 1748 he became engineer to Deeping Fen. Among numerous reports written in the 1750s two at least led to major works. From 1760 the scale of his operations expanded notably, and by 1771 when Yeoman, Smeaton, Grundy, and some others founded the Society of Civil Engineers[4] he occupied a senior position near the top of his profession. After 1776, however, we know of only two reports and no works by him.

Before the canal 'boom' of the late 1760s few civil engineers in England could earn more than a modest income. To supplement his earnings Grundy began trading in timber from the Baltic and goods from Holland, and later owned a sloop and a sailing barge. Also, from 1762, he took on a part-time job as Collector of the Land Tax for the Spalding District.

In January 1743, he married Lydia Knipe at Spalding. Two of their children, Mary and Lydia, survived infancy. Mary married William Thompson of Spalding, and their descendants represent the family at the present time. Grundy's wife died in 1764. Two years later he married Mrs Ann Maud, widow of the Revd John Maud of St Neots, but there were no children of this marriage. Grundy was buried at Spalding, the town where he and his family had lived for many years in a house facing the River Welland. His portrait[5] was painted probably about 1770.

SOURCES

Grundy kept manuscript copies, and any printed copies, of his reports and plans, the originals having been sent to the clients. A complete short-title list is given in his 'Catalogue of Books and Papers' compiled in 1767 with additions to 1776, which also includes a list of his father's papers.[6] This is referred to here as *Catalogue*. At some later time he copied selected reports and drawings (on a reduced scale) into a series of probably ten quarto 'report books'. These and various other engineering documents were purchased by Sir Joseph Banks in 1793 from Grundy's son-in-law William Thompson, and eventually sold at Sotherbys in 1918.[7] Vol. 2 of this set is now in the Institution of Civil Engineers[8] referred to as *Report Book*. The present whereabouts of the others is not known. A few original manuscript reports and plans exist in libraries and county archives, and all of his printed reports and plans have been found.

The other type of source material consists principally of Minute Books of the various authorities for whom Grundy acted as engineer. Most of these books are now in either the Lincolnshire or Humberside county record offices.

V

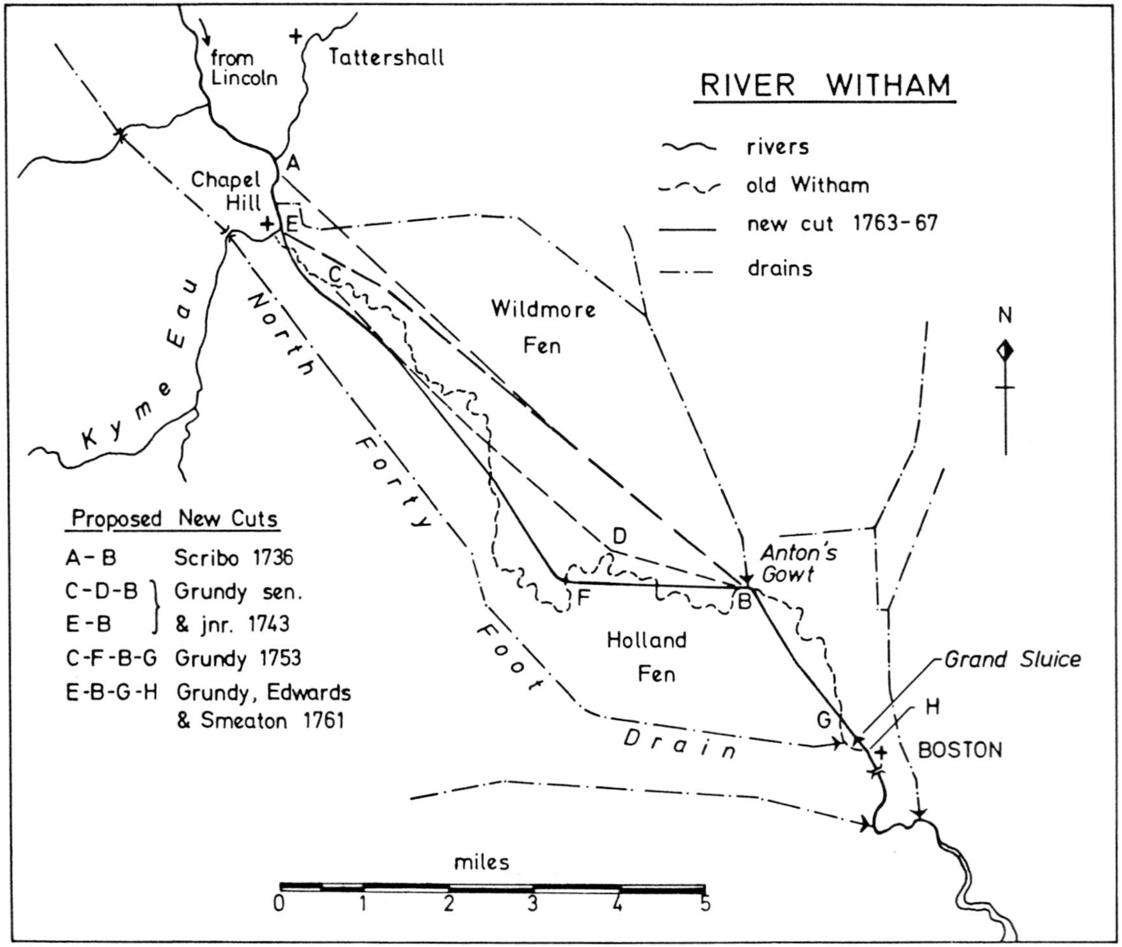

Fig. 2 The River Witham

THE NOVICE ENGINEER, 1739–44

Pinchbeck Sluice

An outfall sluice was built between October 1739 and April 1740 on the Blue Gowt Drain for the Pinchbeck area, near Spalding. It had a 12ft waterway with pointing doors 7½ft high. The contract document is signed by John Grundy (senior) and he drew up the estimate, but the drawing accompanying the contract is by Grundy junior.[9] We can assume that he acted as his father's assistant during construction.

Sea Wall at Palling and Waxham

Between August 1742 and January 1743 Grundy executed repairs to the breached sea wall embankment on this part of the Norfolk coast, in accordance with plans prepared by his father. His report gives details of the work with sketches, including groynes or 'jetties', and sheet piling.[10]

River Witham, 1743–44

The Grundys *père et fils* produced a survey of the Witham from Lincoln to the estuary below Boston, with numerous accurate levels. This was issued as an engraved plan dated 1743[11] and accompanied a 48-page printed report published in 1744;[12] both signed 'John Grundy Sen. and Jun. of Spalding, Engineers'. The report is chiefly from the pen of Grundy senior; the survey is perhaps mainly by the son.

The object of their investigations was to devise a scheme for improving navigation on the Witham and drainage of the adjacent fenland. They identified the three main causes of trouble as the bad state of the river embankments, the 'prodigious meandering Course of the River' between Anton's Gowt and Chapel Hill (see Fig. 2), and the diversion of the upland waters north of Kyme Eau through a tunnel into the North Forty Foot Drain (opened *c.* 1720). The authors then gave eighteen 'theorems' or guiding principles and finally put forward their proposals, with estimates. First: the existing river channel between Anton's Gowt and Chapel Hill could be scoured and widened, but a preferable scheme was greatly to shorten the course by a new cut about 7 miles long as shown at CDB in Fig. 2 or by a new cut of similar length entirely north of the river in Wildmore Fen as shown at EB; this being a modification of a proposal already made by James Scribo.[13] The cut was to be properly embanked, the banks set back at least 30ft from the water's edge, and the cut to have a top width of 70ft with an average depth of 7ft. Second, to scour, widen, and bank the river from Chapel Hill to Lincoln. Third: to build two stanches of flash locks in this length to maintain an adequate depth of water in dry seasons. Fourth: to extend the drain west of the river (and north of Kyme Eau) almost up to Lincoln and turn its water into Kyme Eau, stopping the North Forty Foot Drain just south of this stream. Fifth: all drains, banks, and sluices to be reinstated.

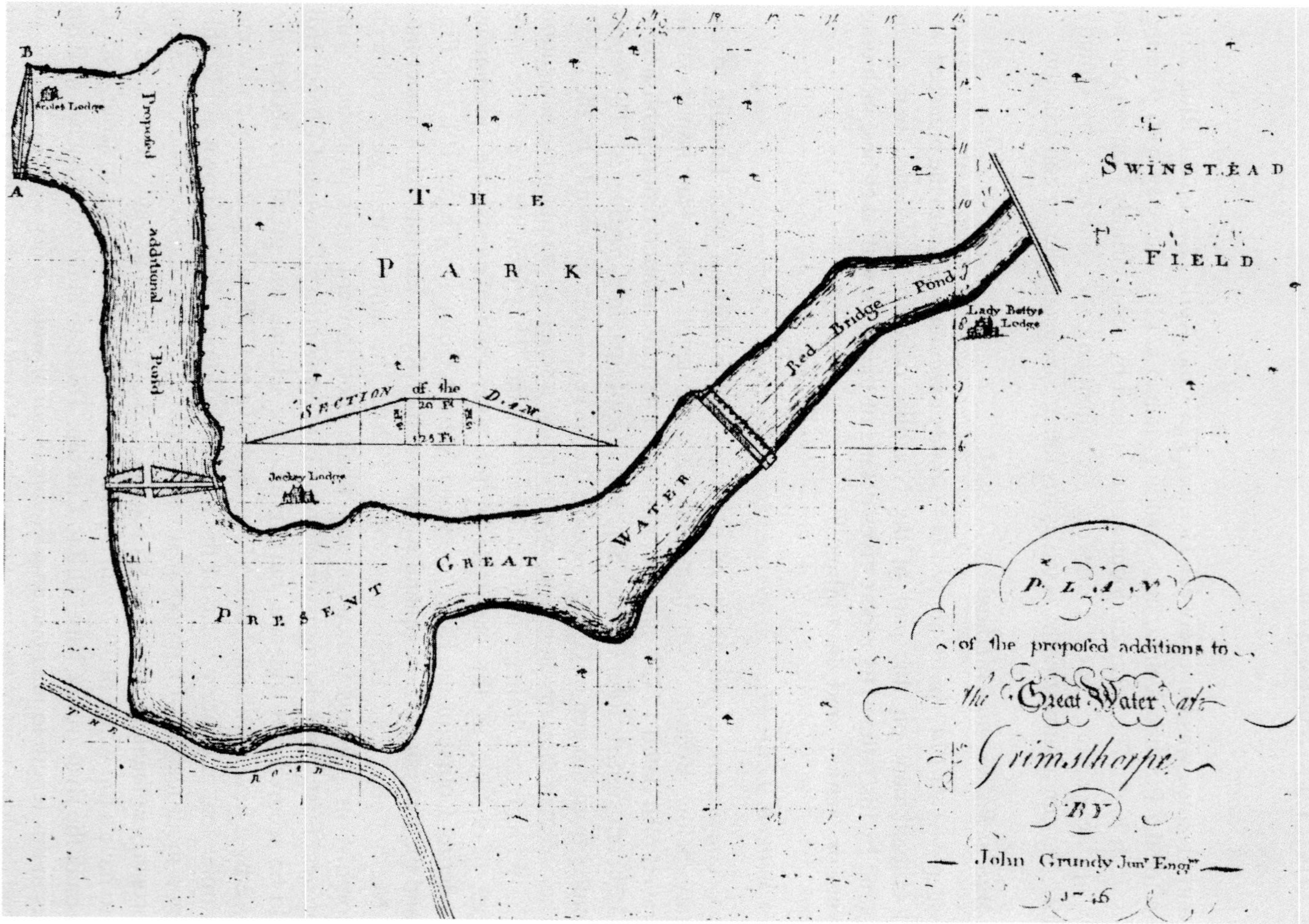

Fig. 3 Grimsthorpe Great Water Plan, 1746

The scheme proved too expensive for the landowners to contemplate at that time, but it served as a basis for future plans and indeed most of its features, with some modifications, were eventually put into effect as we shall see.

EARLY WORKS, 1745–59

Grimsthorpe Dam

Between 1745 and 1752 Grundy carried out several commissions for the 3rd Duke of Ancaster on his estate at Grimsthorpe in Lincolnshire. In approximate chronological order, these include an earth dam to impound a large ornamental lake 36 acres in extent called the Great Water (see Fig. 3), regrading a slope in the park to allow a better view of the lake, a reservoir for topping up the lake, a piped water supply to the house (Grimsthorpe Castle), and new Stew Ponds for fish.

Completed *c.* 1748 the Great Water dam is the earliest known example of an earth dam with a clay core-wall, a type of structure used for some late 18th-century canal reservoirs and widely adopted in the 19th century for water supply reservoirs.

The dam has a maximum height of 18ft, a length of 420ft and a crest width varying from 15ft at the ends to 25ft in the middle. The downstream slope, modified by more than 200 years of soil creep, was probably formed originally at 2 : 1 and from what can be seen of the upstream face it slopes at approximately 3½ : 1. The original outlet sluice, near the middle of the dam, has been replaced by a spillway on the right bank.

In the *Report Book* there are two relevant plans, both by 'John Grundy Junr. Engr.' and both dated 1746. The first of these[14] shows two small lakes or 'ponds' above and one below Red Bridge and, further downstream, the Mill Dam Pond of 17 acres, the latter being impounded by an earth bank 10ft high. This is clearly the layout before Grundy began his alterations. Three sketch sections show what he proposed to do, namely (i) to raise the little dam at Red Bridge to enlarge and combine the two upper ponds into one, shown as 'Red Bridge Pond' in Fig. 3, (ii) to raise and strengthen the Mill Pond dam, and (iii) to build a new dam 12ft high near Jockey Lodge impounding a pond of about 4 acres. A section of this dam shows a clay core. Moreover, a report[15] by Grundy, dated 1741 in the *Catalogue*, mentions 'clay for the wall within the bank' when referring to a small dam for a proposed 3-acre pond at Osbaston in Leicestershire; so the idea of using a clay core was not new (so far as Grundy is concerned) even in 1746.

Evidently the raising of Red Bridge dam went ahead, more or less as planned, but instead of proposals (ii) and (iii) the Mill Pond dam was removed and a new (i.e. the present) dam built near Jockey Lodge, much larger than in the first proposal and some 50yds further downstream.

As the Ancaster account books are missing from 1741 to May 1747 the cost and exact date of construction of the Great Water dam are not known. Mr Binnie, who first realised the significance of this pioneer structure[16], suggests 1748. This is an entirely acceptable date as the dam could not possibly have been finished before 1747 and certainly it must be earlier than mid-1749 since Grundy's final payment (in a total of £346) for 'Pond Works and Levelling' dates to 15 June 1749.[17]

The dam undoubtedly had, and still has, a clay-core wall. Not only is one shown in what may be considered as the preliminary design (i.e. the first porposal of 1746), but it is specifically mentioned in a report of 1758 when investigations were in progress to trace the course of leakage from the Great Water.[18] In fact the dam was sound in all its parts including the 'Clay Wall'; leakage occurred through a swallow hole in the lake bottom.

The second plan[19] (Fig. 3) shows a 'Proposed additional Pond' to be formed by a dam at Scoles Lodge on the line A–B. An undated report[20] gives a selection and specifications for this dam. Leading dimensions and some other details are shown in Fig. 4. Except for a greater height it probably represents closely the Great Water dam as built. The clay is specified to be 'well Rammed and Watered', and the estimated cost comes to around £800.

There were several swallow holes in the valley downstream of the Great Water

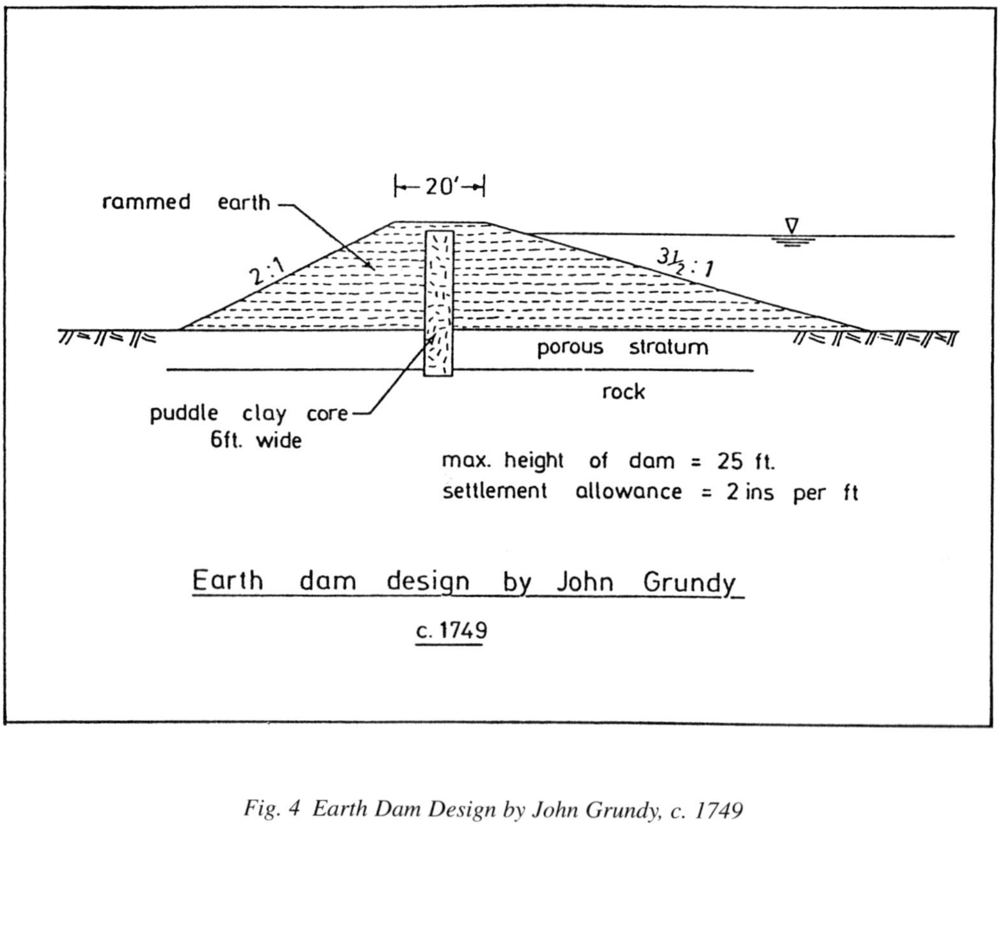

Fig. 4 Earth Dam Design by John Grundy, c. 1749

(indeed, one near the site of Scoles Lodge is clearly visible today), and in his report Grundy suggests that after trying to stop (i.e. to plug) them, a trial bank should be built to impound a temporary lake as an experiment to check on watertightness, the bank then to be incorporated in the dam. The trial bank was constructed, a short distance upstream of the line A–B and of the existing swallow hole. Traces of the bank can still be seen and it is marked on a plan of 1767, but the experiment must have proved to be a disappointment.

However, in 1766 Grundy revived the notion of a dam at Scoles Lodge, this time with a proposal to construct an impervious clay blanket over the entire valley bottom.[21] He also gave a revised design for the dam, shown in sketch profile added to Fig. 3 and in more detail in the plan of 1767 previously mentioned,[22] again with a central clay core but now with a maximum height of 30ft, somewhat flatter slopes, and the water face protected by stone pitching. The upstream clay blanket, which he specified with great care, is technically sound and of historical interest. But neither the dam or the blanket was constructed; the cost may have been too high or perhaps the Duke had lost interest in the scheme.

The idea of a dam at Scoles Lodge might have originated in 1746 as the date on Fig. 3 suggests, but the design in Fig. 4 is probably later. The report is listed in the *Catalogue* between Grimsthorpe items dated 1747 and 1751. The trial bank could have been built in the first batch of work, i.e. before the mid-1749 payment (though obviously subsequent to completion of the Great Water), or it may be the 'work at the New Pond' referred to in the Accounts between November 1750 and February 1751, under the direction of Grundy and William Hodgson. It cannot be later than the payment (totalling £142) to Grundy in 1752 for the second batch of work at Grimsthorpe which includes the water supply scheme, the new Stew Ponds, and 'Stopping Swallow Pitts'. A date of *c.* 1749 for the design (Fig. 4) is therefore unlikely to be wrong by more than a year or two.

Grimsthorpe Water Supply

The scheme involved pumping from a deep well into a cistern and taking the water in lead pipes from the cistern down to Red Bridge and up to Grimsthorpe Castle, a distance of one mile with a difference in head of 10ft, the water to be raised from the well by three pumps in series operated by a triple crank driven from a 'horse engine' or gin.[23] An elegant circular engine house was erected next to a square building for the cistern beside the well. The accounts record payments in June 1751 to Joseph Smith, engineer, for 'Pumps and Engine' and from June to September 1751 to labourers for digging trenches and laying pipes. Grundy supervised the work, which must have cost well over £300.

Deeping Fen, 1748–64

Grundy was appointed 'Agent' to the Adventurers of Deeping Fen on 18 May 1748 at a salary of £50 *p.a.* Featherstone continued as assistant at £40 *p.a.* Grundy held this post until 1764, when Thomas Hogard succeeded him, but he then became consulting engineer to the Adventurers with a retaining fee equal to his previous salary.

As Agent he had charge of all engineering and maintenance work on the banks, drains, sluices, and the River Welland, and also kept the accounts. His account book from 1753 still exists.[24] This and the Adventurers' minute books[25] give a good idea of the work involved, though no technical details are available.

A continuing charge was the raising, strengthening, and maintenance of the Welland banks for which the Adventurers had responsibility: Deeping Bank, 13 miles in length, and the 5 mile Country Bank, on the west and east sides of the river respectively, upstream of Spalding. A particularly demanding job must have been the almost complete rebuilding of the Six Doors Sluice on the Welland. Originally constructed *c.* 1731 by John Perry as a scouring sluice, Grundy extended the floor, raised the existing doors and provided another set of doors pointing seaward to stem the tide. A little later, in 1754, he rebuilt the navigation lock. Next year three new wind drainage engines were erected on Hills Drain and in 1758 a new sluice on the Forty Feet Drain, while in 1759–61 he deepened the Welland by more than 2ft, between the Six Doors and Spalding bridge, using a 'hedgehog': a spiked roller dragged along the river bed to loosen the silt so that it could more easily be sluiced away to sea. These are some of the works accounting for an annual expenditure averaging about £1400 during the years 1748–64.

Stallingborough Sluice, 1751

In November 1749 Grundy prepared preliminary estimates for a new outfall sluice in the south bank of the Humber near Stallingborough, about half way between Grimsby and Immingham, the purpose being to improve the drainage of low grounds belonging to Mr Bouchrett.[26] By 1751 the decision had been made to go ahead, and early in July of that year Grundy spent several days taking levels of the land and drains in relation to the floor of the existing old sluice. On the assumptions (i) that there should desirably be a fall of 1ft per mile between water level in the drains and the floor of an outfall sluice, and (ii) that water in the drains should be at least 1½ft below ground level, he concluded that the floor of the new sluice must be laid 4ft lower than the old one. The drains would have to be deepened correspondingly.

This having been agreed, contracts were let, and on 19 August Grundy visited the site to find the carpenters already making the dovetail piles and preparing timbers for the floor. A fortnight later work was well under way, with William Hodgson as foreman or resident engineer. No doubt the job was completed by the end of that year.

The *Report Book* includes a sketch elevation and a foundation plan. The sluice is in the usual 18th-century style with counterforted brick walls, a pair of self-acting pointing 'sea doors' 8½ft high and a draw door on the land side to maintain water level in the drains during dry seasons and also to be used occasionally for scouring the outfall. The waterway has a clear width of 6½ft. The planked floor and the walls rest on cross beams or 'ground sills' 10 x 8 inches in section, with square piles 8ft long under the walls. Three rows of dovetailed sheet piling between 7 and 10ft run right across the structure, protecting the foundations against the effects of underseepage. The floor of the sluice was 9ft above low water in the Humber and about 6ft below ground level in the drainage area.

River Witham, 1753

Landowners interested in the drainage of land adjacent to the Witham held meetings in 1752 and 1753 to consider further the proposals put forward by Grundy senior and junior in 1744, and a later scheme by Daniel Coppin involving the North Forty Foot Drain. Details of the latter need not concern us, as a preference emerged for the plan of shortening the river course along the line CDB (in Fig. 2) and Grundy was asked to reexamine this scheme.

In fact his report, written in October–November 1753, with a sketch plan,[27] came up with an improved line CFB striking a balance between land in Wildmore and Holland Fens, an extension of this line from B to G (near Boston), and a proposal for a large outfall sluice, with an adjacent navigation lock, just upstream of Anton's Gowt at B. The cuts, and the river from C to Chapel Hill, were to have a bottom width of 56ft and depth of 6 to 8ft, with banks kept 40ft from the water's edge, and the 'Grand Sluice' had to provide a 50ft waterway. The rest of his report followed closely the scheme for river and fens above Chapel Hill as proposed in 1744, including stopping off the North Forty Foot south of Kyme Eau and taking up the tunnel there. The total cost of the works was estimated by Grundy at about £23,000.

A meeting held at Lincoln on 15–17 November 1753 approved his report, with one exception: the sluice should be placed near point G rather than B.

The inclusion of an outfall sluice on a tidal river is an interesting point, as such a structure had specifically been ruled out in the 1744 plan in accordance with the opinion that tidal flow must be restricted as little as possible, even at the expense of higher river banks having to be built miles further upstream. By 1750 or thereabouts Grundy had departed from this view, strongly held by his father, and may well have been influenced by the successful reconstruction of Denver Sluice on the Great Ouse completed in that year.

THE YEARS 1760–71

During this period Grundy directed construction on four major schemes, and work was carried out on two others for the plans of which he was largely responsible.

River Witham, 1761

It will be recalled that in 1753 the position of the proposed Grand Sluice was still undecided. Grundy suggested in 1757 that it might be built at Boston bridge. Langley Edwards, asked to comment on this idea, reported in favour of placing the sluice on a short cut (GH in Fig. 2) just upstream of Boston.[28] In other respects he agreed with Grundy's 1753 plan, though recommending that the cut or channel for the Witham should be 10ft wider.

Encouraged by what seemed to be the emergence of an acceptable scheme, the landowners asked Smeaton to join Grundy and Edwards in a reexamination of the whole question. The three engineers took 'a fresh View of the River and Fens' in October 1761 and reported on 23 November.[29] An engraved plan by Grundy appeared early in 1762.[30]

Their report, one of the classics of 18th-century engineering literature, states briefly and logically the principles involved, and gives detailed recommendations with a summary of the estimated costs. In essence (i) the best line for the new channel is EBGH (Fig. 2), (ii) the channel should have a regular declivity along its length (about 5 inches per mile), a bottom width of 50ft, and banks 40ft back from the edge of the cut, (iii) the sea sluice to be on the cut GH, with pointing doors giving a 50ft waterway, the adjacent lock to have three pairs of gates, (iv) the depth of the new channel and elevation at the top of the sluice draw doors to be such that water level is not less than 2ft below ground, and (v) there should be three pound locks, not stanches, on the river between Chapel Hill and Lincoln. Proposals for scouring and widening the river and improving drainage of the fens followed much the same pattern as in the 1744 report. The estimated cost was rather more than £54,000.

The choice of the line EB was based strictly on engineering principles. For more practical reasons the line CFB remained preferable, and at a meeting on 13 January 1762 this was agreed,[31] with the modification that the cut should be extended almost up to point E.

All three engineers gave evidence on the Bill in Parliament and the Act was obtained in June 1762. Work began in April 1763 with Langley Edwards as engineer in charge.[32] Grundy's involvement with the scheme now came to an end. A measure of his contribution is that, when the Witham Commissioners paid the engineers' fees, he received £254; Edwards charged £90, and Smeaton £73.

Louth Navigation

In October 1756 Grundy was asked to survey and produce a scheme for a navigation or canal from Louth to the north-east Lincoln coast. He selected Tetney Haven as the best point of entry, and submitted a detailed report, with a plan, for which he was paid £30 16*s*. in April 1757. A copy of his report is in the Louth Navigation minute book.[33]

Nothing much happened during the next three years, but from January 1760 the minute book gives a rather full account of events, starting with the decision to open a subscription list. In March the Town Clerk of Louth wrote to Smeaton, saying 'An actual Survey hath already been taken by Mr. Grundy and he is of opinion the making this intended Navigation is very practical. But yet the Gentlemen are not willing to proceed till they have Grundy's opinion confirmed to you.' A fortnight later Grundy's 'Survey and Plan' were delivered to Smeaton in London. The two engineers met on site in August 1760, some letters subsequently passed between them and, with minor revisions and corrections, their reports were printed in August 1761[34] along with the engraved plan (Fig. 5).[35] As would be expected, Smeaton made several useful and perceptive comments, but in all essentials he agreed with his colleague's proposals.

It is interesting to note that in 1759–60 Grundy's fee was one guinea a day plus 7*s*. 6*d*. for expenses. This contrasts with the fee of 5*s*. or 6*s*. per day charged in 1757 by the well-known land surveyor William Fairbank of Sheffield.[36]

After further delay the Act was obtained in March 1763. Probably at about that time Grundy prepared 'Plans of Sluices, Locks, Bridges and Tunnels for Louth Navigation', as listed in the *Catalogue*, and in February 1765 he was appointed engineer at a salary of £300 *p.a.* out of which he had to pay the resident engineer. The latter, James Hogard, took up his duties in the following month and work began.

From March 1765 to June 1766 the principal jobs were to execute a cut 40ft wide, 9ft deep and ³/₄ mile long from the old sea bank towards low-water mark, and to build the outfall sluice and adjacent navigation lock. Similar to, but smaller than, the combined structure of this type on the Market Weighton scheme (Figs. 13 & 14), the sluice had a 16ft waterway in two arches with pointing doors and draw doors, and the lock, 100ft in overall length, had four pairs of gates: two pairs to land and two to sea. The sea doors of the lock were 15ft 9ins high.[37]

By the summer of 1767 the canal had been cut for the greater part of its 7 miles from the sea lock through flat fenland, the depth being such that water level was nowhere less than 2ft below ground surface. This ensured that the canal could act as a main drain: one of Grundy's guiding principles.

Hogard now took over as engineer at a salary of £140 *p.a.* It was under his direction, then, that the seven locks from the edge of the fenland up to Louth were built, in a length of four miles, and also that the decision was made in 1769 to locate the canal basin further out of the town than originally planned, in order to save the cost of two further locks.

Remains of six of the seven locks still survive, from which it can be seen that they have plank floors on a timber grillage. The walls are in brickwork, with a stone coping, rising to a height of 13 to 15ft above the floor. The leading dimensions are: 86ft between gates, 15ft 3ins clear width, and 5ft 4ins of water on the sills, to accommodate Yorkshire Keels. The falls range from 5 to 8ft, totalling 45ft in the seven locks. Above

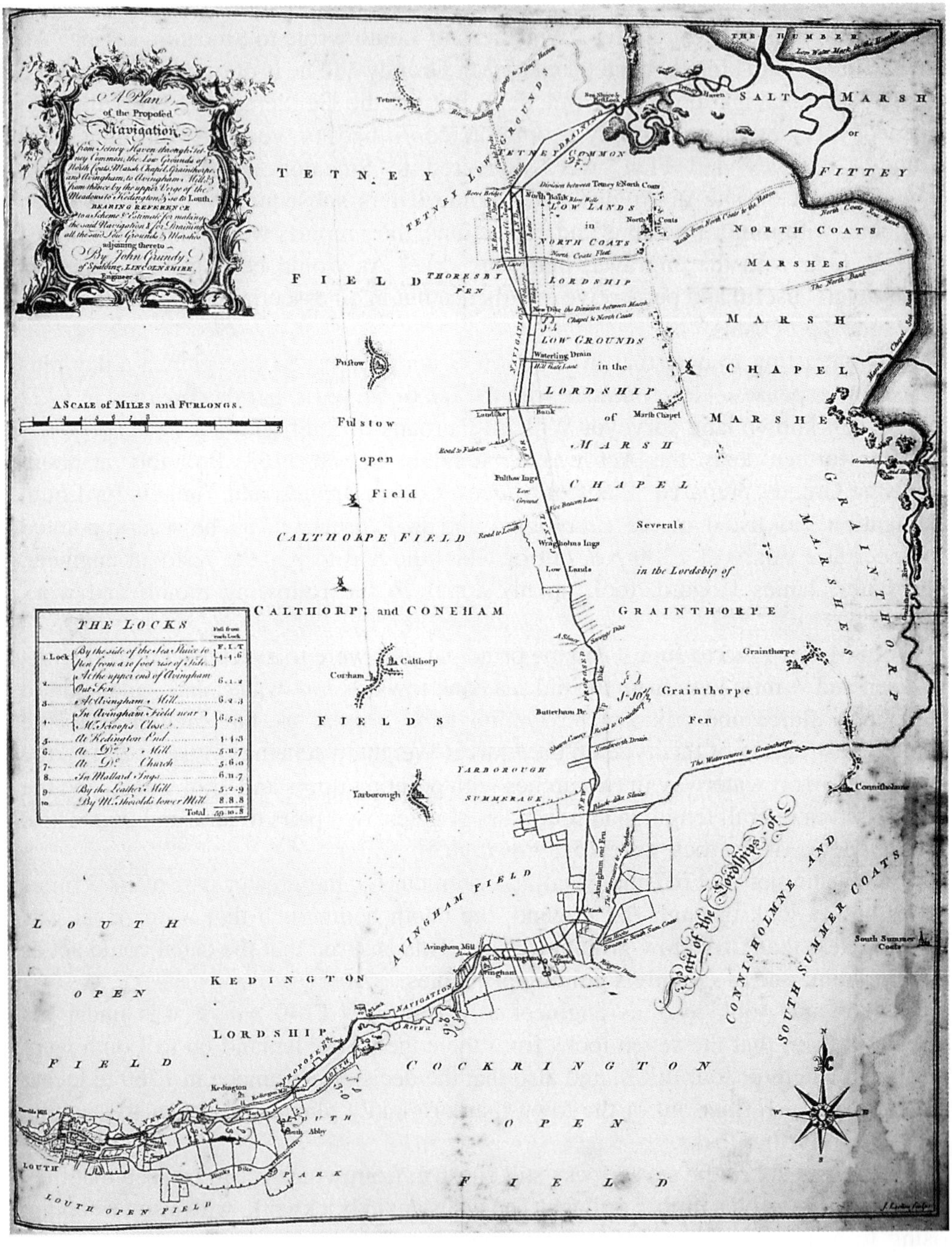

Fig. 5 Plan of Louth Navigation, 1761

Fig. 6 Remains of Salter Fen Pound Lock on the Louth Navigation

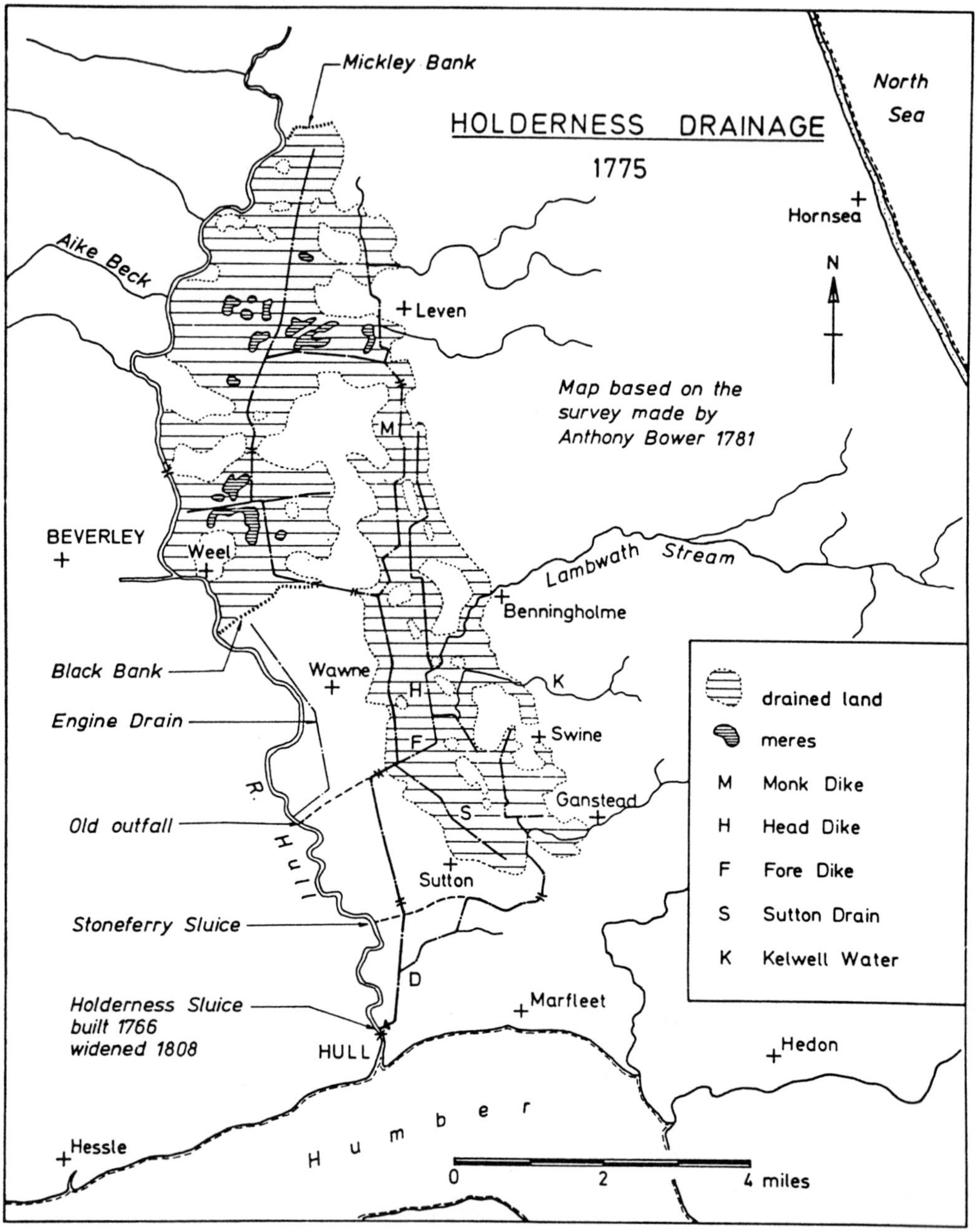

Fig. 7 Holderness Drainage, 1775

the top lock (now demolished) the canal opens out into a basin 66ft wide and 700ft long.

The first lock up from the fen has straight walls. But the walls of the six other locks were built in four segmental arches with vertical timber posts at the springings held by land ties (Fig. 6). The locks appear to be exactly as described by Padley in 1828, apart from later repairs and their now ruinous condition, and he mentioned that one of them is 'getting old and weak'. Moreover the minute book reveals no programme of rebuilding prior to 1828, so it seems certain that this very unusual form of construction was adopted in the original works. Clearly the intention was to increase the resistance of the walls to lateral earth pressure, presumably as an alternative to counterforts.

The navigation was opened throughout in May 1770 at a total cost of £27,500. It had two fixed bridges and four timber swing bridges. The River Lud, running off the Wolds, provided an ample water supply.

Holderness Drainage

Up to the mid-18th century the 11,200 acres of low ground in Holderness shown in Fig. 7 drained into the River Hull at a point marked 'Old outfall' about four miles north of the Humber. This outfall also received upland waters from Lambwath Stream, Kelwell Water and two streams near Leven. Further south, Ganstead Stream flowed into the river at Stoneferry Sluice. The first important works to achieve a much-needed improvement were carried out under an Act of 1764.

Grundy gave evidence on the Bill, and the Act authorised the greater part of a scheme proposed by him in a printed report written in December 1763,[38] supported by Smeaton in an independent report dated 12 February 1764,[39] and illustrated in an engraved plan by the land surveyor Charles Tate.[40]

The works included (i) remodelling of the internal drains to the pattern in Fig. 7, (ii) raising and strengthening the flood banks on the east side of the river, (iii) cutting a new main drain 1½ miles long from Fore Dike, and another from Ganstead, leading into a Mother Drain one mile long from point D to (iv) an outfall sluice just above the bridge in Hull (see Fig. 17). This new Holderness Sluice (Fig. 8) had a 20ft waterway in two arches separated by a central pier 6ft wide, each arch having a pair of pointing sea doors 9½ft high and a draw door on the land side. The foundations included five rows of sheet piling each running across the entire width of the structure and the walls were in brickwork with an ashlar stone facing. The sills were 1ft below low water in the river which, according to Grundy's levels, lay 6ft below Sutton Carrs and 9ft below the carrs near Leven. The fall from the water level in the drains in these distant grounds to low water in the river at Hull River was around 6 inches per mile, a figure also adopted as the gradient for the bottom of the main drain, and quite acceptable so long as the drains and sluice had an ample capacity. In addition, however, the drains carried the upland waters flowing from a catchment far

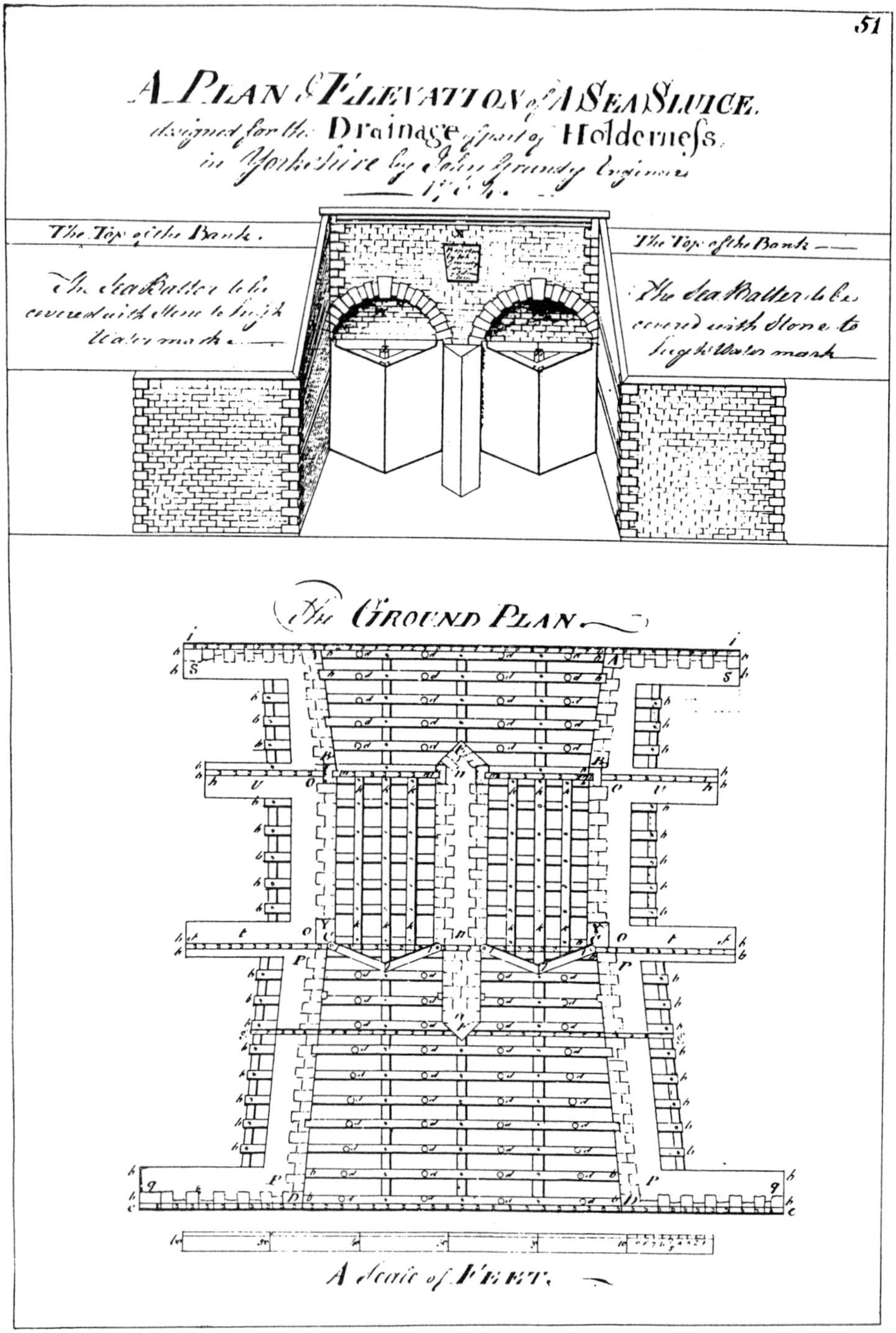

Fig. 8 Holderness Sluice, 1764

larger than the drainage area itself.

In his report Grundy recommended for the western main drain an average depth of 6ft and a bottom width of 14ft (which was adopted). For the Mother Drain he proposed a depth of 10ft and 24ft bottom width; this width was also to be the waterway of the sluice. Smeaton, commenting on the sluice, suggested that given total protection by the river banks a waterway of 18 or 20ft might be sufficient. As we have seen, the latter dimension was adopted but the Mother Drain, from false economy, was made with the narrower width of 18ft.

The Trustees of Holderness Drainage, appointed in accordance with the Act, at their second meeting[41] ordered 'that the two Engineers Mr. Grundy and Mr. Smeaton be wrote to desiring they'll come over immediately to view the low Grounds and Carrs and attend the next Meeting to give such Information relating to this Drainage as may be wanted'. This they did, attending at Beverley in July 1764, and agreed to provide working drawings of the sluice. Grundy produce the drawings, one of which has survived (Fig. 8),[42] and sent them to Smeaton who made some valuable comments (mostly on the foundations) and prepared a very detailed bill of quantities with estimates totalling nearly £1,800. This second report by Smeaton is dated December 1764.[43]

Progress of the work can be followed in the Trustees' minute books. Grundy was engineer in charge, with Tate as surveyor, John Hoggard supervising the river banks, and Joseph Page of Hull as resident engineer on the sluice and main drains, at a salary of £80 *p.a.* Construction of the sluice began in March 1765 and excavation of the drains started a few months later. Grundy made frequent site visits and one of his several progress reports is extant.[44] The sluice, Mother Drain, and the two main drains were completed by October 1767. Grundy and Page then left the scene, but work on the river banks and the drainage north of Fore Dike continued for five or six years under Hoggard's direction. Finally, Commissioners were appointed to conduct an exact survey and valuation of land within the drainage area, and their Award is dated 1775.[45] The total cost of the scheme came to about £24,000.

A very substantial improvement had undoubtedly been achieved, but further works were necessary to perfect the drainage. Apart from widening the drains and sluice back to or even beyond Grundy's original proposals, the principal feature was the separation of the upland waters from the internal drainage; this was achieved first by William Jessop in the years 1787–95 and secondly under an Act of 1832 following Edward Page's scheme of leading the drainage water to a new outfall on the Humber at Marfleet. Grundy's Holderness Sluice and western main drain thereafter continued (well into the present century) to carry the upland waters of Lambwath and Leven into the river Hull.

Driffield Navigation

Works carried out during the years 1767–70 to Grundy's design and under his

V

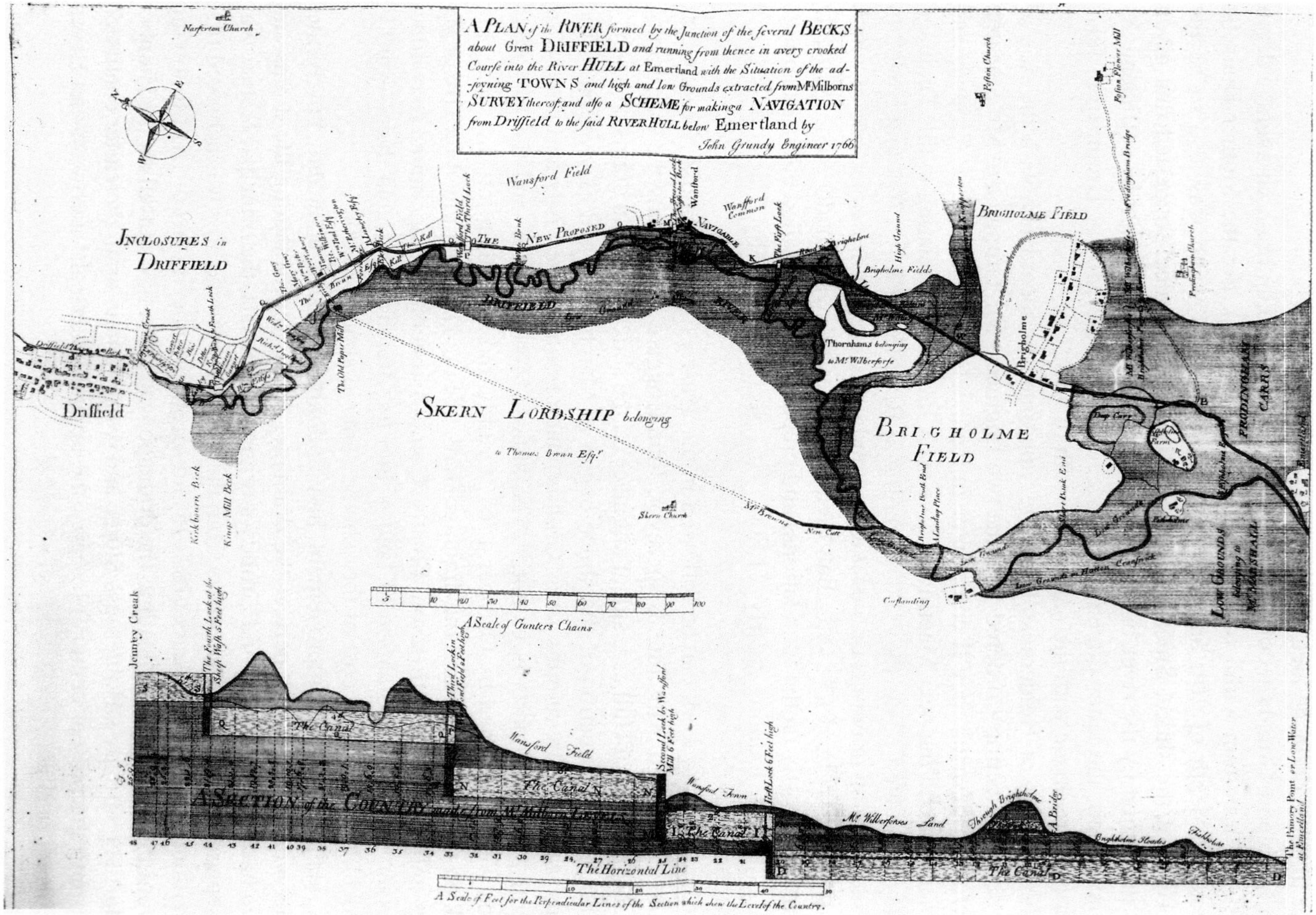

Fig. 9 Plan and Section of Driffield Navigation, 1766

direction enabled Yorkshire Keels to navigate between a canal basin in Great Driffield and the River Hull at Emmotland Cottage, whence the river was navigable to the Humber.

Driffield Beck, culverted into the head of the basin, provided a sufficient supply of water. From Driffield to Wansford a fall of 25ft in three miles was accommodated in four locks, thence the canal ran on a dead level for two miles past Brigham to Fisholme where it entered Frodingham Beck, a tributory of the Hull, and the Beck was deepened along its ¾ mile length to Emmotland. Without a sluice at Fisholme the two mile level is effectively part of the river. It is embanked for most of its length, and its depth (typically 8ft) was such that it could serve both for navigation and as a drain for the adjacent low ground.

The survey, by Isaac Milbourn under Grundy's direction, was published as an engraved plan and section (Fig. 9)[46] and Grundy's printed report on the scheme is dated 18 December 1766.[47] He gave evidence in Parliament, the Act was obtained in May 1767, and in August he became engineer in charge.[48] His working drawings are listed in the *Catalogue* as 'Plans of Driffield Navigation, the Locks and Swing Bridges'.

Samuel Allam of Spalding arrived in October as resident engineer; work began on the first lock soon afterwards, and the firm of James Pinkerton & John Dyson contracted to excavate the canal in May 1768. The navigation was opened in May 1770 at a cost of about £13,000.

The locks were slightly smaller than on the Louth Navigation, accommodating vessels up to 61ft in length and 14½ft wide with a draught of 4ft. They had timber floors, sheet piling under the gates, bearing piles supporting the structure, and straight brick walls with masonry coping. A horse-engine was used to drain the excavations. Timber swing-bridges carried the roads at Wansford and Brigham over the canal.

Adlingfleet Drainage

Smeaton in 1755 and Grundy in 1759[49] had been consulted on the drainage of Haldenby and Eastoft respectively, parts of the 5,000 aces of low ground known as Adlingfleet Level (Fig. 10). In those days a somewhat haphazard system of drains led to an inadequate outfall sluice at Ousefleet, and little in the way of local improvement could be done until a more logical drainage scheme was introduced for the whole area. At the request of the principal landowners Smeaton prepared such a scheme in December 1764, based on his own levels and a land survey by Charles Tate. This chiefly involved a new main drain 5.8 miles long and a 10ft waterway sluice at Trent Falls where low water mark is almost 2ft lower than at Ousefleet.

The necessary Act was obtained in March 1767, Commissioners were appointed, and in June they asked John Grundy to take charge of design and construction, with David Buffery as resident engineer (salary £50 *p.a.*), and Tate continuing as

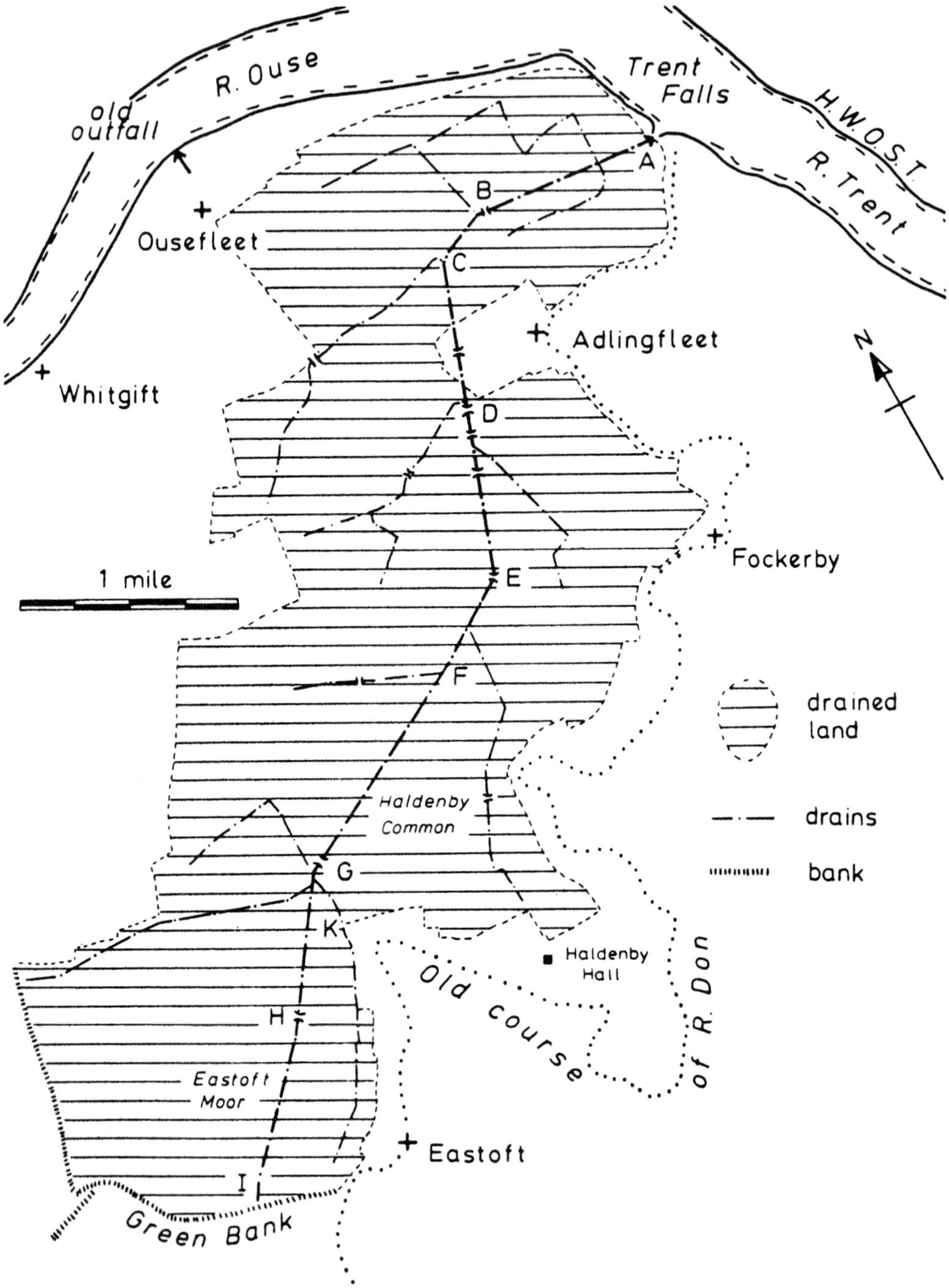

Fig. 10 Adlingfleet Drainage, 1772

surveyor.[51]

Grundy followed Smeaton's plan with only minor variations. He produced drawings and specifications for the sluice on 17 July, contracts for building it were let four weeks later; soon afterwards work began on raising the southern barrier bank (Green Bank in Fig. 10) and in October Pinkerton & Dyson contracted to excavate the main drain 'agreeable to Mr. Grundy's dimensions'.

By June 1768 the sluice had been finished and Grundy issued instructions for opening the outfall into the Trent. In December of that year he wrote a report[52] detailing the small 'stop sluices' to be built on each of the side drains, where they entered the main drain. The report also gave an exact account of the work, now completed, by Pinkerton & Dyson on the main drain (170,000 cu yds of excavation), backfilling against the sluice walls, cutting through the Trent embankment, and excavating for the foundations of bridges over the drain.

In June 1769 'all the material works of the Drainage being completed', Grundy was paid £302 for his fees and expenses. Tate likewise received £275, and Buffery was retained on an *ad hoc* basis to continue letting small contracts and certifying work done on some further bridges, deepening of existing drains, additional raising of barrier banks, and so on. Finally, the Commissioners made their Award in June 1772. The total cost was approximately £7,000.

Leading dimensions and some technical details may be summarised as follows. Main drain: bottom width 10ft, cut on a dead level from A to E the depth varying from 15ft at the sluice to about 8ft at the lowest ground around D and E; further south the bottom rises at a slope of 10 inches per mile, ending at a depth of 6ft; average fall of water level from I to A just over 12 inches per mile; the drain is crossed by eight brick arch bridges. Sluice: 10ft waterway, counterforted walls in stone-faced brickwork 5ft thick at base, timber floor with cross beams, four rows of dovetailed sheet piling 10 to 12ft deep and bearing 9 to 10ft long; a pair of oak pointing doors 12ft high and a draw door, also in oak, 8ft high at the ends of an arch 10ft long; sills about 1ft above low water in the Trent. Side drains: typically 5ft bottom width, 6ft deep, totalling 12.3 miles in length. Barrier bank: 7ft high, 6ft top width with side slopes of 1.75 : 1.

The drainage was admirably planned and executed. Up to 1929, when a new sluice was built nearby to the original one (the remains of which can still be seen), the only significant alteration was widening the main drain by 2ft from A to the bridge between C and D, and the consequent rebuilding of the bridge at B, in 1839.

Laneham Drainage

An engraved plan of a scheme for draining low grounds in the flood plain of the River Trent, on the west side of the river north of Laneham, in Nottinghamshire, was published in 1769 by Grundy and George Kelk.[53] From the following entries in the *Catalogue* it is clear that Grundy prepared estimates and drawings and that the

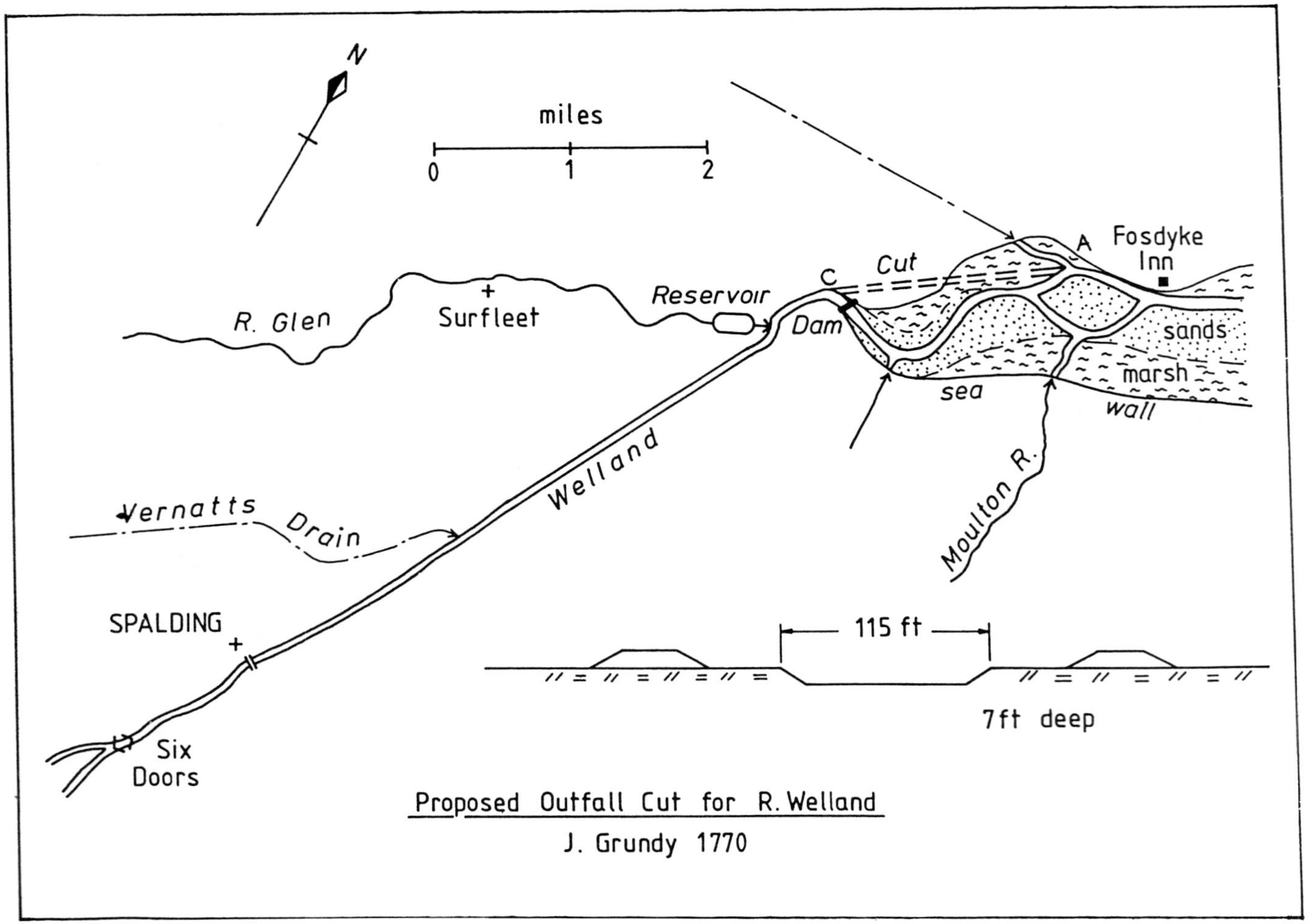

Fig. 11 Deeping Fen and Welland Outfall

scheme was carried out: 'Laneham Drainage, Scheme, Estimates as projected in 1764 and Executed 1770, 1771, 1772', 'Plans of Laneham Drainage 1769 & 1770', and 'Plans and Estimates for the Sluices for Ditto'. The minute book for this period has not been found, nor is it known whether Grundy was engineer in charge of construction.

Deeping Fen, 1769–70

Ordered in March 1769 to produce a comprehensive scheme for 'perfecting' the drainage of Deeping Fen, Grundy completed in February 1770 a report of 151 pages with a plan and an additional 14 pages of estimates (dated April 1770).[54] The most important of his proposals was the provision of an improved outfall for the Welland by making a cut about 100ft wide, 7ft deep, and 1½ miles long, partly behind the sea wall and partly across salt marshes, from the head of the estuary at C (Fig. 11) to the main channel at A, which runs along the edge of the sands to Fosdyke. A dam would be required across the existing channel just below point C to divert flow into the cut. Grundy also recommended that a sluice and navigation lock be built at point A to prevent silt being brought up the rivers in high tides.

His other recommendations included widening and deepening the River Glen, and the Welland for 3 miles downstream of Spalding, and several of the principal drains, as well as the construction of a catchwater drain west of Market Deeping proposed already in 1768 by Thomas Tofield.[55]

Finally, Grundy adopted another of Tofield's proposals, for extending Vernatts Drain alongside the Welland, under the Glen in an inverted syphon and on to a new outfall sluice at A. But whereas Tofield had suggested that the extension and the drain itself should be cut with a bottom width of 40ft on a dead level from the sluice right up to the Fen at a depth of 2ft below low water mark, Grundy thought this extra depth was not justified technically or in terms of cost.

This difference of opinion emerged at a meeting between the two engineers at Retford, and they agreed to seek Smeaton's advice. In reply (27 March 1770) Smeaton outlined the principles of flow in open channels, showing that water below the level of the outlet is not 'dead' but, on the contrary, the extra depth is beneficial; and inferring that where, as in this particular case, the fall has a very low gradient (less than 4 inches per mile) the additional cost is justified.[56] In other words, Tofield was right.

The Adventurers, having considered all these matters, decided in 1772 to enlarge Vernatts Drain and extend its outfall, but only so far as the Reservoir (a proposal made nearly twenty years earlier by Grundy senior), to construct a catchwater drain, and to enlarge the Glen.[57] Further action on the Welland outfall was postponed, and Grundy's idea of a sluice at point A seems never to have attracted serious attention. But after more than thirty years and the publication of several reports a cut was eventually made from point C to Fosdyke, with great benefit to the drainage of

Deeping Fen.

Grundy's bill for his services came to £250, paid in 1772. His daily fee must by now have increased to three or even five guineas.

WEIGHTON DRAINAGE AND HULL DOCK

The last years of Grundy's active professional career, from 1771 to 1776, are distinguished by two of his most important projects: Market Weighton Drainage and Navigation, and the planning of Hull's first dock. During this period he also reported on at least fifteen other schemes.

Weighton Drainage and Navigation

An Act was obtained in May 1772 'for Draining and Preserving certain Commons, low Grounds, and Carrs in the Parish of Market Weighton and other Adjacent Parishes . . . and for making a Navigable Cut or Canal from Market Weighton to the River Humber'. The land to be drained, shown in Fig. 12, amounted to 16,400 aces; a cut 6 miles long from the Humber to point C served as the Mother Drain and lower part of the canal; from C up towards Market Weighton the canal was to be completely separate from the drainage system. Foulney River needed scouring and deepening, numerous side drains were required, and the Act stipulated that water level in the canal must be kept not less than 3ft below ground surface. On the other hand a minimum depth of 4ft of water was required for navigation.

The scheme had been drawn up by John Smith, and in June 1772 he was appointed engineer in charge with Samuel Allam as resident engineer (at £54 *p.a.*).[58] Three months later, however, the Trustees called on Grundy for a second opinion. He reported without delay on 2 September,[59] proposing a better site for the outfall sluice and lock, a different method of building it, a lower level for its sill, and 9 locks instead of 4 to take the fall of about 60ft. The report was accompanied by a longitudinal section and drawings of the sluice and lock.[60] His revised scheme found acceptance; he became chief engineer, new contracts were signed 'according to the plans of Mr. John Grundy', and construction started almost immediately. The works were carred out in three well defined stages.

(i) 1772–75. Sluice, lock and Mother Drain. Grundy in charge (his fees to December 1775 being £160), Smith as assistant (until 1773), Allam as resident; James and John Pinkerton contractors for excavation, Joseph Smith for masonry and brickwork, John Everingham for carpentry. Cost about £26,000.

(ii) 1776–80. Cutting the drains (total length 43 miles), deepening Foulney River (length 12 miles) and other streams (length 6 miles). Allam in charge, Pinkertons principal contractors. Cost about £14,000.

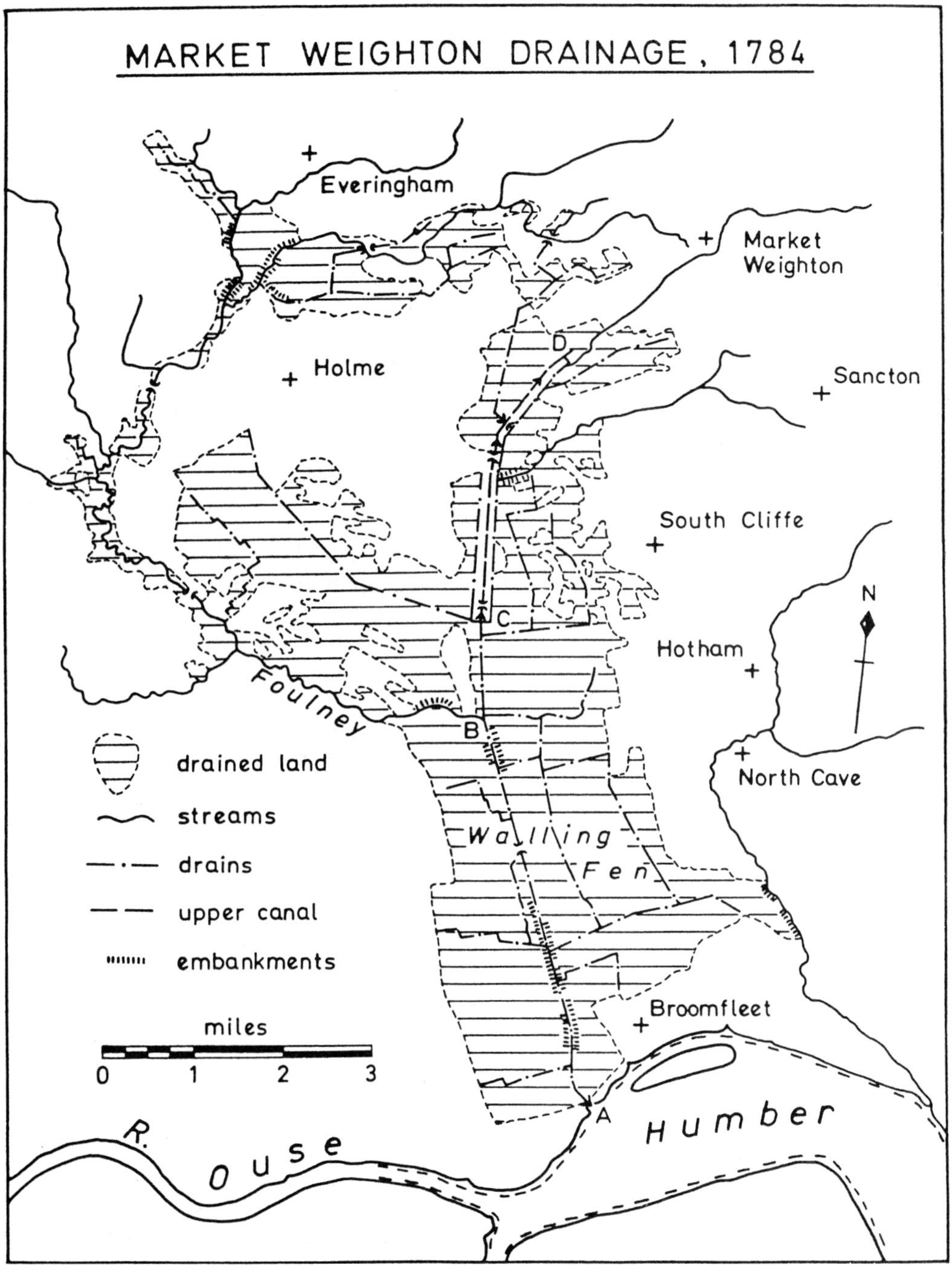

Fig. 12 Market Weighton Drainage, 1784

(iii) 1781–83. Canal above point C, with 3 locks in 3½ miles to the canal head 1½ miles from Market Weighton; this being a revised plan by Luke Holt,[61] who took over from Allam in 1781. Contractor for excavation George Vickers, for masonry Rowland Myers and for carpentry John Harrup. Cost about £8,000.

Several land surveyors were employed, the principal ones being Joseph Jewitt, William Hildyard, and Charles Tate. The Commissioners made their Award in November 1784. One-quarter of the total cost of £48,000 was attributed to navigation and three-quarters to drainage.

The Mother Drain was cut with a bottom width of 30ft and side slopes of 1½ : 1, on a dead level at low water level in the Humber. This gave slightly more than the minimum required depth of 7ft (3ft freeboard and 4ft of water) below the lowest ground in Walling Fen, and an average depth along the 6 mile length of about 14ft. The total volume of excavation in this drain amounted to 870,000 cu yds.[62] It is banked where it passes through very low ground to avoid flooding.

A measured section and general plan of the sluice and lock are shown in Figs. 13 & 14. The sluice has a 30ft waterway in two arches; the lock has a clear width of 15ft and a length of 70ft between each of the two pairs of gates. Sill level of the sluice and lock is at LWST. The wall over the arches (apart from the parapet), the side

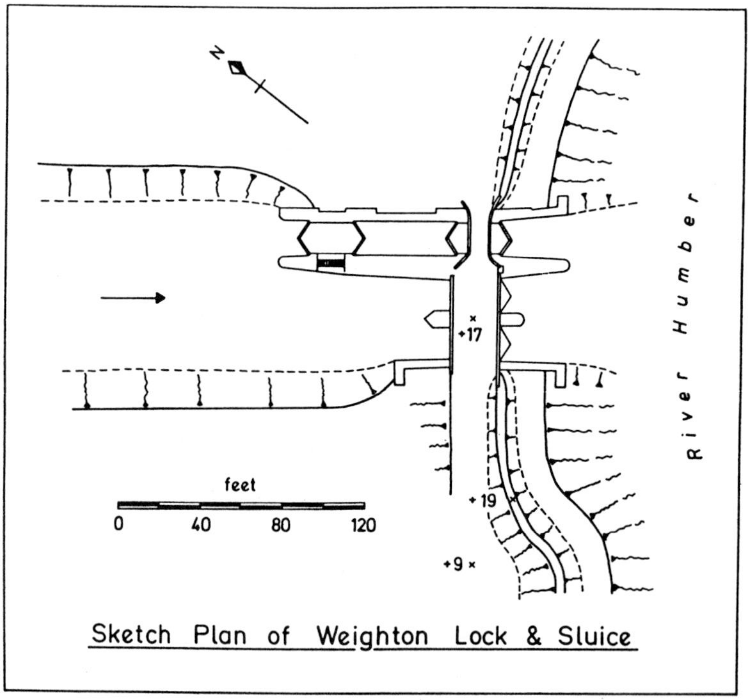

Fig. 13 Sketch Plan of Weighton Lock and Sluice

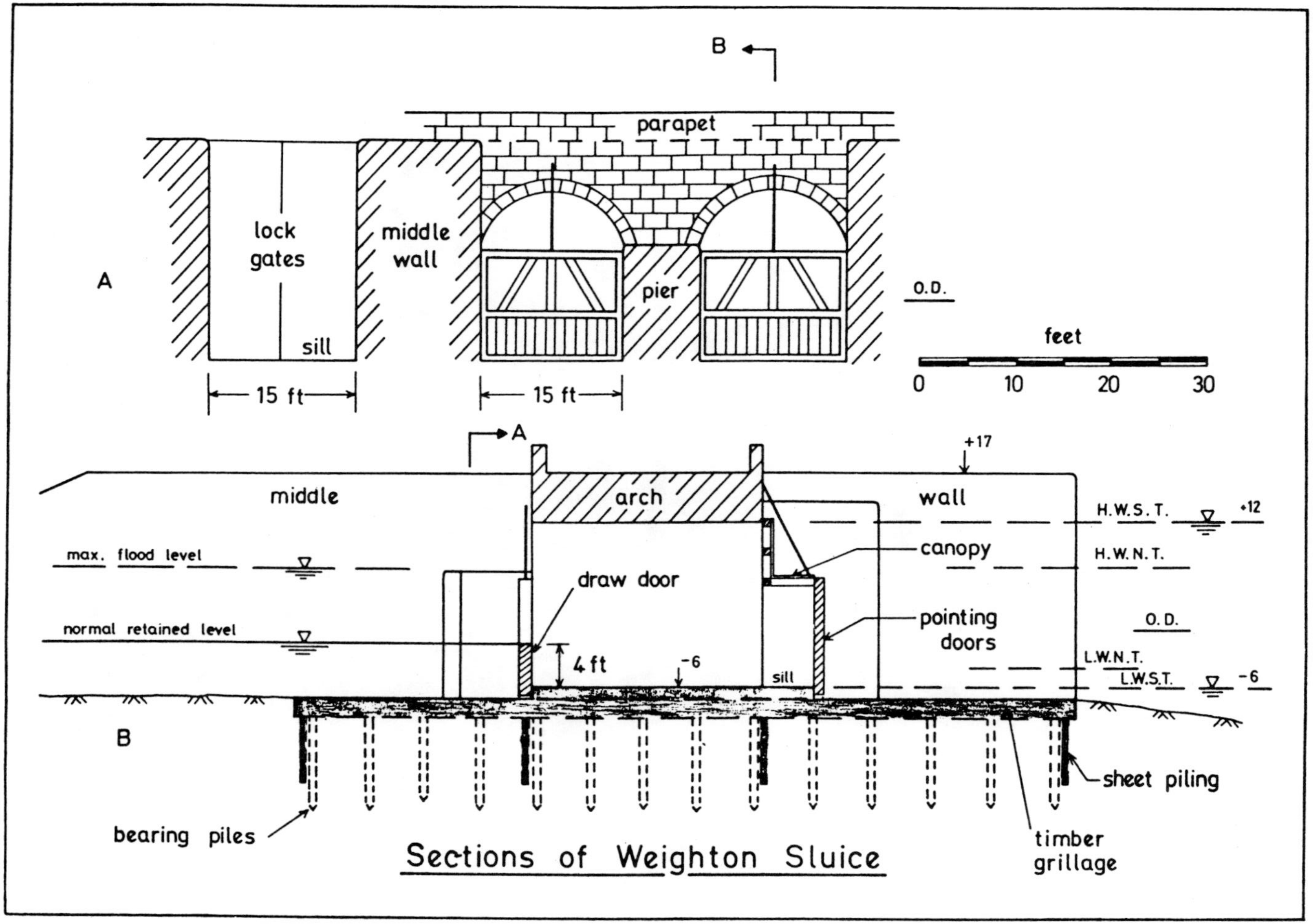

Fig. 14 Sections of Weighton Sluice

Fig. 15 Weighton Outfall Sluice

Fig. 16 Weighton Mother Drain

walls, and middle wall rise 5ft above HWST which is 18ft above LWST; the lock walls are therefore 23ft high above sill level. The pointing doors (Fig. 15) are 12ft high, with a canopy. The whole structure rests on a timber grillage supported by bearing piles, and there are four rows of dovetailed sheet piling.

The great pit for the foundations, 18ft deep with a bottom area of 12,600 sq ft, was excavated behind the Humber bank. After completing the structure new banks were built curving back to the side walls of the lock and sluice, and the old bank removed down to floor level. In this way the use of a cofferdam was avoided.

Repairs were carried out in 1826 and again in 1855–56, and very recently the draw doors have been replaced by roller gates. In most other respects the lock and sluice remain essentially as completed to Grundy's design in 1775. A view from the sluice looking back along the Mother Drain is shown in Fig. 16.

Hull Dock

The need for extra space for shipping in Hull harbour and for a Legal Quay had long been recognised, but it was not until 1773 that a satisfactory solution to the problems emerged.[63] Asked for his opinion on the effects of building the Quay as a solid structure projecting 40ft into the harbour on its west side, Grundy reported in October 1772 that it would oppose too much obstruction to flood water coming down the river, that a flood relief channel could however be provided by enlarging the old moat or 'Town Ditches' (Fig. 17) and that, with great advantage, the north part of the moat would provide a suitable site for a large 'wet dock' (i.e. a dock with lock gates to maintain water at high tide level) to accommodate unladen ships, thereby relieving pressure on space in the harbour.[64]

This idea was quickly taken up and elaborated: the dock should be of sufficient depth to accommodate fully laden ships so that the Legal Quay could be located on one side of the dock, leaving the need only for a 15ft quay on open piled foundations in the harbour itself. Acting for the Board of Customs, John Wooler in January 1773 prepared outline plans and estimates. In April, Smeaton reported favourably on the 15ft quay in the harbour. Further to satisfy those who still feared the flood threat, Smeaton and Grundy jointly examined the problem again and wrote (separate) reports in December.[65]

The way now lay open for an application to Parliament. Grundy drew up detailed plans and in an estimate covering four folio sheets dated 31 March 1774 worked out the cost at £67,832 exclusive of ground to be purchased, parliamentary charges, fees, etc.[66] The Act was obtained in May and in November Grundy received £300 for his work.

His plans and report on the dock are unfortunately lost, but recently discovered entries in the *Catalogue* include:

(i) Rough Plan of the Town and Harbour of Hull in Yorkshire, with the Designs for a Wet Dock, Quays, &c. done in 1774.

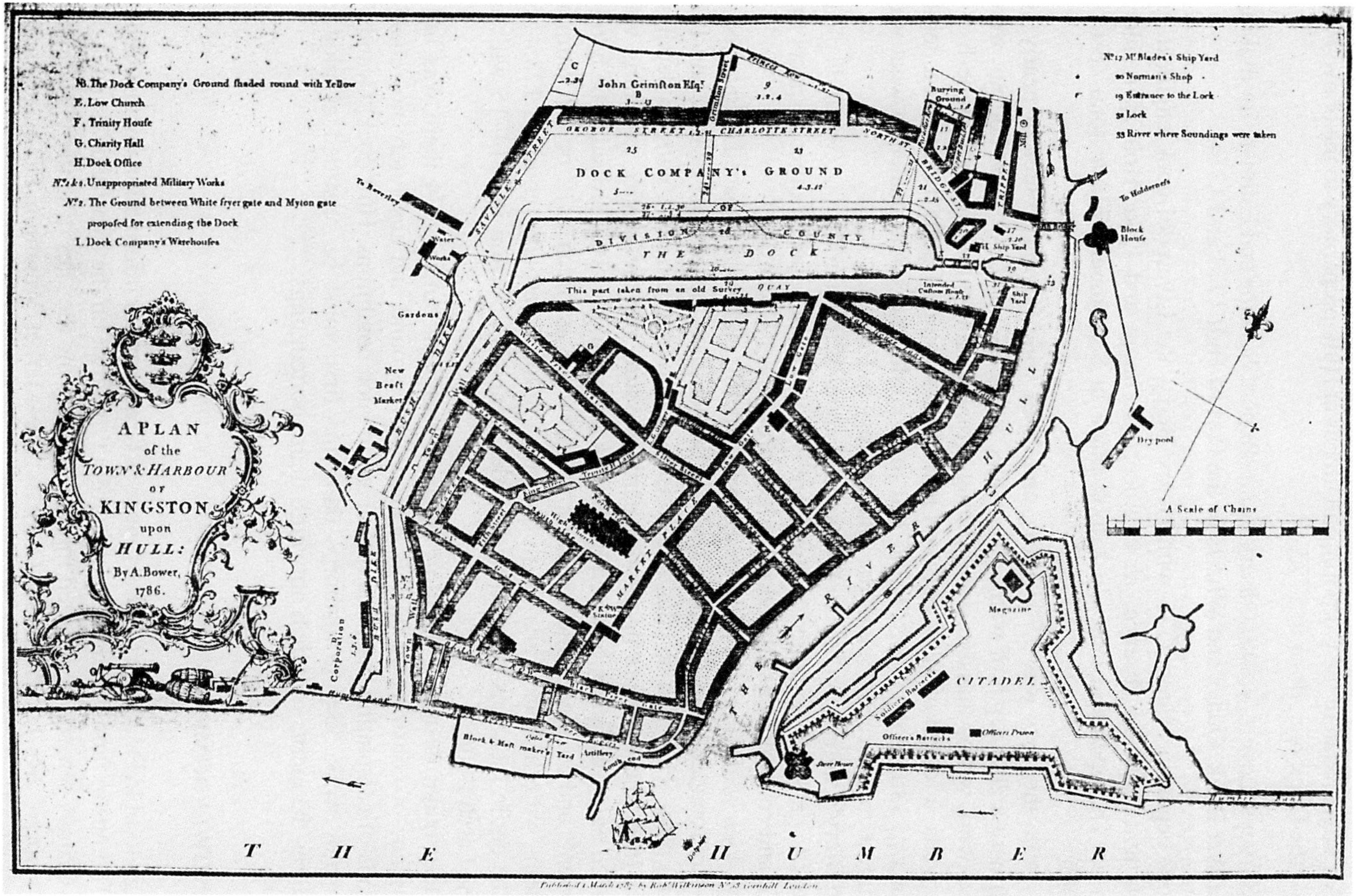

Fig. 17 Hull Dock Plan

(ii) Mr Page's Plan of the Corporation Ditches at Hull, preparitory to the above Plan of the Dock &c., done in 1773.

(iii) Surveys, Levels, Views, Observations, Report, Scheme and Estimate for making a Wet Dock, Quays and Wharfs in the Port of Hull 1773 and 1774.

Work began in March 1775 with Henry Berry, the Liverpool dock engineer on intermittent leave of absence, as engineer in charge and Luke Holt as resident engineer. The opening of the dock took place in September 1778 at total cost of £73,000.

As built, the dock was 1,703ft long and 254ft wide (giving an area of nearly 10 acres) with a 20ft depth of water. The walls, rising 24½ft above dock bottom, were in brickwork with a massive stone coping. The lock had a timber floor, stone-faced brick walls, a length of 120ft between gates and a width of 36ft; it retained a depth of water of 18ft on the sills.[67] Up to 100 ships could be accommodated in what was the largest commercial dock of the 18th century in England.

On Berry's suggestion the dock was made about 17ft wider than in Grundy's plan, and the Legal Quay enlarged.[68] But there is no evidence that in other respects the layout and construction differed materially from the 1774 design.

The dock walls (Fig. 18) had a based width of 10ft and widely spaced counterforts, and were founded at dock-bottom level on timber planking, sleeper beams, and 10ft bearing piles. They had an adequate resistance to overturning but, with no depth of soil in front and therefore no passive earth pressure, their resistance to sliding was provided only by friction between brickwork and planking. This proved to be barely sufficient, and in places the walls moved forward as much as 2 or 3ft. Similar problems with the lock walls led to their having to be partially rebuilt in 1785–86, and the entire lock was reconstructed under Rennie's direction in 1814–15 with an inverted arch floor.

The original design is therefore open to criticism, but in 1774 there was little experience of construction on this scale in soft ground and absolutely no reliable methods of calculating the forces involved in such conditions. Hull dock, Grundy's last major work, like Grimsthorpe dam, his first, must be regarded as pioneer structures the merits of which outweigh their shortcomings.

ACKNOWLEDGMENTS

I am grateful to Dr and Mrs J. M. Hopper of Grantham for information on the Grundy and Thompson family and for the loan of a typescript on the two John Grundys. Mr P. B. Grimes generously devoted time to showing me the sites of Grundy's works at Grimsthorpe. Mr L. C. Sands, Clerk to the Market Weighton Drainage Board, provided facilities for my wife and myself to study plans and Minute Books etc., as did Mr G. W. Haywood, Clerk to the Adlingfleet & Whitgift Drainage

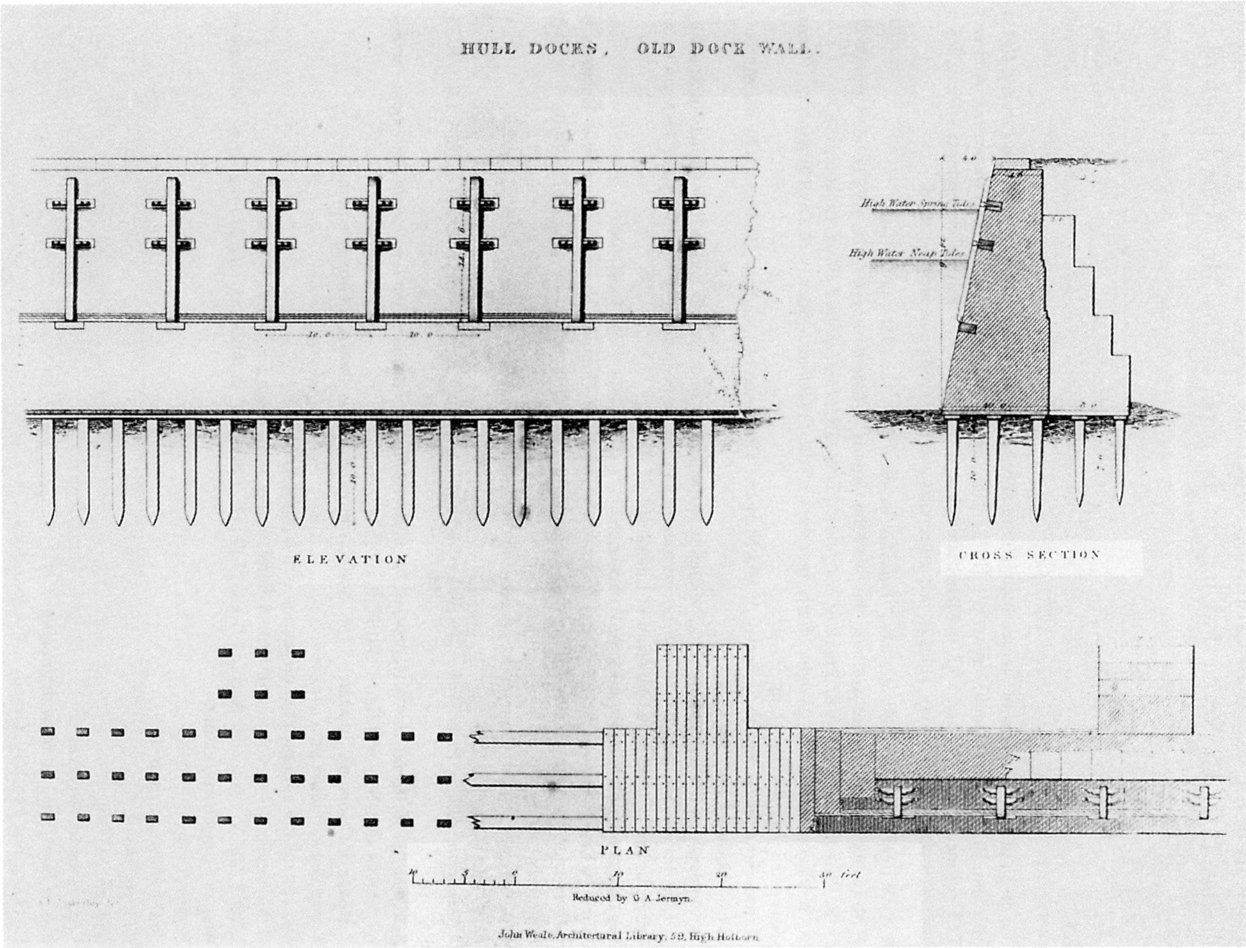

Fig. 18 Hull Dock Wall Section

Commissioners. Mrs Catherine Wilson kindly allowed me to spend a day examining documents (including Grundy's Catalogue of Books and Papers) temporarily withdrawn from the excellent exhibition which she had arranged at the Museum of Lincolnshire Life. The seven drawings in the paper are by Mrs Anne Langford of Imperial College.

NOTES

Where not given in full, locations of source material are abbreviated as follows: BDL = Bodleian Library, BM = British Museum (British Library and Map Room), HRO = Humberside Record Office, LAO = Lincolnshire Archives Office, LCL = Lincoln City Library, MCL = Manchester Central Library, RGS = Royal Geographical Society, RS = Royal Society (Smeaton Collection). Titles of printed reports and engraved plans are shortened.

1 E. C. Wright, 'The early Smeatonians,' *Trans. Newcomen Soc.*, 18 (1938), pp. 101–10.
2 J. M. Hopper, 'Lincolnshire worthies: the two John Grundys', *Lincolnshire Life*, 20 (1983), no. 3, pp. 24–7.
3 N. R. Wright, *John Grundy of Spalding, Engineer, 1719–1783. His Life and Times*, Lincoln, 1983.
4 A. W. Skempton & E. C. Wright, 'Early members of the Smeatonian Society of Civil Engineers', *Trans. Newcomen Soc.*, 44 (1971), pp. 23–42.
5 Portrait (artist unknown), the property of Mrs Margaret Thompson of Langham. Photograph by the Museum of Lincolnshire Life.
6 'A Catalogue of Books and Papers . . . belonging to the late Mr John Grundy and to John Grundy of Spalding in Lincolnshire, taken in 1767', MS bound quarto, 112 pp. In possession of Mrs Margaret Thompson and loaned by her in 1983 to the Museum of Lincolnshire Life.
7 Personal communication from Mr A. P. Woolrich.
8 'Surveys Levels Reports Schemes and Estimates in Works of Draining Navigation and other Business in Engineering. By John Grundy of Spalding Enginr. Vol. II', MS bound quarto, 353 pp., Institution of Civil Engineers Library.
9 LAO Measure 3/1/23. Contract documents and drawing for Pinchbeck Sluice.
10 *Report Book*, pp. 77–130.
11 *A Map of the Antient River Witham* (with Levels) by John Grundy Sen. & Jun. 1743; BDL, BM, LCL, RGS, RS.
12 *A Scheme for Restoring the Navigation of the River Witham and draining the Low Lands contiguous thereto*, by John Grundy Sen. and Jun. 1744; BM.
13 *J. House of Commons*, 30 March 1736. Scribo's proposed cut is shown in Grundy's plan.
14 *Report Book*, folding plan at p. 184.
15 *ibid.*, pp. 59–65 with plan but no section of the proposed dam.
16 G. M. Binnie, 'The evolution of British dams', *Trans. Newcomen Soc.*, 47 (1976), 207–24.
17 LAO 2 Anc 6/5, Ancaster Account Book 1749–52. 'Levelling' in this entry refers to regrading the slope in front of Grimsthorpe Castle.
18 *Report Book*, pp. 176–82.
19 *ibid*, folding plan at p. 151.
20 *ibid*, pp. 143–9.
21 *ibid*, pp. 187–97.
22 LAO Ancaster Archives.
23 *Report Book*, pp. 151–65 with drawing.
24 LAO Deeping Fen 1/10, Grundy's Account Book.
25 LAO Deeping Fen 1/1 and 2, Adventurers' Minute Books 1669–1755 and 1756–1776.
26 *Report Book*, pp. 275–303. 1749 estimates, report and site diary of 1751, and drawing.
27 Printed report of meeting, held at Boston 19 October 1753, with Grundy's report and plan. Institution of Civil Engineers Library.
28 The reports by Grundy and Langley Edwards are summarised in W. H. Wheeler, pp. 149–50. *A History of the*

Fens of South Lincolnshire, Boston, 1897, pp. 149–50.

29 *The Report of John Grundy, Langley Edwards and John Smeaton on the River Witham and the Fen Lands on both sides of the said River*, Lincoln, 1761; BDL, BM, LCL.

30 *A Plan of the River Witham and adjoining Fens and Low Grounds with the New Works proposed to be Executed for Draining the said Fens and Restoring the Navigation*, by J. Grundy, 1762; BDL, BM, LCL, MCL, RGS, RS.

31 As noted and shown on the engraved plan.

32 LAO, LRA 1/1, Minute Book of the Witham Drainage Commissioners 1762–75. BS 4/1/1 Minute Book of the Commissioners for Navigation 1762–86.

33 LAO, LRA 5/1 Louth Navigation Minute Book 1760–1828.

34 *A Scheme for executing a Navigation from Tetney Haven to Louth. By John Grundy. To which is added the Report of John Smeaton*, Nottingham, 1761; LCL.

35 *A Plan of the Proposed Navigation from Tetney Haven . . . to Louth*, by John Grundy; BM, LCL, MCL, RGS.

36 T. W. Hall, *The Fairbanks of Sheffield*, Sheffield, 1932.

37 J. S. Padley, *Louth Canal. A Description of the Locks, Bridge, &c.* [in 1828], Louth, 1832; LCL.

38 *The Report of John Grundy, of Spalding, Engineer. Concerning the Drainage of the Low Ground and Carrs . . . in Holderness*, 1764, HRO.

39 *Reports of the late John Smeaton*, London, 1812, Vol. 1, pp. 88–9.

40 *A Plan of an Actual Survey of part . . . of Holderness. By C. Tate, with a Scheme for Draining by Messrs. Grundy & Smeaton*, 1764; BM, MCL, RS.

41 HRO, DCMO 1/1, Minute Book of Holderness Drainage Trustees, 1764–68.

42 RS, Smeaton collection 5.f.51. The drawing of Holderness Sluice by Grundy 1764, sent by him to Smeaton for comment.

43 *Smeaton's Reports, op. cit.*, pp. 90–98.

44 *Report in Consequence of a View of the Works for the Drainage of Holderness Level*, by J. Grundy, 1765; HRO.

45 Holderness Drainage. Commissioners' Award. Printed 1781 with a folding engraved map by Anthony Bower, HRO.

46 *A Plan of the River . . . from Mr. Milborn's Survey and also a Scheme for making a Navigation from Driffield to the River Hull by John Grundy*, 1766, MCL.

47 *Mr Grundy's Observations . . . and his Report, Scheme and Estimate for making a Navigation between Great Driffield and the River Hull*, 1766; HRO, MCL.

48 HRO, DDX 40/1, Driffield Navigation, Minute Book of Commissioners 1767–1824.

49 *Report Book*, pp. 338–53.

50 *Smeaton's Reports, op. cit.*, pp. 199–206.

51 Adlingfleet & Whitgift Drainage Commissioners, Crowle, Minute Book 1767–1823.

52 MS report by John Grundy, 1768. Institution of Civil Engineers Library.

53 *A Plan of the Ings, Meadows, Marshes and other Low Grounds in Laneham, etc.* Surveyed by John Grundy and George Kelk 1769. LCL.

54 BM, MS 54784. MS report by John Grundy, 1770.

55 *Mr Tofield's Report on a Survey of Deepening Fen*, 1768; LAO.

56 The correspondence is printed in *Min. Proc. Inst. C.E.* 4 (1845), 205–9.

57 LAO, Deeping Fen 1/2, Adventurers' Minute Book 1756–76.

58 Market Weighton Drainage Board, Pocklington, Weighton Drainage & Navigation Trustees' Minute Books 1772–1838, Treasurer's Accounts 1772–84, Contracts 1772–80.

59 *The Report of John Grundy respecting the Drainage and Navigation proposed for Walling Fenns, &c.* 1772, with annexed engraved section. HRO, Hull University Library.

60 A copy of the lock and sluice foundation plan is in Yorkshire Water Authority, Beverley.

61 HRO, DDMW/7/404, MS Section of Weighton Canal, June 1781; Luke Holt.

62 Hull University Library DDLA/16/3, MS report by Grundy of site visit July 1773.

63 M. W. Baldwin, 'The engineering history of Hull's earliest docks', TS in Science Museum Library. An abridged version in *Trans. Newcomen Soc.*, 46 (1973) pp. 1–12.

64 *The Report of John Grundy, Engineer, respecting Quays or Wharfs on the West Side of the River Hull*, 1772, HRO, Hull City Archives.

65 *The Report of John Grundy, Engineer* [on Hull Harbour and proposed Dock], 1773, Hull City Archives. *Smeaton's Reports*, Vol. 3, pp. 157–62.

66 This information is given by the Dock Company Chairman, William Hammond, in *Remarks on a Publication intitled 'The Case of the Merchants of the town of Kingston-upon-Hull' with Additions*, London, 1787. Hull City of Archives.

67 J. Timperley, 'An account of the harbour and docks at Kingston-upon-Hull, *Trans. Inst. C.E.*, 1 (1836), pp. 1–52.

68 Hammond, *op. cit.*, gives the area of the dock and quay as built and as in 'Mr. Grundy's Plan, which was that directed by the Act'.

POSTSCRIPT

Several years after publication of this paper the missing report books (see p.3) were discovered in the Brotherton Collection at the University of Leeds. The story is told by O.S. Pickering (*Trans. Newcomen Soc. 60* (1988–89) 137–143) who gives a brief analysis of their contents. Together with vol. II in the Institution of Civil Engineers library, it is now seen that the complete set constitutes fifteen quarto volumes, totalling some 5000 pages, of fair copies of reports, estimates, drawings, observations and letters dating from 1741 to 1776 plus two volumes of reports, with separate pagination, up to 1779. This valuable source awaits detailed study; but Grundy's *Catalogue*, which was available when writing the paper, fortunately gives titles of all the main reports in vols I to XV, so the list of works (p.39) will require little change.

JOHN GRUNDY: Works & Selected Reports

	Planning		Works		
	Report	Plans	Construction	Engineer in charge	Resident Engineer
Pinchbeck Sluice [design by Grundy sen.]		1739	1739–40	Grundy sen.	? Grundy
Newark Navigation	*1741*	*1741*			
Osbaston dam	1741	1741			
Norfolk sea wall [scheme by Grundy sen.]	1743	1743	1742–43	Grundy	
R. Witham (with Grundy sen.)	*1744*	*1743*			
Grimsthorpe Great Water dam		1746	1746–48	Grundy	
Grimsthorpe water supply		*c.*1748	1751	Grundy	Jos. Smith for engine & pump
Deeping Fen: various works			1748–64	Grundy	Joseph Featherstone
Stallingborough Sluice	1749, 1751	1751	1751	Grundy	Wm. Hodgson
R. Dee, Chester	*1750–59**				
R. Witham	*1753*		*1753*		
Romney Marsh, sea wall	1755		1755		
Louth Navigation	1756	1756*		see below	
Eastoft Drainage	1759				
R. Witham (report with Edwards and Smeaton)	*1761*	*1762*	[1763–67	Langley Edwards	?]
Louth Navigation	*1761*	*1761*	1765–67	Grundy	Jas. Hogard
Fossdyke (with Smeaton)	1762				
Holderness Drainage	*1763*	*1764*	1765–67	Grundy	Jos. Page
Driffield Navigation	*1766*	*1766*	1767–70	Grundy	Sam. Allam
	1767	1767			
Swale & Cod Beck Navigations	*1767*	*1767*	[not completed	John Smith	?]
Thorney Fen bank		1767			
Adlingfleet Drainage [scheme by Smeaton]			1767–69	Grundy	Dav. Buffery
Misterton, Gringley & Everton Drainage	1768*				
Laneham Drainage	1769*	*1769*	1770–72	?	?
Deeping Fen & Welland outfall	1770	1769			
Chesterfield Canal variation	*1770*	*1770*			
Weighton Drainage & Navigation	*1772*	*1772*	1772–75	Grundy	Sam. Allam
Hull Quay	*1772, 1773*				
Hull Dock	1774*	1774*	[1775–78	Henry Berry	Luke Holt]
Hedon Haven	1774	1773			
East Fen Drainage	*1774*	*1774*			
South Holland Fen Drainage	1774*	1775			
Wells Harbour (with Thos. Hogard and Jos. Hodskinson)	1782				

italic: printed report, engraved plan

* not found

VI

The Eleventh Dickinson Memorial Lecture

William Chapman (1749-1832), Civil Engineer

INTRODUCTION

Chapman's first biographer is quite well informed on his work in Ireland and mentions his various mechanical inventions, but has extraordinarily little to say about his civil engineering works in England, preferring, for some reason, to devote much attention to the unrealised project for a canal from the Tyne to Solway Firth.[1] Partly as a result of this lack of balance, which is reflected later in the short memoir by Richard Welford,[2] and partly because harbours and land drainage are still rather unfashionable subjects, most of Chapman's achievements as a civil engineer remain practically unknown to engineering historians.

Briefly stated his works include: two major land drainage schemes in Holderness and another in the Vale of Pickering; important harbour works at Scarborough, Shoreham, Leith and Seaham; improvements in the navigation of the rivers Shannon, Barrow, Hull and Tees; the introduction, so early as 1787, of the correct method of building oblique arch bridges; and the introduction of steam dredging in the improvement of tidal river navigations, on the Orwell in 1806. In association with William Jessop he worked on the Grand Canal of Ireland and during the early stages of the Ringsend Docks at Dublin, with John Rennie he was joint engineer of the Humber Dock at Hull; and with Daniel Alexander he planned the East London Dock. Chapman also drew up the original proposals for the Circular Line Canal in Dublin, and, in association with Benjamin Thompson, for the Newcastle & Carlisle Railway.

All the above mentioned civil engineering schemes are described in the paper. Chapman was, in addition, concerned with many projects including: the Tyne-Solway canal, improvements in the outfalls of the Welland and the Witham, an early version of the Manchester Ship Canal, and a proposed suspension bridge over the Tyne. His reports on these and some other projects, and some papers and books or tracts on various topics, are listed in the Appendix,

In mechanical engineering it is well known that Chapman invented the self-acting machinery for lowering coal waggons from the staiths into ships, and that he and his brother Edward Walton Chapman took out a patent in 1812 which included the first description of a pivoted bogie for locomotives. The brothers were also responsible for improvements in rope-making machinery and, with Edward as manager, they ran for more than thirty years a successful rope manufactory at Willington.

WILLIAM CHAPMAN (1749–1832), CIVIL ENGINEER

PART 1 LIFE

William Chapman was born on 7 March 1749 at Whitby, into a Quaker family well known in the North East as ship owners, sea captains and, later, as bankers.[3] His father, Captain William Chapman (1713–1793), communicated two papers to the Royal Society, one on a method of converting salt water to fresh, and the other describing a Jurassic fossil crocodile, discovered in 1758 jointly with his friend John Wooler the Whitby harbour engineer. Captain Chapman left Whitby about 1766 to live at Barnes near Sunderland, the home town of his wife, the former Miss Hannah Baynes. In the 1780s he moved to Newcastle where, it is said, he established the rope-works at Willington Quay.

Fig. 1. William Chapman (1749-1832): from the portrait, artist unknown, reproduced in Weale's Quarterley Papers, 1843.

As a young man William Chapman followed his father's profession and for about three years had command of a ship trading with west and northern Europe. By 1778, however, he and his brother John Chapman (1750–1814) had taken the lease of Byker Colliery and also of Wallsend Colliery. Already in the previous year William was writing to James Watt about a steam engine for pumping at Byker. This, the earliest Boulton & Watt engine on Tyneside, was installed in 1778 at the St. Anthony pit. All seemed to go well there, but the sinking of the first shaft at Wallsend led to financial difficulties, the mortgagor foreclosed, and the Chapmans were declared bankrupt in 1782.

John Chapman managed to set up in business as a coal merchant. William decided to try his fortune as an engineer; he had recently developed an improved method of coal-winding and had been in correspondence with Watt on steam engine matters other than those arising from the Byker engine. Moreover in 1779 he had entered an open competition for designs to improve the pumping engine at Rotterdam, and won the second prize.

Anxious no doubt to get away from Newcastle, for the time being, he took the opportunity of visiting Ireland in company with Matthew Boulton in the autumn of 1783 and as things turned out he stayed there, apart from short trips home, for the next eleven years. To begin with he acted as agent for Boulton & Watt, supervising the erection of an engine at a colliery in County Tyrone, among other activities; but from his letters to Watt it is clear that his inclinations were moving towards civil engineering. Watt gave very practical help, for example by sending plans and specifications, supplied by courtesy of the contractor John Pinkerton, for work on the Birmingham Canal; and in 1785 Chapman wrote a report, on his own initiative, proposing a canal around the south part of Dublin. This was later built with some modifications under the direction of William Jessop. But the publication of Chapman's report brought his name and ability to public attention, and there can be little doubt that it led to his first civil engineering appointment: as engineer to the Kildare Canal. It was there, during the construction of the canal (1787–89), that he designed and built the earliest oblique arch bridges.

From 1789 to 1794 Chapman was fully occupied as a civil engineer in Ireland. During these six years he acted as assistant to Jessop on the Grand Canal and in the early stages of building the large Ringsend Docks in Dublin. In addition works were carried out to his plans and under his direction on the River Barrow navigation and on the Limerick-Killaloe Canal. His brother Edward Walton Chapman (1762–1847), joined him in Ireland in 1786 and probably worked as assistant on the Kildare Canal; certainly he was resident engineer on the Ringsend Docks right through to their completion in 1796 and did an excellent job in that capacity.

William was elected a Member of the Royal Irish Academy in 1794, and would seem to have had an assured future in Ireland, but he now felt the need to return to his native country, and family. Towards the end of 1794 he settled in Newcastle, having been commissioned to report on a projected canal from the Tyne to Solway Firth. Almost immediately he was elected into the Smeatonian Society of Civil Engineers, on the proposal of Jessop, seconded by John Rennie and Robert Mylne.[4] He attended meetings of this Society for over thirty years. Rather surprisingly he never joined the Institution of Civil Engineers, but in 1837 his widow donated a practically complete set of his printed reports to the Institution. This valuable and unique collection is still at the Civils and provides one of the main sources for the study of Chapman's professional career. The other primary sources are chiefly Minutes of the Commissioners of the various public works for which Chapman acted as consulting or chief engineer.

Nothing came of the Tyne-Solway canal project, but in 1796 Chapman was asked to prepare plans and estimates for two large land drainage and flood protection schemes in east Yorkshire. This proved to be the beginning of a period of great activity in which, after drawing up his reports and plans, work began on the following undertakings: Beverley & Barmston Drainage 1799, Muston & Yedingham Drainage 1800, Scarborough Harbour 1801, Driffield Navigation 1802, and Keyingham Level Drainage 1803. Also in 1803 construction began on the Humber Dock at Hull, for which Rennie and Chapman were joint engineers.

All these works were well under way by 1806, when Chapman decided to embark again on coal mining. This was a difficult time for civil engineers. Canal construction had come almost to a standstill; railways, on a large scale, were still in the future; the country was involved in the Napoleonic Wars; and Telford and Rennie had, between them, the direction of most of the larger public works. Presumably, therefore, as a second line of financial support Chapman took a share in the lease of East Kenton and Coxlodge Collieries jointly with Robert Knowsley. There he was responsible for winning the Jubilee Pit, and installing a 24 H.P. pumping engine, and probably it was also he who laid out the 4 mile Kenton & Coxlodge railway. The railway and the pit were opened in 1809. Chapman also designed the first 'coal drop', in 1807, at Benwell staith; an ingenious machine for lowering the coal waggons direct into the ships instead of tipping the coal down the 'spouts'. He disposed of the colliery leases in 1811. In the following year he and his brother Edward jointly patented a chain-haulage steam locomotive The patent included the first description of a pivoted bogie. A locomotive of this type was tried out at Heaton Colliery in 1813. The chain system appears not to have been very effective, but the bogie proved to be a useful idea as it enabled heavier and therefore more powerful locomotives to run on the existing rails with the weight distributed by six or eight wheels instead of four An eight wheel loco had successful trials on the Lambton Colliery line in December 1814, and not long afterwards both of Hedley's Wylam locomotives were rebuilt as eight-wheelers, i.e. with a pair of four-wheel bogies.

WILLIAM CHAPMAN (1749–1832), CIVIL ENGINEER

During the years 1806–14 Chapman of course had not given up civil engineering. He completed the Yorkshire land drainage works and constructed an extension of the Keyingham Level scheme, and work continued at Scarborough, while in 1806 construction began on a new channel, near Ipswich, for improving the navigation of the River Orwell. Chapman planned the entire project, which included several other cuts or channels, and introduced on the cut below Ipswich the first steam dredger to be used on a tidal river navigation and indeed one of the earliest steam dredgers ever built. This, like the Coxlodge engine and the Heaton locomotive of 1813, was made for him at the Butterley ironworks. In this period, also, he and Telford did some preliminary planning of the Carlisle Canal and work began in 1809, under his direction, on improvements of the River Tees navigation below Stockton.

Chapman had an interest in the rope manufactory at Willington, near Newcastle, of which his brother Edward was manager. They installed a Boulton & Watt engine at Willington in 1799 and the brothers took out several patents for improvements in rope-making machinery. Chapman's book on cordage, published in 1808, gives perhaps the best historical account of the early developments of such machines. This, it may be noted, is only one of his publications (apart from the reports). Others include a book on canals (1797), another on the preservation of timber (1817) and an important tract on the necessity of keeping accurate mine records for the prevention of future accidents (1815). Chapman also read an interesting paper to the Royal Society in 1816 on the formation of coal. He correctly deduced that this substance originated as peat, formed on the surface of the earth in remote geological time, and that its transformation into coal resulted from compaction under the weight of subsequently deposited overlaying strata. If not the first, this must be a very early statement of the principles involved.

His reports of 1813 and 1814 on Laneham Drainage in Nottinghamshire and on the Sheffield Canal mark the prelude to the third phase of Chapman's civil engineering career, corresponding with the post War period. Construction of the Sheffield Canal began in 1816. Shoreham Harbour followed in 1817. Work at Scarborough, which had been proceeding slowly, was notably accelerated and enlarged in scope after 1817. Construction of the Carlisle Canal started in 1820 and at Leith Harbour, following prolonged negotiations, in 1827. All these schemes were completed in accordance with Chapman's designs and under his personal direction. His last work, Seaham Harbour, was also completed according to his plans but he lived to see only about a half of the harbour actually built. This was constructed for the Marquis of Londonderry, for whom he had in 1821 designed a floating crane for transferring coal from barges into ships lying in Sunderland harbour. Of Chapman's project reports in this period special mention may be made of his proposals for a railway from Newcastle to Carlisle and his planning of the East London Dock in 1824.

Apart from a short spell in York in 1804-5 and at Morton in Co. Durham, the home of his father-in-law, around 1811-12, Chapman lived at Newcastle from his return to England in 1794 until his death; which occurred on 29 May 1832. He had an office in Saville Row, and his residence in Ridley Place. He left about £8000 (say £120,000 in present-day money) to his wife Elizabeth, his brother Edward and sister Hannah, and to his daughter and only child Elizabeth Hannah. Chapman's library, consisting of 535 volumes and numerous maps, plans etc., was sold by auction in April 1833. As previously mentioned, his printed reports were presented by Mrs. Chapman to the Institution of Civil Engineers in 1837.

Of Chapman's pupils the most distinguished was Thomas Elliot Harrison (1808-1888) who became President of the Institution in 1873. Of his resident engineers James Leslie, F.R.S.E. (1801-1889) is outstanding. He was selected by Chapman for his first responsible post as resident engineer at Leith, and went on to become the leading engineer in Scotland on harbour works and water supply.

Chapman appears as a somewhat reserved and studious person, patiently devoted to his work, ready to change his mind when new evidence emerged but a little too ready, perhaps, to take offence when subject to what he felt to be unjustified criticism. He was respected by his fellow engineers, with whom his relationship seems invariably to have been courteous and friendly. With clients he naturally did all in his power to further their interests, even, as in the Leith job, undertaking delicate negotiations with the Admiralty when the Harbour Commissioners themselves had failed to make progress; but on occasion his patience was tried beyond endurance by the clients' treatment of a resident engineer (as on the Carlisle Canal) or by their reluctance to pay his own scrupulously honest fees (as at Shoreham). Of his personal life little is known. He visited Mr. and Mrs. James Watt at Heathfield House on several occasions. Later he was on friendly terms with Sir George Cayley, John Buddle and other leading figures in the northern scientific world. One imagines him spending much of his leisure time with his family and in his library at Ridley Place, attending meetings of the Newcastle Lit. & Phil. Society, and, on visits to London, going to a dinner at the Smeatonians; in other words, leading a quiet, useful life, the main results of which are to be seen in his writings and engineering work.

PART II: WORKS

In this part of the paper I give a catalogue *raisonné* of the principal works on which Chapman was actively engaged This is an unusual form to adopt in a biographical memoir but it serves to display a good deal of information as briefly as possible; and during my historical studies I have often felt the lack of such details, which do not exist for any other engineer.

The works are set out in chronological order, with a few exceptions where a more convenient grouping is obvious and the dates in each heading are those for the period of construction or, if in brackets, the period during which Chapman was involved in construction. In a few instances the dates refer to the year in which the design was carried out. Costs are given for the engineering works usually without the sums for puchasing land and obtaining parliamentary Acts, but in some cases the items cannot be separated. Chapman generally charged his fees on a daily basis including the time spent 'giving instructions or Drawing plans at home', As quoted here the fees include travelling expenses which, in his own words, 'are never unnecessarily high, as both by inclination and habit I am disposed to economy'.

PROPOSALS FOR THE CIRCULAR LINE CANAL AT DUBLIN (1785)

Chapman's first civil engineering report concerned a topic of major importance to the city of Dublin. Its title is

'Observations on the Advantages of bringing the Grand Canal round
by the Circular Road into the River Liffey.
By William Chapman, Engineer'. Dublin, 23 May 1785.

At that time the Grand Canal still terminated in a basin at St. James's Street in the west part of Dublin, without a connection to the Liffey, but in 1784 a survey had been made by the canal company for a short link from the basin to the river. Chapman's better and more imaginative plan, based partly on environmental considerations, involved a new cut 4 miles long with eight locks, passing around the south of Dublin and entering the Liffey through a dock, downstream of the city bridges, either at Ringsend or opposite the new Customs House.

The canal to Ringsend, with some modifications of the original proposals, was built 1790-92 by Archibald Miller with Jessop as consulting engineer. (See also, Ringsend Docks).

REFERENCE

1. Ruth Delany *The Grand Canal of Ireland* (Newton Abbot 1973).

KILDARE CANAL (1787-1789)

LOCATION: Sallins to Naas, Co. Kildare (Fig. 2).

CLIENTS: County of Kildare Canal Company.

ACT: 8 May 1786.

ENGINEER: William Chapman (salary £200 p.a.).

COST: £10,500 (Irish) = £9,700.

PRINCIPAL WORKS: (i) Canal cut 2½ miles long from the Grand Canal to Naas (ii) Rise of 44 ft taken in 5 locks, the locks 14 ft wide (iii) Canal basin at Naas (iv) Three bridges over canal. Work continued from 1789 for a year or two on an extension towards Corbally but was temporarily abandoned for lack of funds; it was completed in 1810 by John Killaly.

WILLIAM CHAPMAN (1749–1832), CIVIL ENGINEER

SPECIAL FEATURES: Three skew or oblique arch bridges at Osberstown, Oldtown and Naas (Finlay bridge). The latter (Fig. 3) had an obliquity of 51°, span 25 ft, rise 5 ft 6 ins; a geometrical diagram of the winding courses is given in Chapman's article in Rees's *Cyclopaedia*. The design dates from 1787, about ten years before March Barn Bridge was built on the Rochdale Canal, generally acknowledged as the earliest example of this type of construction in England.

REFERENCES:

1. Watt to Chapman 15 Nov 1785. Chapman to Watt 25 July 1788, 19 May 1790. Boulton & Watt Collection (Birmingham Library).

2. William Chapman 'Oblique arches' Rees's *Cyclopaedia* vol. 25 (London, 1813).

3. Ruth Delany 'The County of Kildare Canal' *Journ. Co. Kildare Archaeol. Soc.* vol. 15 (1972) 122–35.

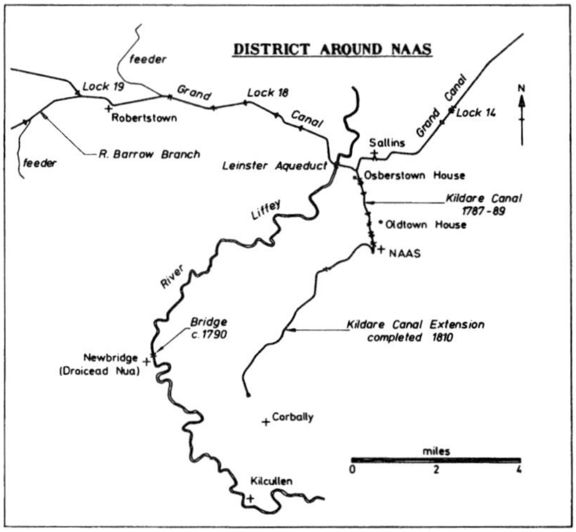

Fig. 2.

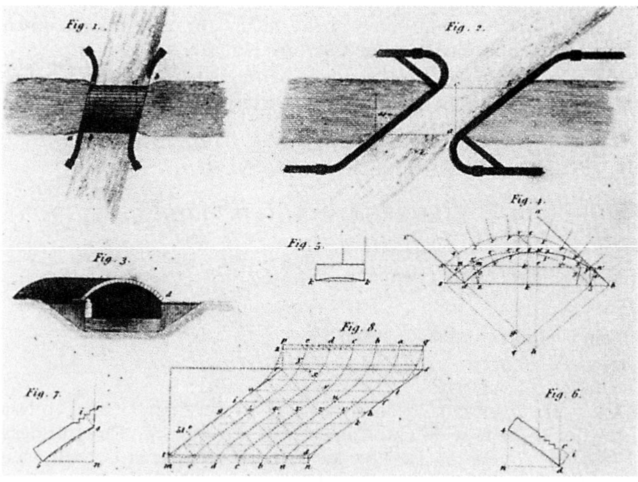

Fig. 3. Finlay Bridge, Kildare Canal. Designed by Chapman, 1787, built 1789. Engraving by W. Lowry, published in Rees' *Cyclopedia,* 1814.

WILLIAM CHAPMAN (1749–1832), CIVIL ENGINEER

BRIDGE OVER THE RIVER LIFFEY (c. 1790)

LOCATION: Newbridge (Droicead Nua), Co. Kildare (Fig. 2).

ENGINEER: Chapman designed the bridge at the request of the Duke of Leinster.

DESCRIPTION: A 5-arch masonry bridge, total length across the river about 480 ft. It replaced, on a slightly different site, the mediaeval bridge which had recently been destroyed by a flood. Difficult foundation conditions are reported to have been successfully overcome under Chapman's direction.

REFERENCES:

1. Chapman to Watt 19 May 1790 (B. & W. Collection).
2. John Weale *Memoir of William Chapman* 1843.
3. Personal communication from Ruth Delany.

GRAND CANAL OF IRELAND (1789-1794)

LOCATION: Dublin to Banagher on R. Shannon.

CLIENTS: Grand Canal Company.

ENGINEER: Construction began in 1756. William Jessop was chief consulting engineer 1789–1802. Chapman acted as his assistant 1789–1794.

PRINCIPAL WORKS BY CHAPMAN: Apart from his report of 1785, proposing the Circular Line around the south of Dublin, Chapman (i) surveyed the entire length (53 miles) of the canal west of the summit level to the Shannon between 1789 and 1792 (ii) reported on the works in progress (1791) on the canal where it crosses the Bog of Allen (iii) rebuilt (1793–94) sixteen of the original eighteen locks between Dublin and the summit level near Robertstown (iv) designed several bridges.

REFERENCES:

1. Jessop's reports, in Grand Canal Company Minute Books. I am indebted to Mrs. Delany for copies of these.
2. William Chapman 'Remarks on the General Repairs of the Grand Canal' Dublin, 17 Nov 1795 (I.C.E. Library).
3. Ruth Delany *The Grand Canal of Ireland* (Newton Abbot 1973).

RINGSEND DOCKS (1792-1796)

LOCATION: Dublin.

CLIENTS: Grand Canal Company.

ACT: 5 May 1791.

ENGINEER: William Jessop; with Chapman as assistant in early stages of design and construction.

RESIDENT ENGINEER: Edward Walton Chapman (salary 300 gns. p.a.).

CONTRACTORS: Messrs. Myers, until 1794; works completed probably by direct labour.

COST: £112,000 (Irish) = £105,000.

PRINCIPAL WORKS: (i) Two linked wet docks, total area 26 acres (ii) Three dry docks, the largest 230 ft in length (iii) Two ship locks, one 148 ft long and 37 ft wide with 18 ft of water on the sill at spring tides the other 120 ft X 26 ft X 16 ft (iv) a barge lock 16 ft wide and 135 ft long overall built as a staircase pair with three sets of gates.

FORMAL OPENING: 23 April 1796.

WILLIAM CHAPMAN (1749–1832), CIVIL ENGINEER

RESULTS: Extensive dock accommodation at junction of the Circular Line of the Grand Canal with the River Liffey.

SPECIAL FEATURES: The largest wet docks built in the British Isles before 1800. One of the earliest steam engines ever used on civil engineering works, employed in pumping the excavation.[5] Chapman was in correspondence with Boulton & Watt about the engine 1791-92; it began work probably in 1793 and by 1794 'is now doing well'.

REFERENCES:

1 Jessop's reports, in Grand Canal Company Minute Books.
2. Chapman to Boulton & Watt 5 Dec 1791, 27 Dec 1791, 20 Jan 1792, 31 March 1792.
3. Chapman to Boulton 17 Aug 1793 (Birmingham Assay Office).
4. Ruth Delany *The Grand Canal of Ireland* (Newton Abbot, 1973).

RIVER BARROW NAVIGATION (1790-1795)

LOCATION: Athy, Co. Kildare, to the tide-lock at St. Mullins, Co. Carlow.

CLIENTS: Barrow Navigation Company.

ACT: 5 May 1790.

ENGINEER: William Chapman, from 1790 to 1795 (salary £200 p.a.).

RESIDENT ENGINEER: Humphrey Mitchell.

CONTRACTORS: not known.

COST (1790-1800): £63,000 (Irish) = £58,000.

REPORT: 'Report on the Means of Perfecting the Navigation of the River Barrow from St. Mullins to Athy. By William Chapman, Engineer'. Naas, 9 October 1789. (This is exceptionally detailed, the estimates alone run to 33 pages).

PRINCIPAL WORKS: (i) Four of the eight pre-1789 locks rebuilt; (ii) Ten new locks, 80 ft X 16 ft, and the associated weirs; (iii) Several lateral cuts.

RESULTS: In 1789, after 30 years work, the river was navigable by 15 ton boats. The improvements were designed to allow navigation throughout the 42 miles from St. Mullins to Athy by 80 ton boats. This aim only partially achieved by 1800. Five more locks were built subsequently, bringing the total to 23, four short of the number planned by Chapman.

REFERENCES:

1. Chapman to Watt 17 May 1790 (B. & W. Collection).
2. *Statistical observations relative to the County of Kilkenny made in the years 1800 and 1801* (Dublin, 1802).
3. V.T.H. & D.R. Delany *The Canals of the South of Ireland* (Newton Abbot, 1966).

LIMERICK–KILLALOE CANAL (1791-1794)

LOCATION: Between Killaloe, at south end of Lough Derg, and Limerick.

CLIENTS: Limerick Navigation Company.

ACT: Work carried out under Act of 27 May 1768.

WILLIAM CHAPMAN (1749–1832), CIVIL ENGINEER

ENGINEER: William Chapman, from 1791 to 1794.

RESIDENT ENGINEER: Mr. Lascelles.

CONTRACTORS: Browning, Clark & Carrol.

COST: (1791-99): £24,500 (Irish) = £22,500.

PRINCIPAL WORKS: (i) Rebuilding several locks on the 5½ mile Cloonlara cut, including conversion of the triple Erinna Lock as a staircase pair (ii) Setting out the 2 mile Moys cut from Killaloe to Cusane (iii) Foundations of Cusane Lock using inverted arch construction (1794).

REFERENCES:

1. Chapman to Boulton 3 Feb 1791 (Birmingham Assay Office).
2. Chapman's printed letter 'To the Subscribers of the Limerick Navigation'. Dublin, 26 Nov. 1795 (I.C.E. Library).
3. John Brownrigg's report to the Directors-General of Inland Navigation, 1 Sept. 1800.-(P.R.O. Dublin). I owe this reference to Mrs. Delany.
4. V.T.H. & D.R. Delany *The Canals of the South of Ireland* (Newton Abbot, 1966).

WILLINGTON ROPERY

According to tradition the ropery at Willington Quay, near Newcastle (Fig. 16) was established by Captain William Chapman in the 1780s. It was certainly carried on by his sons William and Edward after their return from Ireland. Between 1797 and 1799 they were engaged with developments in rope-making machinery and took out the following patents, singly or jointly: 2191 (1797), 2219 (1798), 2265 (1798), 2326 (1799). They installed an 8 H.P. Boulton & Watt engine at Willington in 1799, the second to be used in a rope manufactory (the first being John Grimshaw's at Sunderland in 1797). Patent 2513 followed in 1801, for the pre-treatment of tar to render it more suitable as a preservative for ropes immersed in water, and in 1807 the brothers patented (3078) a method of making belts or flat ropes. The Chapmans' machinery was used not only at Willington but also, under licence, by Messrs. Hood, Errington & Co. of Newcastle, from 1799. In these inventions William Chapman played the leading role, but Edward managed the Willington works and lived in a house near the gateway office. Ropes of excellent quality were produced for mines, inclined planes on railways and other purposes, and ships cables up to 24 ins circumference.

Though he retained a financial interest in the concern, William's active participation probably ceased after about 1816, when his civil engineering practice greatly expanded. John Paterson became a partner before 1818, and in 1827 Robert Rawlinson joined the firm which, in the 1820s, was known as Edward Chapman & Co. Edward himself retired about 1830, moving to Northumberland Street, Newcastle, and the ropery came under the management of Mr. Washington Potts. Finally the business was acquired c.1840 by Robert Hood Haggie. It continued to flourish well into the 20th century.

REFERENCES:

1. James Watt jnr. to Chapman, 11 June 1799. Engine drawings in Portfolio 186 (B. & W. Collection).
2. Trade card of 'Wm. Chapman & Co. Willington Ropery' 1802 (Goodrich Collection, Paper 14. Science Museum).
3. W. Chapman *A Treatise on the Progressive Endeavours to improve the Manufacture and Duration of Cordage* (London 1808).
4. Circular issued by Edward Chapman & Co., Willington Ropery. Jan. 1827 (Durham C.R.O. NCB I/JB/1843).
5. W. Richardson *History of the Parish of Wallsend* (Newcastle, 1923).

VI

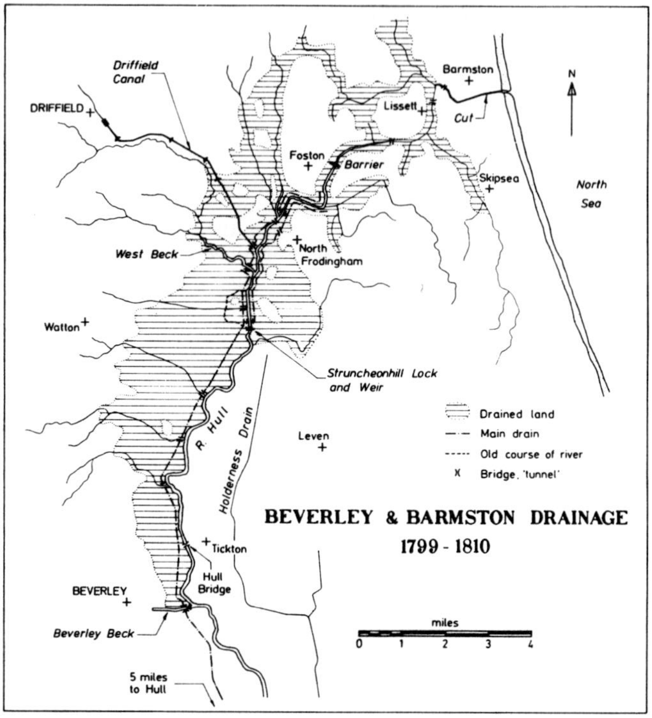

Fig. 4.

BEVERLEY & BARMSTON DRAINAGE (1799–1810)

LOCATION: Valley of the River Hull. Holderness, east Yorkshire (Figs. 4 & 5).

CLIENTS: Beverley & Barmston Drainage Commissioners.

ACT: 21 June 1798.

ENGINEER: William Chapman (fees £2158).

RESIDENT ENGINEER: William Settle (salary £175 p.a.).

SURVEYORS: Francis Haigh, Samuel Dickinson, John Dalton, John Foster.

CONTRACTORS: for earthwork: Robert Dickinson, Francis & James Pinkerton, Thomas Dyson, Robert Nicholson & James Smith. For outfall sluice, 'tunnels' and culverts: Thomas Dyson. For bridges: James Smith, Thomas Dyson.

COST: £115,000

WILLIAM CHAPMAN (1749–1832), CIVIL ENGINEER

REPORT: 'The Report of Mr. Chapman, respecting the Drainage of the Low Grounds lying below the Wolds, on the West Side of the River Hull, and in Frodingham Carrs, and at Lissett'. Newcastle, 8 October 1796.

PRINCIPAL WORKS (i) Barmston Cut 2 miles (1799–1801) (ii) Main drainage cuts, 23 miles (iii) River embankments, 20 miles (iv) Outfall sluice at Hull (1800–2) (v) Eleven tunnels[6] leading drainage cuts under Beverley Beck, Driffield Canal, River Hull and its tributaries (vi) Twenty-seven road bridges, some of oblique construction, and numerous culverts and occupation bridges. (See also, Driffield Navigation). Two of Chapman's drawings for the Beverley 'tunnel' are shown in Figs. 6 and 7. This structure remains unchanged to the present day (Fig. 8).

RESULTS: Drainage and flood protection of 12,600 acres lying between Beverley and Lissett. The previously often flooded low grounds and water-logged carrs were transformed into some of the richest agricultural land in England.

SPECIAL FEATURES: Chapman's scheme was more imaginative and thoroughly worked out than the proposals submitted earlier in 1796 by Richard Hodgman and by Jessop. Two basic principles were involved (i) Diversion of upland waters away from the area to be drained; achieved by diverting streams from the neighbourhood of Skipsea and Burton Agnes eastwards to the sea through the new Barmston Cut, with a 'barrier' or small earth dam across the river at Foston (ii) Separation of land drainage from the 'living waters' of the river and its tributaries; achieved by constructing a system of drainage cuts independent of, and where necessary passing under, the watercourses. In this way the river, suitably embanked, was left free to discharge its own flood water (in any case reduced by the Barmston diversion) without at the same time being encumbered by extra flow from the land drains – as in the traditional system of fen drainage.

The Beverley & Barmston scheme deserves to be recognised as a classic civil engineering work.

REFERENCES:

1. Beverley & Barmston Commissioners Minute Books, Proceedings, Annual Accounts, Schedule and Award (East Riding Record Office).

2. Chapman's reports 1796–1809. (I.C.E. Library).

3. Survey by John Dalton and John Foster, 1810. (E.R.R.O.).

4. June Sheppard, *The Draining of the Marshes of East Yorkshire*. Unpublished Ph.D. Thesis. University of London, 1956.

Fig. 5. General view of Beverley and Barmston drain and embanked River Hull.

Photo. by author.

VI

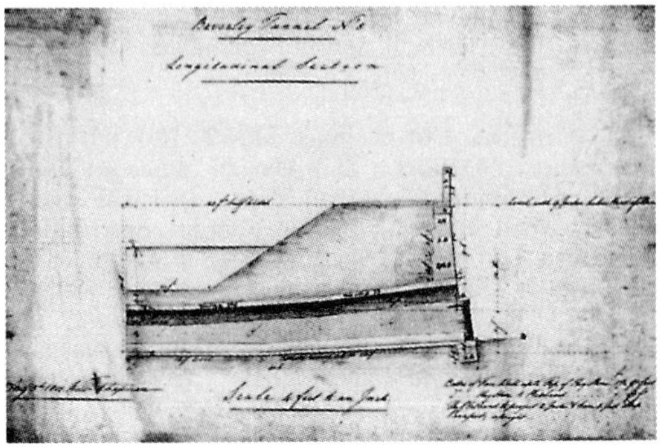

Fig. 6. Longitudinal Section of Beverley 'Tunnel'. Drawing by Chapman dated 3 Aug. 1801. East Riding Record Office.

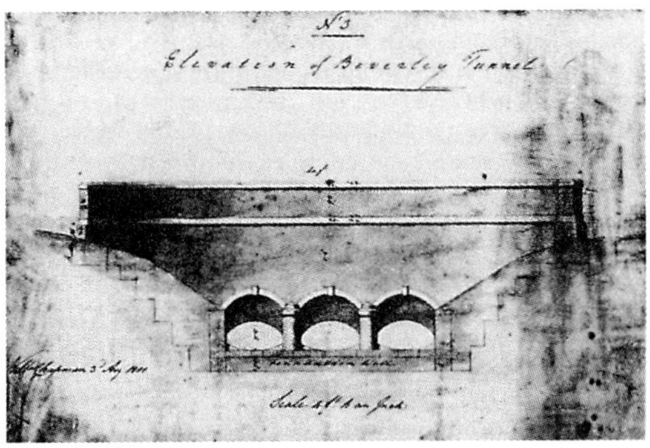

Fig. 7. Elevation of Beverley 'Tunnel'. Drawing by Chapman dated 3 Aug. 1801. East Riding Record Office.

Fig. 8. Beverley 'Tunnel', leading the main drain under Beverley Beck. Completed to Chapman's designs, 1802.

Photo. by author.

WILLIAM CHAPMAN (1749–1832), CIVIL ENGINEER

DRIFFIELD NAVIGATION (1802–1805)

LOCATION: River Hull, east Yorkshire (Fig. 4)

CLIENTS Driffield Navigation Commissioners.

ACT: 2 July 1801.

ENGINEER: William Chapman (fees £320).

RESIDENT ENGINEER: William Settle.

SURVEYOR: Samuel Dickinson.

CONTRACTORS: for river works, lock and weir: Thomas Atkinson and Thomas Dyson. For Hull Bridge: Appleton Bennison.

COST: £6,000.

REPORT: 'Estimate for the Expence of Perfecting the Driffield Navigation. By William Chapman'. Newcastle, 30 November 1796.

PRINCIPAL WORKS: (i) New channel ¾ mile long, with embankments, by-passing loop in river (ii) Lock and weir at south end of channel, near Struncheon Hill (also known as Hempholme lock) (iii) deepening of the river from the entrance to the Driffield Canal (built 1767–70 by John Grundy) to Frodingham bridge (iv) Hull Bridge at Tickton (v) The Beverley & Barmston drain was taken under the new river channel in a 'tunnel' just upstream of the lock.

REFERENCES:

1. Driffield Navigation Commissioners Minute Books (E.R.R.O.)
2. Charles Hadfield *The Canals of Yorkshire and North East England, vol. 2* (Newton Abbot, 1973).

MUSTON & YEDINGHAM DRAINAGE (1800–1808)

LOCATION: Vale of Pickering, Yorkshire (Fig. 9)

CLIENTS: Muston Drainage Commissioners (Sir George Cayley, chairman of Directors).

ACT: 28 July 1800.

ENGINEER: William Chapman (fees £1380).

RESIDENT ENGINEER: Robert Wilson (salary £130 p.a.)

ASSISTANT RESIDENT ENGINEER: Barnham Taylor.

SURVEYORS: Isaac Milbourn, Ralph Burton.

CONTRACTORS: for Sea Cut: Francis & James Pinkerton, Thomas & Simon Hamer. For Hertford channel: James Metcalfe, William Shaw. For Derwent channel: Thomas & Simon Hamer. For Everley Weir: Edward & Robert Coates. For bridges over Sea Cut and Scalby Beck: John King. For bridges over Derwent channel: William Outhwaite. For Yedingham sluice: George Willoughby & Jonathan Booth. For 'tunnels' and culverts: Thomas Dyson.

COST: £42 000

REPORT: 'Report of William Chapman, Engineer, on the Means of Draining the Low Grounds in the Vales of the Derwent and Hertford'. Newcastle, 16 January 1800.

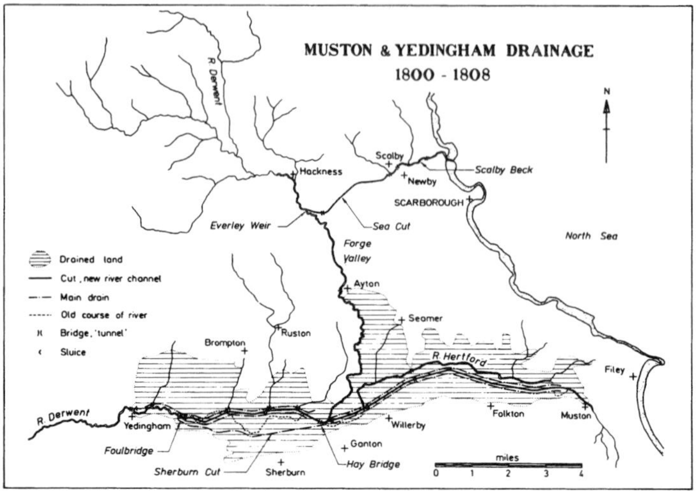

Fig. 9.

PRINCIPAL WORKS: (i) Sea Cut 3 miles and (ii) Everley Weir (Fig. 10), completed 1804. (iii) New Hertford channel 7 miles. (iv) New Derwent channel 4 miles. (v) Main drains 22 miles. (vi) Eight masonry bridges. (vii) Yedingham sluice. (viii) Seven 'tunnels'.

RESULTS: 10,500 acres of land drained and protected from flood.

SPECIAL FEATURES: A beautifully designed work carried out under Chapman's close personal direction. The weir and sluice at Everley diverted upland flood water of the Derwent along the Sea Cut and Scalby Beck to the sea whilst allowing ample flow down the river for the mills and the navigation from Malton to the Ouse. The rivers Hertford and Derwent given new channels (totalling 11 miles in length) between Muston and Yedingham with flood embankments formed by soil excavated from the channels and the main drains or 'delphs' running alongside. Rigorous separation of land drainage from the rivers and tributaries.

Fig. 10. Everley Weir, near Hackness. Completed to Chapman's designs in 1804. The River Derwent, in foreground, flows south through the sluice; excess flood waters pass down the cut to the sea near Scalby. *Photo. by author.*

Chapman was retained as consultant after completion until 1830 and Wilson remained as Surveyor or engineer until his retirement in 1836.

REFERENCES:

1. Correspondence up to 1800 (E.R.R.O., DDSY/11).

2. Muston & Yedingham Commissioners Minute Books, Accounts, and Award. Robert Wilson's Account Book (Muston & Yedingham Drainage Board).

3 'A Plan of the Low Grounds between Muston and Yedingham which adjoin to the Rivers Derwent and Harford. By Ralph Burton. 1808'. (M. & Y. D.B.).

KEYINGHAM LEVEL DRAINAGE (1803–1807)

LOCATION: South Holderness, Yorkshire (Fig. 11).

CLIENTS: Keyingham Level Drainage Commissioners.

ACT: 7 May 1802.

ENGINEER: William Chapman.

RESIDENT ENGINEER: not known.

SURVEYOR: Robert Stickney.

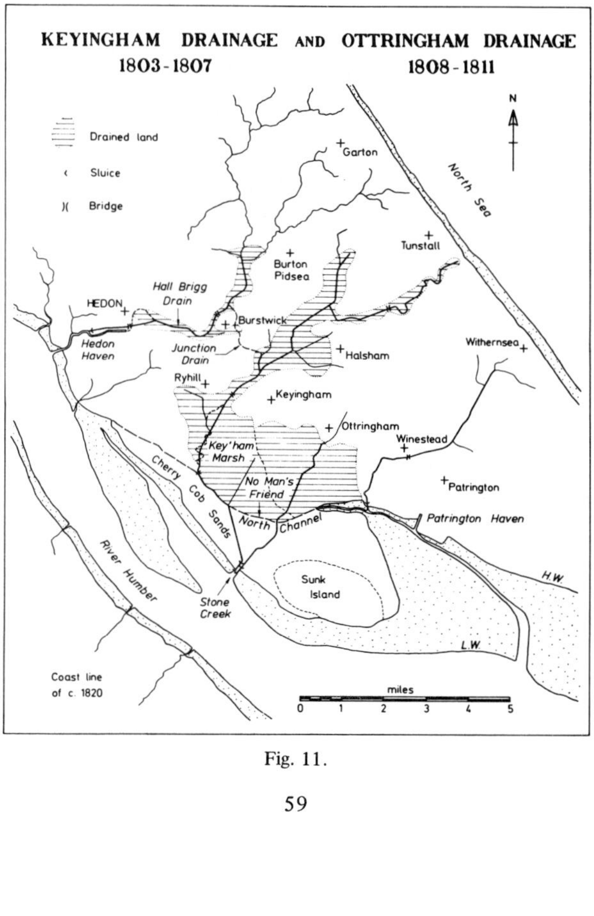

Fig. 11.

VI

WILLIAM CHAPMAN (1749–1832), CIVIL ENGINEER

CONTRACTORS: not known.

COST: £30,000.

REPORT: 'Report of William Chapman, Engineer, on the Drainage and Navigation of Keyingham Level, in Holderness'. Newcastle, 30 June 1797.

PRINCIPAL WORKS: (i) New cut to Hedon Haven 1½ miles and (ii) outfall sluice (iii) New cut across Keyingham Marsh 2 miles (iv) New cut across Cherry Cob Sands 1½ miles and (v) outfall sluice at Stone Creek (vi) All the existing drainage channels enlarged and deepened.

RESULTS: Improved drainage and flood protection of 5 500 acres.

SPECIAL FEATURES: Flood water, which used to flow over Keyingham Level, diverted to Hedon Haven. A permanent, non-silting outfall secured at Stone Creek.

REFERENCES:

1. 'A Map of Part of the Wapentake of Holderness . . . comprising the Levels of Keyingham & Burstwick with the Drainage thereof and projected improvements. Robt. Stickney, Surveyor. 1802.' (E.R.R.O.)
2. H. E. Strickland *A General View of the Agriculture of the East-Riding of Yorkshire* (York, 1812)
3. June Sheppard 'The Drainage of the Marshes of South Holderness and the Vale of York' *East Yorks. Local History Series* No. 20 (1966).

N.B. The Keyingham Commissioners Minutes before 1811 no longer exist.

OTTRINGHAM DRAINAGE (1808–1811)

An extension of the Keyingham Level scheme, carried out under an Act of 14 August 1807 with Chapman as nominated engineer. The drainage of 2500 acres improved by a new cut 1½ miles long to new outfall sluice (built 1809) at Stone Creek adjacent to the Keyingham sluice (Fig. 11).

REFERENCES:

1. Proposals outlined in a printed letter 'To the Proprietors of Lands within the Keyingham Level Drainage' by William Chapman. Hedon, 24 July 1800.

HUMBER DOCK (1803–1809)

LOCATION: Hull, between Myton Gate and the Humber.

CLIENTS: Hull Dock Company, Hull Corporation and Trinity House.

ACT: 22 June 1802.

JOINT ENGINEERS: John Rennie for the Dock Company, William Chapman for the Corporation.

RESIDENT ENGINEER: John Harrop.

ASSISTENT RESIDENT ENGINEER: George Miller.

CONTRACTORS: not known

COST: £230 000.

REPORTS: The principal planning reports are two letters (i) by Rennie and Chapman c. August 1802 (ii) by Chapman 5 October 1802.

PRINCIPAL WORKS: (i) Wet dock 7.2 acres in area (ii) Entrance lock 42 ft wide, 158 ft long, 26 ft of water on sills at spring tides (iii) Entrance basin 2.7 acres with (iv) two jetties projecting 100 ft in front of (v) new Humber Wall.

FORMAL OPENING: 30 June 1809.

WILLIAM CHAPMAN (1749–1832), CIVIL ENGINEER

RESULTS: Dock accommodation provided for 70 vessels of 180 tons burden.

SPECIAL FEATURES: Lock of inverted arch construction on piled foundations. Dock walls curved in profile, and piled. Cofferdam segmental in plan spanning 280 ft built to withstand 30 ft head of water. Two 6 H.P. steam engines began operations in 1804, pumping, pile-driving and mixing mortar. Steam dredger started work in 1812. Broadly speaking Chapman was planning engineer, and Rennie carried out the structural designs. They jointly supervised construction from 1803 to 1806; Rennie seems to have acted alone 1806–9.

REFERENCES:

1. Rennie and Chapman's letters 1802-9. Ms copies in Rennie's Reports vols. 2–5 (I.C.E. Library).
2. Plan of the intended Humber Dock at Kingston-upon-Hull, 1803. Drawn by John Buckton. (P.R.O. HDC 4/1/3/ C.8639).
3. John Timperley 'An account of the harbour and docks at Kingston-upon-Hull'. **Trans. I.C.E.** vol. 1 (1836) 1–52.
4. M. W. Baldwin 'The Engineering History of Hull's Earliest Docks'. The present volume, P. 1.

SCARBOROUGH HARBOUR (1801–1831)

LOCATION: Scarborough, Yorkshire (Fig. 12).

CLIENTS: Scarborough Harbour Commissioners.

ACT: A new Act obtained 23 May 1806.

ENGINEER: William Chapman (fees £1270).

RESIDENT ENGINEERS: Matthew Shout,[7] 1801–4; Roger Nixon, 1804–22; William Barry, 1822–38. (Salary £105 p.a.).

CONTRACTORS: Work carried out chiefly by direct labour until 1818. Thereafter for East Pier: Messrs. Whittal and Coulson & Tisseman. For West Pier: Barry & Hovington. For rebuilding end of Vincent's Pier: Whittal. For Spur: Coulson.

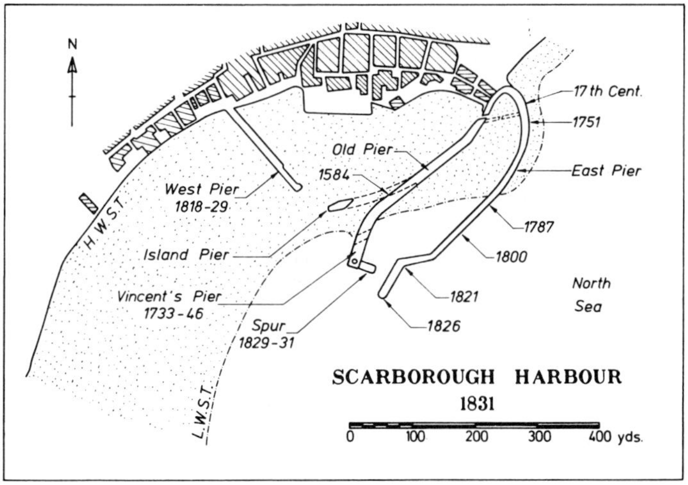

Fig. 12.

COST: about £65,000 (1801–31).

REPORTS: 'Report on the Harbour of Scarborough; and on the Means necessary for its Improvement. By William Chapman, Engineer'. Hull, 20 September 1800.

'Mr. Chapman's Report on the Antient, the Intermediate, and Present State of the Harbour of Scarborough, and on the Measures requisite for its Final Improvement'. Newcastle, 4 August 1829.

PRINCIPAL WORKS: (i) Extension of the great East Pier 500 ft, 1800–26 (ii) West Pier 480 ft 1818–20, widened 1827–29 (iii) Rebuilding end of Vincent's Pier 1823–26 (iv) Spur extension from Vincent's Pier 85 ft 1829–31. All these works in massive masonry construction (Fig. 13).

Fig. 13. Scarborough Harbour Entrance, as completed by Chapman in 1831. Lithograph by George Hawkins, after W. Beverley, published 1845.

RESULTS: In 1801 the harbour works comprised a single pier (an Elizabethan structure extended in the mid 18th century by William Vincent) and about half of the East Pier. By 1831 the harbour had been transformed practically into the state as seen today, providing one of the best havens and small ports on the north-east coast. Inner Harbour 8½ acres, Outer Harbour 5½ acres.

SPECIAL FEATURES: The East Pier founded on *pierres perdues* at a depth of at least 15 ft below L.W.S.T. The line of the East Pier fixed as experience increased, notably after 1817, following a period of uncertainty going back as far as Smeaton's report of 1765. Sluicing basin built c.1808 at head of outer harbour but eventually (1829) abandoned.

REFERENCES:

1. Chapman's reports 1800–31. (I.C.E. Library).

2. Scarborough Harbour Commissioners Minute Books (Scarborough Library).

3. 'Plan of Scarborough Harbour 1831. From Surveys by T. O. Blackett and John Barry, under the Directions of Wm. Chapman Esq. C.E., M.R.I.A.'

WILLIAM CHAPMAN (1749–1832), CIVIL ENGINEER

RIVER ORWELL NAVIGATION (1806–1808)

LOCATION: River Orwell from Ipswich to Freston Reach, Suffolk (Fig. 14).

CLIENTS. River Orwell Commissioners.

ACT: 10 July 1805

ENGINEER: William Chapman (fees £304).

RESIDENT ENGINEER: William Shaw[8] (salary £120 p.a.).

SURVEYOR: John Bransby.

CONTRACTORS: Work carried out mainly by direct labour

COST: £9,700 (1806–8) excluding cost of dredger.

REPORTS: 'Report on the means of Improving the Navigation from Downham Reach to Ipswich. By William Chapman, Engineer'. Newcastle, 15 May 1797.

'Mr. Chapman's Report upon Plans and Estimates for deepening and improving the River Orwell'. 5 November 1803.

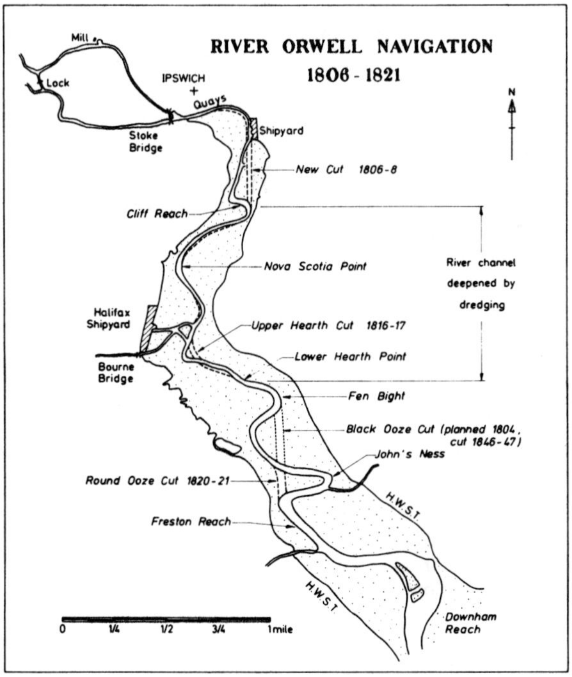

Fig. 14.

WILLIAM CHAPMAN (1749–1832), CIVIL ENGINEER

PRINCIPAL WORKS: (i) New cut 550 yards in length from Ipswich shipyard to Cliff Reach, made 1806–8 (ii) Widening and deepening river channel up to Ipswich Quays and at various places downstream of Cliff Reach to Lower Hearth Point (iii) New cut at Upper Hearth Point 1816–17 (iv) New cut at Round Ooze, 1820–21. These works, executed under the superintendence of William Shaw, completed Chapman's plans; but Chapman himself only directed work on the first cut.

STEAM DREDGER: The use of a steam dredger was proposed by Chapman in his 1797 report. Within a fortnight of the Act being passed in July 1805 he placed an order with Boulton & Watt for a 6 H.P. bell-crank engine. The bucket-ladder machinery was built by the Butterley Company, and the vessel by William Cordingley of Ipswich, to Chapman's designs. Total cost about £3000. The dredger began regular work in November 1806 and remained in service until replaced by a 10 H.P. machine, built by Hunter & English, in 1839.

RESULTS: Greatly improved navigation for 2½ miles below Ipswich, to deep water in Freston Reach; the basis for the growth of Ipswich as an important river port.

SPECIAL FEATURES: The first use of steam dredging for improvement of a tidal river navigation[9].

REFERENCES:

1 Ms copies of Chapman's reports 1797–1806 (Ipswich Port Authority, and East Suffolk Record Office).

2. 'Plan of the Proposed Improvements of the River Orwell from Ipswich to Freston Reach. Sept. 20, 1804'. (I.P.A.).

3. River Orwell Commissioners Minute Books (I.P.A.).

4. Boulton, Watt & Co. to Chapman 8 Aug. 1805, 25 Oct. 1805. Foundry Letter Book, 7 Aug. 1805. Small Engine Book. (B. & W. Collection).

5. Butterley Co. Ledger Book B (Derbyshire Record Office).

6. George Hurwood 'On the River Orwell and the Port of Ipswich'. *Min. Proc. I.C.E.* vol. 20 (1860) 14–25.

RIVER TEES NAVIGATION (1809–1810)

LOCATION: River Tees between Stockton and Portrack (Fig. 15).

CLIENTS: Tees Navigation Company.

ACT: 27 May 1808.

ENGINEER: William Chapman.

RESIDENT ENGINEER: not known.

CONTRACTOR: John Wilkinson.

COST: £9,300.

REPORTS: 'Report of the Committee appointed to Enquire into the Expediency of making a Cut across the Neck of Land between Stockton and Portrack (26 November 1802). Also Mr. Chapman's Report and Estimate, with Plans'. York, 20 December 1804.

PRINCIPAL WORK: Mandale Cut 220 yards in length, 250 ft top width and about 16 ft deep, with stone revetment on concave side.

FORMAL OPENING: 18 September 1810.

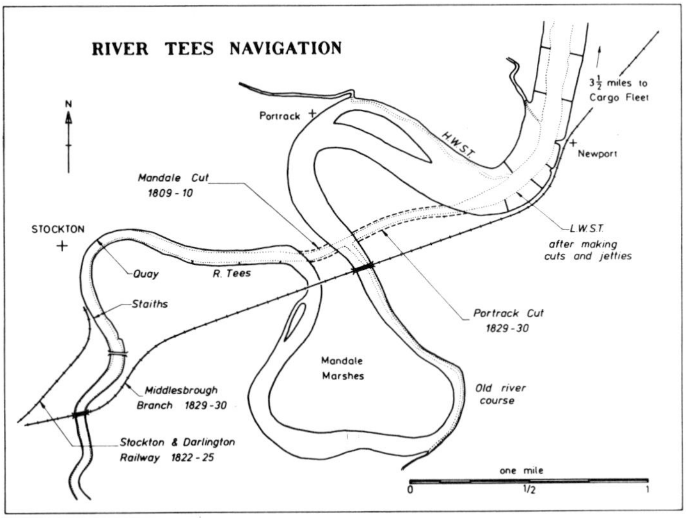

Fig. 15.

RESULTS: Minimum depth of water in the reach upstream increased from 9 ft to 11 ft at ordinary spring tides. 2½ mile loop in river by-passed. Ships enabled to sail up to Stockton from the mouth of the estury, at Cargo Fleet, in a single tide. Tonnage handled at Stockton Quays rose from about 24,000 tons p.a. before making the cut to 43,000 tons in 1812.

LATER WORK: Chapman envisaged a second cut, downstream of Portrack. This was eventually constructed 1829–31 under the direction of Henry H. Price, together with several river-training jetties.

REFERENCES:

1. John Brewster *History and Antiquities of Stockton-upon-Tees* (Stockton, 1829).
2. John Sykes *Local Records* vol 2 (Newcastle, 1833).
3. Tidal Harbours Commission. Parl Papers 1846, vol.18.

KENTON AND COXLODGE COLLIERIES (1806–1811)

From 1806 to 1811 Robert Knowsley and William Chapman were leaseholders of East Kenton and Coxlodge collieries near Newcastle (Fig. 16). Already in November 1805 they and a partner, James Tindall, negotiated a wayleave over ground in East Kenton, and in May and September 1808 further wayleaves were granted to Knowsley and Chapman across land in Gosforth, Long Benton and Wallsend. These evidently refer to the Kenton & Coxlodge waggonway, which is known to have been completed in 1809, with a length of about 4 miles, and is said to be one of the earliest iron railways on Tyneside. It had a gauge of 4 ft 7½ ins.

Rather more than 2000 iron rails were invoiced to 'William Chapman Esq. of Newcastle' in May and June 1806 by the Butterley Company, at a total cost of £352. There can be little doubt that these were destined for the western part of the railway at Kenton. A further 875 rails were invoiced to Kenton & Coxlodge in December 1808, but far more would have been required for the entire length of the waggonway.

Plans for winning a new pit at Coxlodge were drawn up by Chapman in 1807, and approved in November of that year by John Buddle, the well known coal-viewer. The partners purchased a 24

H.P. pumping engine from Butterley, which was invoiced the 'The Owners of Kenton & Coxlodge Collieries' in March 1808 at £765. William Bourne spent eleven weeks (from July to October 1808) erecting it and Chapman later expressed great satisfaction at his work on the Coxlodge engine. Much of its cost was offset by the Butterley Company purchasing rope, presumably from the Willington ropery.

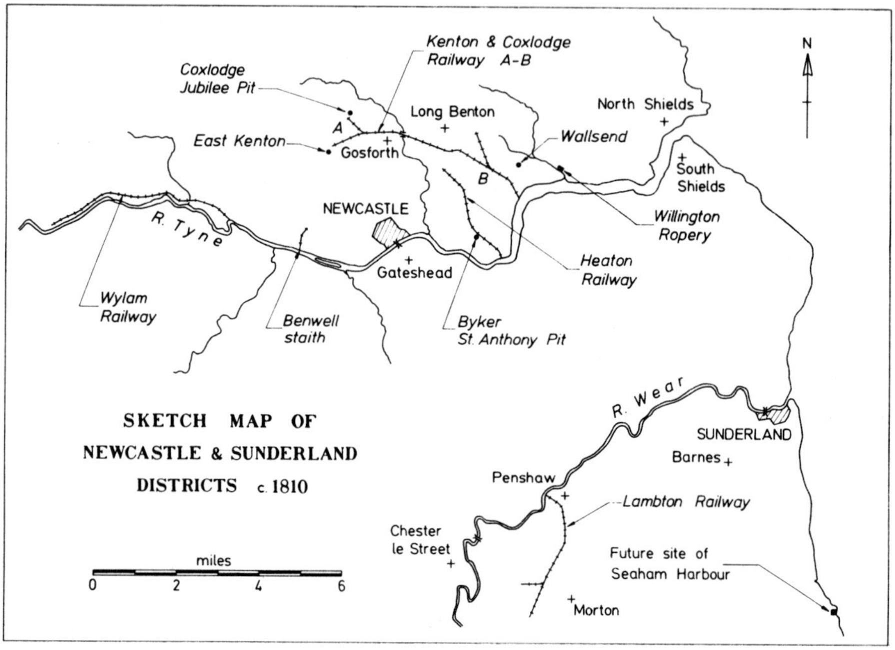

Fig. 16.

The Jubilee Pit, as it was called, proved to be successful, but Knowsley and Chapman sold their shares to Tindall in November 1811.

REFERENCES:

1. Coxlodge & Kenton Arbitration of 1811, and View Book of 1807. Watson Collection (Neville Hall, Shelf 9 No. 14, Shelf 8 No. 8).

2. William Chapman, 'Terms for leading Kenton & Coxlodge Coals, and Expences of Carriage'. 18 December 1809. Buddle Papers vol 3 (Neville Hall).

3. Butterley Co., Ledger Book B (Derbyshire Record Office).

4. Charles E. Lee 'Tyneside Tramroads of Northumberland' *Trans. Newcomen Soc.* vol. 26 (1949) 199–224.

BENWELL COAL DROP (1807)

In 1807 a machine was installed at Benwell Colliery staiths, on the Tyne, by means of which the coal waggons could be lowered vertically from the end of the staith directly into the ship's hold; thereby avoiding the breakage inevitably associated with the traditional method of tipping the coal down the shutes or 'spouts'. It was known as 'Mr. Chapman's lowering down Machine'. The waggons, on a detachable platform, descended by their own weight under the control of a brakesman and, when empty, returned to their original position by the action of a counterpoise. The machine at

Benwell cost £270. It was the first of the kind ever erected. Chapman took out a patent (No. 3030) in May 1807 to cover this idea and several variants.

Five years later, at Bewicke Main staiths, Benjamin Thompson introduced the rotating jib mechanism. In this form the 'coal drop' was adopted almost universally on the Tyne banks. However, Chapman's perpendicular lowering system, as improved by George Leather, proved to be better suited for staiths at a lower height above the water as at Port Clarence and Middlesborough on the Tees.

REFERENCES:

1 Benwell Colliery papers Buddle Collection (Neville Hall, Shelf 45 vol.4).

2. *Inventions, Improvements and Practice of Benjamin Thompson* (Newcastle, 1847).

3. George Turnbull 'Account of the drops used for shipment of coals at Middlesboro' on Tees' *Min. Proc. I.C.E.* vol.5 (1847) 248–54.

HEATON LOCOMOTIVE (1813)

DESIGN: Based on patent No. 3632 of 30 December 1812 by William and Edward Walton Chapman for a chain-haulage locomotive with a pivoted bogie.

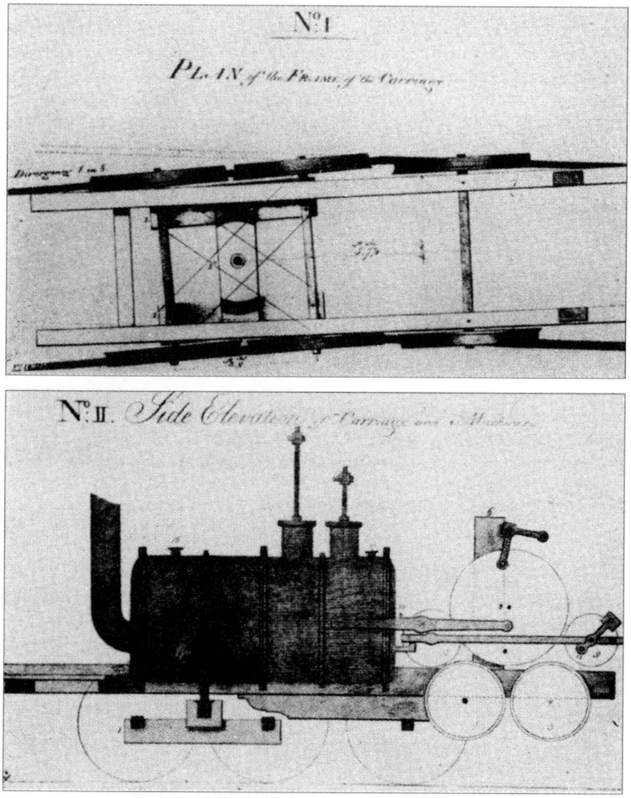

Fig. 17. Copies of two of the original drawings of the Heaton Locomotive, dated 3 Feb. 1813. Reproduced in *The Engineer* (1936) from lithographs in Derby Museum.

WILLIAM CHAPMAN (1749–1832), CIVIL ENGINEER

BUILT: By the Butterley Co., Derbyshire for Messrs. Chapman & Buddle of Newcastle.

DRAWINGS Lithographed copies exist in the Derby Museum of the original drawings, with accompanying descriptive details, dated 3 February 1813 (Fig. 17).

DESCRIPTION: Cast iron boiler 7 ft long, 3 ft 6 ins. diameter with two safety valves designed to operate at 60 psi Two vertical cylinders 8 ins. diameter 2 ft stroke. Piston rods connected to side levers working the crankshaft which, through gearing, drives the band wheel. A continous chain laid along the railway passes over the band wheel. The engine is mounted on a timber frame carried on a pair of leading wheels, with a fixed axle, and a 4-wheel bogie. The bogie frame is pivoted centrally and supported at the sides by two conical rollers bearing on segmental iron plates, allowing rotation about a vertical axis and a horizontal traverse axis. Gauge of rails about 4 ft 4 ins.

COST: In the Butterley ledgerbook 'a Locomotive steam Engine' is invoiced to 'W. Chapman Esq. Newcastle on Tyne' on 13 August 1813 for £514. On the credit side of the ledger appears £149 'By Chapman & Co. for ropes', £200 'By Cash per J. Buddle', and the balance of £165 was paid 'By draft'.

TRIALS: The engine arrived at Heaton colliery on 21 September 1813. The traditional date of October 1813 for the first trials may be correct, but from a letter written by Chapman to Buddle in May 1814, during a visit to London, it appears that the 'Elephant' (as Chapman nicknamed the locomotive) had only recently been brought into working order. In reply to questions by Sir John Sinclair, Buddle wrote on 5 October 1814 that the engine can 'haul itself and its load up hill, as has been effectually proved under very disadvantageous circumstances on the wooden railway from Heaton colliery to the river Tyne'. (Fig. 16). After serious flooding of the Heaton colliery in 1815 the engine was used for pumping.

SPECIAL FEATURES. The first steam locomotive to be fitted with a bogie, providing a reduction and partial equalisation of wheel load on the rails.

REFERENCES:

1. 'Copy of the Specifications of a Patent, granted to William . . . and Edward Walton Chapman . . . for their invention of a Method . . . of reducing the expence of carriage on railways'. (Newcastle, 1813).
2. C. F. Dendy Marshall 'Links in the history of the locomotive; William Chapman's engines' (the 1813 drawings) *The Engineer* vol. 162 (1936) 223–25.
3. Butterley Co. Ledger Book B (Derbyshire Record Office).
4. Chapman to Buddle, 16 May 1814 (Durham C.R.O. NCB I/JB/263).
5. 'Answers to Sir John Sinclair's queries respecting the conveyance of coals on rail-ways by steam engines' *Repertory of Arts* 2nd Series vol.33 (1818) 19–24.
6. L. G. Charlton *The First Locomotive Engineers* (Newcastle, 1974).

LAMBTON LOCOMOTIVE (1814)

In his letter to Buddle, written from London on 16 May 1814, Chapman mentions 'the new Elephant' the plans for which he had left in Newcastle. Construction was about to begin, or may even have started. There was to be a wrought iron boiler working at 50 psi, 'well protected from the Escape of heat', and provision for chain haulage.

These remarks certainly refer to the locomotive built by Phineas Crowther of Ouseburn Foundry, Newcastle, for J. G. Lambton Esq. at a cost of £483. The trials took place on 21 December 1814. The engine weighed 6 tons and was mounted on eight wheels, doubtless arranged as two 4-wheel bogies. It drew 18 loaded coal waggons (total weight about 54 tons) up a gentle incline of 1:115 at a rate of nearly 4 mph. On this gradient adhesion between the driving wheels and the rails was sufficient, but for steep ascents the chain would be required. The waggonway had cast iron rails, of 4 ft 2 in gauge, and ran from the Lambton collieries to the river Wear at Penshaw (Fig. 16).

Buddle, who was viewer for Lambton at this period, as well as at Heaton, refers in his 1815 view-book to an extension of the way and notes on 19 July 1815 'Tried the Engine – she took 10 Waggons from new Lambton to Pensher. It required the Steam to be 50 lbs. p.In. to take them up the Bank from Beckworths to Ransoms Carr'. This remark tends to confirm what is implied in the published account of the trials, namely that the engine used chain haulage for the steep gradients but after arriving at the head of an ascent the chain was disengaged and the engine then worked as an adhesion locomotive.

Of great practical importance is the fact that the rails did not have to be replaced; the bogies reduced the wheel loads approximately to those of the existing (previously horse-drawn) waggons. The success of the eight-wheeled system at Lambton seems to have led to its adoption on the Wylam colliery waggonway. This had iron plate rails which were of inadequate strength properly to bear the weight of the two-cylinder four-wheeled adhesion locomotives introduced by William Hedley in 1814–15. Rather than lay new rails, at great expense, both of the Wylam engines were remounted on two 4-wheel bogies. In this form they continued in service until about 1828. One of the Wylam eight-wheelers is illustrated by Nicholas Wood.

REFERENCES:

1. Chapman to Buddle, 16 May 1814 (Durham C.R.O., NCB I/JB/263).

2. Crowther's bill (Dec. 1814) and a brief description of the waggonway are included in the Buddle Collection (Neville Hall, Shelf 45 vol. 1). I am indebted to Michael Lewis for this information.

3. 'Account of a locomotive engine executed by Messrs. Chapman of Newcastle-upon-Tyne . . . contained in a letter to the editors'. *Repertory of Arts* 2nd series vol.26 (1815) 161–62. See also the *Newcastle Chronicle* for 24 Dec. 1814.

4. Buddle's Lambton View-Book, 1815 (Durham C.R.O. NCB I/JB/2079).

5. Nicholas Wood *A Practical Treatise on Rail-Roads* (London, 1825).

6. P. C. Dewhurst 'Early British locomotives' *The Engineer* vol.172 (1941), 219–20.

7. L. G. Charlton, *The First Locomotive Engineers* (Newcastle, 1974).

SHEFFIELD CANAL (1816–1819)

LOCATION: Sheffield to the head of the River Don Navigation at Tinsley, Yorkshire (Fig. 18).

CLIENTS: Sheffield Canal Company.

ACT: 7 June 1815.

ENGINEER: William Chapman.

RESIDENT ENGINEER: Henry Buck.

SURVEYOR: Josiah Fairbank.

CONTRACTORS: not known.

COST: £76,000.

REPORT: 'Report of William Chapman, Civil Engineer, on the Proposed Canal, from Castle Orchards, Sheffield to the River Dun below Tinsley'. Newcastle, 6 October 1814.

PRINCIPAL WORKS: (i) Canal cut 4½ miles long including the ½ mile Greenland arm (ii) Rise of 70 ft taken in twelve locks arranged in two flights of 4 and 8 respectively, the locks 64 ft X 16 ft with 7 ft of water over the sills (iii) Carr Brook embankment (iv) Attercliffe cutting (v) Aqueduct over the Worksop road (vi) Seven road bridges (vii) Canal basin at Sheffield about 600 ft X 100 ft, with a warehouse and wharves (viii) Steam pumping engine, near the Sheffield basin, supplying water to the canal from old mine workings.

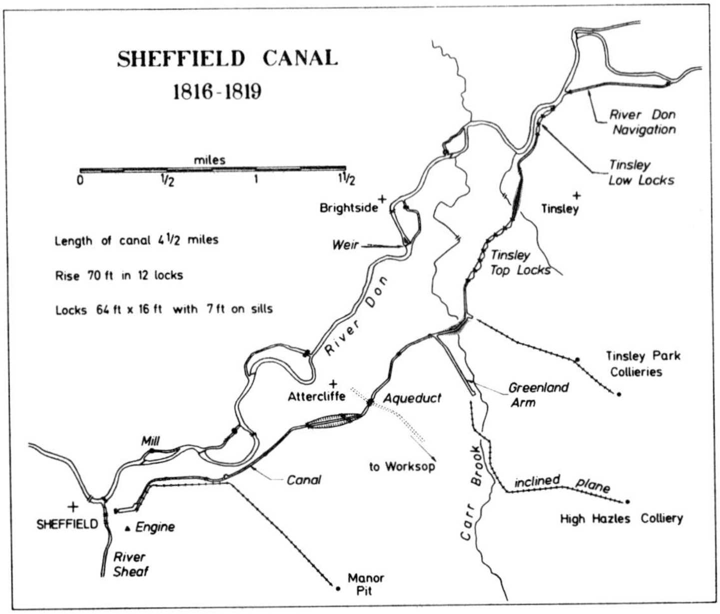

Fig. 18.

FORMAL OPENING: 22 February 1819.

RESULTS: 'the canal, by opening a communication between the River Dunn and Sheffield, is of incalculable benefit to that place, by the facilities it affords of transmitting articles of its extensive and important manufactories, and also of the coal and iron stone which abound in the neighbourhood; and of supplying it with all articles of domestic consumption' (Priestley, 1831).

REFERENCES:

1. 'A Plan of the Intended Canal from . . . Sheffield . . . into the Township of Tinsley . . . By William Chapman Civil Engineer. Surveyed by W. & J. Fairbank 1814'.

2. 'An account of the Opening of the Sheffield Canal' *Sheffield Mercury*, 27 Feb. 1819.

3. Joseph Priestley *'Historical Account of the Navigable Rivers, Canals and Railways throughout Great Britain* (London, 1831).

4. C. M. Butterworth *The Sheffield Canal* (Sheffield 1970).

5. Charles Haffield *The Canals of Yorkshire and North East England* vol.2 (Newton Abbot, 1973).

SHOREHAM HARBOUR (1817–1821)

LOCATION: Sussex coast between New Shoreham and Hove.

CLIENTS: Shoreham Harbour Commissioners.

ACT: 1 July 1816.

ENGINEER: William Chapman (fees charged £1088, at 5 gns. per day plus travelling expenses; fees paid £904).

VI

WILLIAM CHAPMAN (1749–1832), CIVIL ENGINEER

RESIDENT ENGINEER: William Clegram[10] (salary £150 p.a.).

SURVEYORS: William Clegram, for harbour. Richard Buck, for river embankments.

CONTRACTORS: Bellingham & Francis, to November 1817. Hugh McIntosh, December 1817 to completion. McIntosh executed all the works except dredging.

COST: £58,000 including £3,400 for river embankments and £3,500 for dredger

REPORT: 'Report on the Harbour of New Shoreham. By William Chapman, Civil Engineer'. Newcastle, 10 July 1815.

PRINCIPAL WORKS: (1) Harbour entrance 250 ft wide opened through the massive shingle bank, opposite Kingston-by-Sea (ii) Two timber framed chalk-filled piers on each side of the entrance, both piers about 320 ft long (iii) Groynes extending 200 ft seaward beyond the pier ends (iv) Groynes protecting landward ends of the piers (v) Timber framed 'middle pier' 700 ft in length forming 'Light House Point' behind the entrance (vi) Dredging the inner channels leading from east and west towards the entrance (vii) Two sets of sluice gates in eastern arm of harbour, 1 mile from entrance, for scouring this arm and the sand bar (viii) Raising embankments of the River Adur.

DREDGER: Designed by Chapman. Dredging vessel and 4 barges built at Shoreham by Edwards & Balley for £1680. 10 H.P. engine and bucket-ladder machinery built by Butterley Co. for £1930 including £106 charged for William Bourne's time and expenses fitting up. The dredger could operate in 19 ft of water. It began regular work about February 1819, easily raising 60 tons of gravel per hour, under the superintendence of William Clegram.

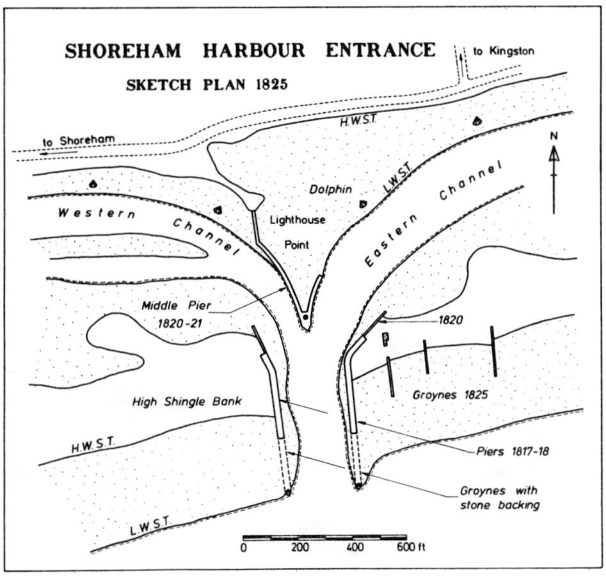

Fig. 19.

WILLIAM CHAPMAN (1749–1832), CIVIL ENGINEER

RESULTS: Creation of a safe haven and useful harbour, with a permanent entrance 1¼ miles from New Shoreham instead of the natural shifting entrance 1½ miles further east. In westerly storms a sand bar tended to form across the harbour mouth from the end of the west pier. The sluices were only partially successful in scouring the bar. They were abandoned in 1828. Harbour dues increased from about £600 p.a. in 1817 to £3200 in 1822 as a direct result of the new works.

SPECIAL FEATURES: An attempt had been made in 1760 to open an entrance opposite Kingston and dam off the eastern arm. This failed. Jessop in 1800 thought little could be achieved other than deepening the channel by means of a bucket-ladder horse dredger. Robert Vazie and Rennie, 1809-10, proposed a (very expensive) wet dock at the west end by Shoreham. Not happy with these proposals, the local inhabitants in 1814 revived the idea of an entrance at Kingston. Chapman was then brought in on the advice of Trinity House. He realised the importance of the eastern arm, which has indeed become the main part of the harbour, and designed the entrance. It was located at a narrow part of the shingle bank opposite the deepest water in the bay. A plan and view of the entrance are shown in Figs 19 and 20.

The Letter Books show that during the period of construction Chapman wrote just over a hundred letters to the resident engineer, averaging about one a fortnight. He made eleven site visits between 1815 and 1821 and visited Butterley in connection with the dredger. He also helped to raise a loan of £15,000 from the Exchequer Loan Bill Commissioners.

The Harbour Commissioners, beset by financial difficulties, paid Chapman's fees up to December 1817 but thereafter, with the exception of £30 on account, left unpaid a bill which by the end of 1821 amounted to over £500. After repeated protests from Chapman, part of this was settled in 1824 but it appears that the balance of £180 was never paid.

LATER WORKS: Three groynes built in 1825 to protect beach beside the east pier. Both piers extended 280 ft by the harbour engineer Thomas Thornton, 1836–39. Subsequent developments include conversion of part of the eastern arm into a wet dock or 'canal' by a lock ½ mile from the harbour entrance in 1854; further extension of the west pier, first by a jetty and then of the pier itself in 1879; and a new lock adjacent to the first in 1933. Extensive remodelling of the entrance took place 1954-57 along with other major improvements.

Fig. 20. Shoreham Harbour entrance, as completed by Chapman in 1821. Painting of c 1830, artist unknown, reproduced by courtesy of Shoreham Port Authority.

WILLIAM CHAPMAN (1749–1832), CIVIL ENGINEER

REFERENCES:

1. 'Memorial from the principal Merchants and Ship Owners to the Commissioners of New Shoreham Harbour' 27 December 1814 (West Sussex Record Office).
2. 'A Plan of the Harbour of New Shoreham, Surveyed by Capt. Clegram in January 1815, showing the Improvements, projected by W. Chapman Esq. Civil Engineer, in July 1815'.
3. Shoreham Harbour Commissioners Minute Books (Shoreham Port Authority).
4. Shoreham Harbour, Letter Books (W.S.R.O.).
5. 'A General Statement of the Receipts and Expenditure relative to the Harbour of New Shoreham from . . . July 1816 up to June 1825'. (S.P.A.)
6. Butterley Co. Ledger Book C (Derbyshire Record Office).
7. 'Plan of Shoreham Harbour [entrance] By W. B. Prichard, 1843' (S.P.A.).
8. H. Ridehalgh 'Shoreham harbour development' *Proc. I.C.E.*, vol. 11 (1958) 285–96.

CARLISLE CANAL (1820–1823)

LOCATION: Carlisle to Fisher's Cross,[11] near Bowness, on Solway Firth, Cumberland (Fig. 21).

CLIENTS: Carlisle Canal Company.

ACT: 6 April 1819.

ENGINEER: William Chapman (fees £1025).

RESIDENT ENGINEER: Henry Buck, to July 1820, followed by Richard Buck.

OVERSEER OF WORKS: Thomas Ferrier.

SURVEYOR: Richard Buck.

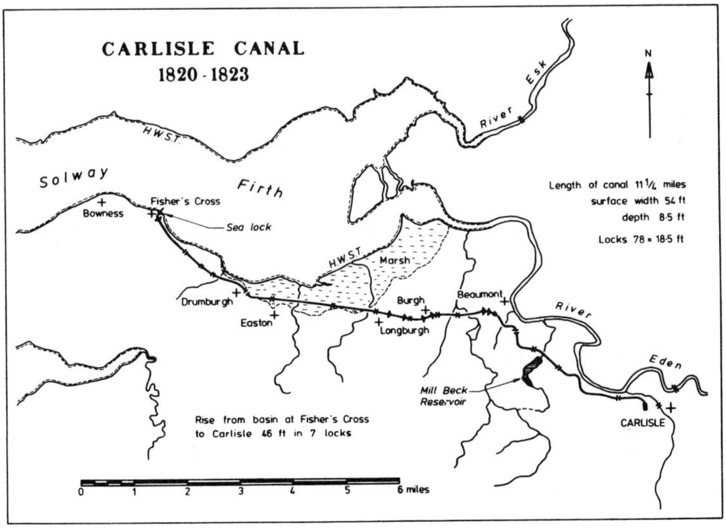

Fig. 21.

WILLIAM CHAPMAN (1749–1832), CIVIL ENGINEER

CONTRACTORS: Hamer & Thompson, Wood & Barker Sanderson & Taylor, W. & W. Gates, Millington & Bainbridge.

COST: about £80,000.

REPORT: 'Mr. Chapman's Report on the Proposed Canal Navigation between Carlisle and Solway Frith'. Newcastle, 8 February 1818.

PRINCIPAL WORKS: (i) Canal cut 11¼ miles long, 54 ft surface width, 8½ ft deep (ii) Basin 250 ft X 80 ft at Fisher's Cross with a sea lock and timber jetty at entrance from the Solway Firth (iii) Rise of 46 ft from the Solway basin to Carlisle taken in 7 locks; the locks 78 ft X 18½ ft with 8 ft of water over the sills (iv) Embankment protecting canal from exceptional high tides where it crosses Easton and Burgh marshes (v) Reservoir on Mill Beck near Grinsdale (vi) Canal basin 450 ft X 100 ft at Carlisle with warehouse and wharves (vii) About 14 double-leaf draw bridges over canal.

FORMAL OPENING: 12 March 1823.

RESULTS: In the year 1824-25 nearly 18,000 tons were carried on the canal, earning gross receipts of £2000. Five years later these figures had increased by 50 per cent. After the opening of the Newcastle & Carlisle Railway in 1839 the canal tolls rose to £6000 p.a. for commercial traffic, but later railways linking Carlisle to the north and south led to closure of the canal in 1853.

SUBMERGED FOREST: The discovery of a pre-historic submerged forest, near Burgh, was described by Chapman in a paper to the Newcastle Antiquarian Society in January 1824. He presented a specimen of well preserved oak wood, and some archaeological finds.

MANAGEMENT COMMITTEE: Relations between Chapman and his resident engineer, on the one hand, and the Company's management committee, on the other, were far from happy. The committee appointed an 'Overseer of the Works' and through him, or on his advice altered various details. Richard Buck the resident engineer, who had carried out the surveys, was therefore placed in a most unsatisfactory position. Chapman described him as 'a young man with great ability and ardour in pursuit of his profession', and strongly objected to his treatment. Chapman was dismissed on 27 December 1822, less than three months before completion of the canal. Richard Buck left in May 1823.

REFERENCES:

1. Canal Company Minute Books. (Scottish Record Office). I am indebted to Charles Hadfield and Roland Paxton for notes from these.

2. William Chapman 'Address to the Subscribers to the Canal from Carlisle to Fisher's Cross'. Newcastle, 31 January 1823.

3. Various newspaper cuttings relating to Carlisle Canal, 1822–29. 'Manuscript History of Canals' c.1835 (Imperial College, Civil Eng. Dept. Library).

4. Charles Hadfield & Gordon Biddle *The Canals of North West England* vol. 2 (Newton Abbot. 1970).

THE 'TRANSFERRER' (1821)

In 1821 Chapman designed a floating machine, in the form of a special type of crane, for transferring coal from keels or barges into sea-going ships. The device, covered by Patent No. 4550, consisted essentially of an upright counterbalanced frame, rotating about a horizontal axis, by means of which a container could be lifted from a barge, swung across the vessel in which the machinery was mounted, and lowered into a ship on the other side.

From a letter written by Chapman in June 1821 to John Buddle, chief viewer of the collieries belonging to the Marquis of Londonderry, it is clear that the prototype was then under construction; and by 1822 several of these machines were in operation, transferring coal brought down the Wear, from the colliery staiths,[1 2] into ships lying in Sunderland harbour. The barges carried eight containers, each holding a chaldron (about 2½ tons), and by the use of a 6 H.P. steam engine fifty tons of coal could be loaded in an hour. The rotating frame was about 50 ft in length.

WILLIAM CHAPMAN (1749–1832), CIVIL ENGINEER

The 'transferrers' on the River Wear were seen and described by two German engineers in 1826 or 1827, but the machines would have been no longer required by the Londonderry collieries after the opening of Seaham Harbour around 1832 (q.v.).

REFERENCES:

1. Chapman to Buddle, 9 June 1821 (Durham County Record Office, NCB I/JB/264).

2. 'Specification of a Patent granted to William Chapman of Newcastle upon Tyne, Civil Engineer, dated 12 April 1821, for a Method of transferring the Ladings of Lighters or Barges into Ships' (1822).

3. *Railways in England 1826 and 1827.* A report by C. von Oeynhausen and H. von Dechen, first published in Germany 1829. Translated by E. A. Forward (Cambridge, 1971).

PLANNING THE EAST LONDON DOCK (1824)

The first London Dock, at Wapping, was built 1801–6 under the direction and to the designs of John Rennie. He envisaged a second dock to the east, communicating with the first and having a new entrance at Shadwell, and land was reserved accordingly, but not until 1824 did the Dock Company resolve to construct this extension. They commissioned Chapman to draw up the engineering plans and estimates, Rennie having died three years earlier.

During August and September 1824 very active discussions took place between Chapman, the Company, and their Surveyor, Daniel Alexander. Chapman produced his plan in a

> 'Report on the most eligible measures to be attended to in the extension of the London Docks' London, 30 August 1824.

His estimates followed two days later. He submitted plans for a slightly smaller dock on 6 September accompanied by estimates drawn up by him for the engineering works (£138,000) and by Alexander for the storage vaults, warehouse etc. (£114,000); giving a total, allowing for contingencies, of around £280,000.

On this basis the Company immediately decided to proceed. Alexander was appointed executive engineer and architect at a salary of £600 p.a. and J. W. Hemingway was appointed resident engineer at £300 p.a. On 21 September the Committee of Treasury and Plans resolved to pay Chapman's bill amounting to £198, and recommended that 'a Gratuity of One Hundred Guineas be presented to Mr. Chapman, accompanied by a Letter of thanks, for his Zeal and active exertions relative to the projected New Works'. His fees were 6 gns. per day, with 1 gn. expenses, plus travelling.

By June 1825 contracts had been let to Hugh McIntosh for excavations and to Thomas Cartwright for the vaults and boundary wall, and in April 1826 to Thomas Smith for the dock walls. Hemingway died in June 1826. He was succeeded by Henry Robinson Palmer, and after Alexander resigned in April 1827 Palmer took over his post while remaining as resident engineer.

The works, completed in 1828, consisted of a dock of 7 acres extent accommodating vessels of up to 700 tons burden; a cut communicating with the Old Dock and equipped with a caisson gate; two iron swivel bridges; a warehouse for 11,000 tons of merchandise, and vaults for storing 10,000 casks of wine or spirits. The total cost, including compensation, came to £302,000.

REFERENCE:

1. London Dock Company. Committee of Treasury and Plans, Minute Books. (P.L.A. Library).

EARLY PLANS FOR THE NEWCASTLE & CARLISLE RAILWAY (1824–25)

In an open letter to Sir James Graham of Kirstall, published in July 1824, Chapman pointed out that the canal line surveyed by him in 1795 between Newcastle and Carlisle would, with a deviation at the eastern end, be suitable for a railway; and that, with the advances recently made in steam locomotives and stationery engines for transporting goods, a railway could be built more cheaply and operated more efficiently than a canal in this terrain. The idea attracted attention, and in the following month Chapman was commissioned to report in detail on the relative merits and costs of the two schemes. As a result he produced

VI

WILLIAM CHAPMAN (1749–1832), CIVIL ENGINEER

'A Report on the Cost and Separate Advantages of a Ship Canal and of a Rail-Way, from Newcastle to Carlisle. By William Chapman, Esq. Civil Engineer'. Newcastle, 27 October 1824.

This came down conclusively in favour of a railway, at an estimated cost of £252,000, in contrast to a canal (with locks of the same size as those on the Carlisle Canal) estimated at over £800,000.

Even in Newcastle, the principal birthplace of steam railways, this report aroused some opposition. Telford's opinion was sought, and Chapman looked forward to going over the line with him, but for various reasons the great engineer could not visit the North and had to content himself with writing a letter in January 1825 agreeing in general with Chapman's opinion about the expense of a canal, though expressing some doubts on the economic advisability of a railway. Josias Jessop was then called in. He reported in March, confirming Chapman's assessments, while advising a stronger foundation bed for the rails, which would add about £40,000 and therefore bring the total estimate to nearly £300,000. Three weeks later a general meeting unanimously resolved to go ahead with the railway.

Surveys were then commenced under the joint direction of Chapman and Benjamin Thompson. The presence of Thompson, one of the directors of the newly formed company, is explained partly by his expert railway experience, and partly by the opinion of James Losh, the chairman, who wrote in his diary in July 1825, 'our engineer, Mr. Chapman (tho' I believe well acquainted with his profession) is by no means an effective person'. Losh probably had in mind a comparison with George Stephenson, and one sees what he meant, but it is only fair to add that Jessop had generously remarked when amending the estimates that he considered 'Mr. Chapman . . . to stand at the head of his profession'.

The line decided upon in December 1835 differed from Chapman's proposals of 1824 chiefly in crossing the Tyne at Scotswood, rather than at Newburn, and in following the valley of the Irthing down towards Carlisle. But when the Parliamentary plans were drawn up in 1828, after detailed surveys by T. O. Blackett and T. Studholme, Thompson reverted to a line passing south of Brampton, similar to Chapman's original route for the section west of the summit.

Construction of the railway began in 1830 with Francis Giles as engineer. It was opened throughout from Newcastle to the canal basin at Carlisle in 1839: total length 61 miles, with several branches.

REFERENCES:

1. 'Observations on the most advisable measures to be adopted in forming a Communication for the Transit of Merchandise . . . to or from Newcastle and Carlisle. In a Letter from Mr. Chapman to Sir James Graham'. Newcastle, 10 May 1824. With a Supplement, 21 July 1824.
2. Telford Mss. vol. 10 (I.C.E. Library).
3. Report of Josias Jessop, Esq. Civil Engineer. Butterley, 4 March 1825.
4. 'The Diaries and Correspondence of James Losh' *Surtees Soc.*, vol.174 (1963).
5. 'Newcastle and Carlisle intended Rail Road. Report of the Directors'. Newcastle, 3 Dec. 1825.
6. J. S. Maclean *The Newcastle & Carlisle Railway.* (Newcastle, 1948).

LEITH HARBOUR (1826–1831)

LOCATION: South shore, Firth of Forth (Figs. 22 and 23).

CLIENTS: Leith Docks & Harbour Commissioners.

ACT: 5 July 1825.

ENGINEER: William Chapman (fees £1080 at 6 gns per day plus travelling expenses).

RESIDENT ENGINEER: James Leslie[13] (salary £176 p.a.).

DOCK SUPERINTENDENT: Patrick Dall.

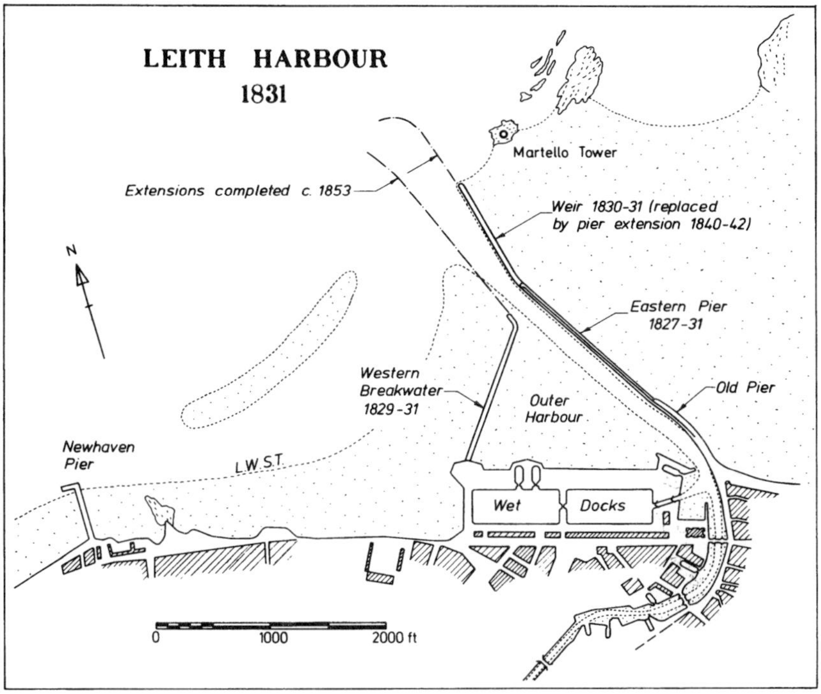

Fig. 22.

Fig. 23. Leith Harbour. View from the East Pier. Engraving by R. Wallis, after W. H. Bartlett, published in Finden's *Views of the Ports, Harbours and Watering Places of Great Britain.* (London, c 1840).

VI

WILLIAM CHAPMAN (1749–1832), CIVIL ENGINEER

CONTRACTORS: for eastern pier: Kenneth Mathieson, to 1830; Barry & Coulson (of Scarborough) from 1830 to completion. For western breakwater: Hemming & Pinkerton.

COST: £40,000 exclusive of dredging.

REPORT: 'Report relative to the Improvement of the Harbour of Leith. By William Chapman, Esq. Civil Engineer', Edinburgh, 1 October 1824.

PRINCIPAL WORKS: (i) Eastern pier 1500 ft long, consisting of a masonry breakwater and timber piled pier (ii) Gravel 'weir' or low embankment extending 1000 ft beyond the eastern pier as a training wall for the approach channel. The weir was replaced in 1840 by an extension of the pier in accordance with the original plan (iii) Western breakwater 1200 ft long surmounted by a timber gangway leading to a pier head (iv) Dredging in the harbour and the approach channel or 'fairway' alongside the eastern pier.

DREDGER: A 16 H.P. bucket-ladder steam dredger was in use from July 1827, employing direct labour under Patrick Dall. Engine and machinery built by Messrs. Girdwood of Glasgow, following design of their Clyde dredger No. 2. Dredging vessel and barges built at Leith by Messrs. Rose.

RESULTS: A large outer harbour with a fairway 19 ft deep at H.W.S.T. providing a safe approach channel to the old inner harbour (mouth of the Water of Leith) and wet docks. The docks had been built 1801–17 to Rennie's designs.

SPECIAL FEATURES: Construction costs kept within a strictly limited budget. Plans capable of future extensions, which were carried out in 1840 and 1853.

PROGRESS OF WORK: The ceremony of driving the first pile took place on 15 August 1826, but construction started in 1827 after further negotiations with the Admiralty who had agreed to pay for the western breakwater, as contributing towards improvements of the naval establishment at Leith, and with the Government for a loan. Sir John Rennie, consultant to the Admiralty, then expressed doubts concerning Chapman's plans. Telford in 1828, on behalf of the Treasury, disaproved Rennie's counter-proposals. The matter was finally settled by Chapman's personal intervention, and work proceeded to a satisfactory completion of his original scheme; the only change being a modification in detail of the east pier cross-section. It is ironical that Chapman's appointment as engineer in 1824 followed a recommendation by Viscount Melville, First Lord of the Admiralty.

Apart from five visits to London in connection with all these negotiations, Chapman attended Parliament in 1825 and, during the period 1824–31, made nine site visits and submitted eight reports.

REFERENCES:

1. Leith Docks & Harbour Commissioners Minute Books (Scottish Record Office).

2. Sir John Rennie's Reports vols.2 and 3 (I.C.E. Library).

3. 'Plan of Leith. Surveyed by P. Dall, Master, R.N., and J. Lelie, Engineer. 1831'.

4. Tidal Harbours Commission. Parl. Papers 1847 vol.32.

5. For Summary of later 19th century developments and modern works, see M. C. White 'The design and construction of the new deep-water quays at Leith' *Proc I.C.E.* Part 2, vol.1 (1952) 55–82.

WILLIAM CHAPMAN (1749–1832), CIVIL ENGINEER

SEAHAM HARBOUR (1828–1835)

LOCATION: 6 miles south of Sunderland, Co. Durham (Figs. 16 and 24).

CLIENT: Marquis of Londonderry.

MANAGER: As chief viewer of the Londonderry collieries John Buddle acted as manager of the entire Seaham project.

ENGINEER FOR HARBOUR: William Chapman.

RESIDENT ENGINEER: Thomas Nicholson.

CONTRACTORS: The harbour seems to have been built chiefly by direct labour.

COST OF HARBOUR WORKS: £165,000.

PLANS: Chapman's first plans are dated 1820 (for Sir Ralph Milbanke) and 1823 (for the Marquis of Londonderry, who purchased the Seaham estate from Milbanke in 1822). Work began in September 1828 in accordance with

'Plan of Seaham Harbour. As projected for . . . the Marquess of Londonderry . . . By William Chapman M.R.I.A., Civil Engineer'. (published December 1828).

A revised scheme for the South Harbour and Dock is given in

'Plan of the Harbour of Seaham and of its Future Extensions. As projected by William Chapman M.R.I.A., Civil Engineer'. (16 January 1832).

PRINCIPAL WORKS 1828–32: (i) North Pier 560 ft long, and (ii) outer 450 ft of South Pier, together forming the Outer Harbour 3 acres in extent, with entrance 110 ft wide between the pier heads (iii) North Basin, or inner harbour, 2½ acres, with 15 ft of water at H.W.S.T. and a minimum depth of 7 ft maintained by a gate at the entrance from the Outer Harbour. Many thousand tons of rock were excavated to form the North Basin, and used as internal filling for the piers. Stone for the massive masonry facing of the piers was brought from Penshaw quarries. (iv) A stone-paved 'glacis' protecting the land from marine erosion near the inner end of the North Pier. (v) Three coal loading berths on North Basin quay.

WORKS 1832–35: (vi) Completion of South Pier (total length 1000 ft) forming the South Harbour of 3 acres extent, opening off the Outer Harbour (vii) Completion of the 2½ acre South Dock with an entrance from the Outer Harbour, provided with a double set of mitre gates, and a second entrance from the South Harbour. These works were carried out with some minor modifications from Chapman's revised plan of January 1832. (viii) Additional coal loading berths.

RAILWAYS: 'Lord Londonderry's Railway' from Rainton to the harbour was an integral part of the scheme. Buddle's original estimates date from 1823. The railway was built 1829–31 by Benjamin Thompson under an agreement to convey the coal from Londonderry's Rainton Colliery at a fixed rate for nine years; at the end of which time the Marquis could acquire the line under named conditions. 'Mr. Braddyll's Railway' from South Hetton Colliery to the harbour was built a little later, probably about 1833. There is no indication that Chapman was directly involved in either of these railways.

FORMAL OPENINGS: The first vessel left the harbour with coal brought down the Rainton-Seaham railway on 25 July 1831. The South Dock was opened 29 July 1835.

RESULTS: 230,000 tons of coal were shipped in 1832. The output rose to about 350,000 tons in 1836 and by 1845 exceeded 700,000 tons.

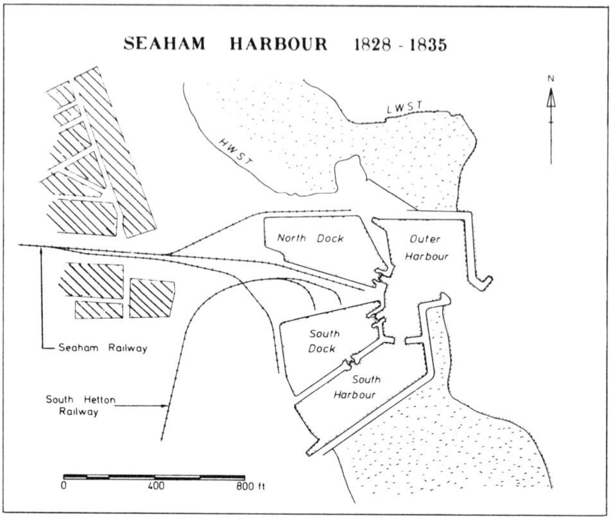

Fig. 24.

PROGRESS: After lengthy but fruitless financial negotiations the Marquis decided in 1828 to build the harbour at his own cost. Chapman was engaged to make plans on 20 August. Preparations on site began on 6 September under Buddle's direction. Chapman arrived on 12 September and worked almost continuously until 27 November, with his pupil-assistant T. E. Harrison. Excavations for the North Basin started on 15 September. Chapman's drawing for the North Pier is dated 2 October and the foundation stone of the pier was laid 28 November 1828.

Work continued with the remarkable expedition already displayed in the preliminary stages. Chapman was setting out 'essential works' at Seaham, probably the South Pier in the summer of 1830. Certainly by the first opening ceremony in July 1831 the outer 300 ft of the South Pier had been constructed; the North Pier was nearly finished, much of the North Basin had been excavated; the masonry jetties forming its entrance had been built, together with the quay wall along its southern side and two coal loading berths; and the north quay wall of the South Dock was completed. Chapman had also published a chart with sailing directions. In January 1832 he produced a revised, and much improved, plan for the South Dock.

By November 1832 Lord Londonderry had spent £118,000 on the harbour. It then became necessary to raise a loan for completing the works from the Exchequer Loan Commissioners. Early in 1833 Telford, on their behalf, approved the plans and estimates, which had been examined by his assistant John Macneill, and work proceeded with Nicholson in charge under Buddle's general direction; an arrangement which seems to have been in operation since Chapman's death.

In a report written at the beginning of 1832 Chapman said

> 'There is no instance of a private Harbour being constructed with such rapidity and to such extent as the present Harbour has been, by the affluence and adventurous spirit of the Marquess of Londonderry, and his advisers, to whom I feel greatly indebted for the confidence placed in me, and I, alone, should have been to blame, had the scheme failed of success, as the whole was left to my decision and opinion, as circumstances should point out . . . Much has also been due to the executive agents employed in constructing the works, and to the regulations of the Harbour master in the admission and departure of vessels.

Clearly Chapman thoroughly enjoyed the last, and one of the most important, of his works.

WILLIAM CHAPMAN (1749–1832), CIVIL ENGINEER

REFERENCES:

1. Various letters, reports and plans in the Londonderry Papers (Durham C.R.O.) particularly in D/Lo/E 593, 600 and 605.

2. W. Chapman 'Chart of the Sea Coast and Plan of Seaham Harbour with Sailing Instructions'. Newcastle, July 1831.

3. 'Report on the Rise, Progress, Present State, and Projected Extension of the Harbour of Seaham. By Wm. Chapman'. (Newcastle, 1832).

4. 'Plan of Seaham Harbour & Town Adjoining. Showing the Lines of Approach to the Harbour by the Several Railways. Surveyed & Drawn by Thomas Nicholson Superintendent of the Works. Nov. 1834'.

5. *Durham Advertiser,* 29 July 1831, 31 July 1835.

6. E. Mackenzie & M. Ross *An Historical, Topographical, and Descriptive View of the County Palatine of Durham.* vol. 1 (Newcastle, 1834).

7. W. Fordyce *The History and Antiquities of the County of Durham.* vol. 3 (Newcastle, 1855).

8. R. Welford 'Benjamin Thompson' in *Men of Mark' twixt Tyne and Tees.* vol. 3 (London, 1895).

ACKNOWLEDGEMENTS

I am greatly indebted to Mrs. Ruth Delany for information on works by Jessop and Chapman in Ireland, and to Miss Joyce Brown for her research in various Newcastle archives. My wife accompanied me on several field trips and gave useful criticism of early versions of the paper. I am also grateful for discussions and correspondence with Mr. Alan Smith, Mr. Philip Riden, Mr. J. G. James, Mr. C. E. C. Townsend, Mr. Charles Hadfield, Mr. Mark Baldwin, Mr. R. Paxton, Dr. Michael Lewis, Mr. P. K. Pepper, Mrs. C. Hiskey and Mr. James Day. Permission and facilities for studying documents in their keeping were kindly granted by Ipswich Port Authority, Muston & Yedingham Drainage Board, Port of London Authority, Birmingham Assay Office, and Shoreham Port Authority. I am obliged to the staff of many libraries and Record Offices for their courtesy and help, and in particular the librarian of the Institution of Civil Engineers, for the loan of Chapman's printed reports. Miss Ursula Schüler made the tracings, Miss Joyce Gurr carried out the photographic work, and Miss Diana Lombard has given valuable secretarial assistance. Mr. L. G. Charlton contributed several useful comments while the paper was in press.

NOTES

1. John Weale (ed.) 'Memoir of William Chapman', *Quarterly Papers on Engineering* vol.1, Pt.1 (1843)

2. Richard Welford, 'William Chapman', in *Men of Mark twixt Tyne and Tweed* vol.1 (London 1895). See also the very brief article on Chapman in D.N.B.

3. Joseph Foster *Pedigrees of the County Families of Yorkshire.* vol. 4 (London, 1874).

4. Chapman was proposed 30 Jan 1795 and elected 13 Feb. 1795. Society of Civil Engineers Minute Books (I.C.E. Library).

5. Another very early example was the engine used by Rennie in building the foundations of the Lune Aqueduct on the Lancaster Canal; it began working in January 1794.

6. Inverted syphons, known also as 'foxes'.

7. Matthew Shout left Scarbough in 1804 to become engineer of Sunderland Harbour.

8. Shaw had been a contractor on the Muston & Yedingham scheme. He remained as engineer of the Orwell Navigation until December 1821.

9. The first bucket-ladder steam dredger came into use in 1803 at Portsmouth harbour, under the direction of Sir Samuel Bentham. Trevithick's dredger, raising ballast from the Thames, began work almost at the same time as Chapman's on the Orwell.

10. Clegram left Shoreham in 1826 to become engineer of the Gloucester & Berkeley Canal.

11. Later known as Port Carlisle.

12. Near Penshaw, 6 miles upstream of Sunderland (Fig. 16).

13. Leslie left Leith in 1832 to become engineer of Dundee Harbour. He was retained as consultant to Leith Harbour (after Chapman's death).

WILLIAM CHAPMAN (1749–1832), CIVIL ENGINEER

APPENDIX

Select List of Chapman's Project Reports additional to those mentioned in the main text, and other Publications
(Short title catalogue)

(A design to improve the pumping engine at Rotterdam). Newcastle, Feb. 1779. *Nieuwe Verhandelingen van het Bataafsch Genootschap der proefondervindelijke Wijsbegeerte te Rotterdam,* vol.1 (1800) 154–78.

'Report on the Improvement of the Harbour of Arklow'. Dublin, Dec. 1791.

'Report on the means of making Woodford River navigable from Lough-Erne to Woodford-Lough'. Limerick, April 1793.

'Report on the Proposed Navigation between the East and West Seas, so far as extends from Newcastle to Haydon-Bridge'. Newcastle, 26 June 1795.

'Second Part of a Report on the Proposed Navigation . . . from Haydon-Bridge to Maryport'. Newcastle, 10 July 1795.

Observations on the Various Systems of Canal Navigation. (London, 1797).

Facts and Remarks relative to the Witham and the Welland. (Boston, 1800).

'Report on the proposed Navigation to Knaresboro'. London, July 1800.

'Report on the means of obtaining a Safe and Commodious Communication from Carlisle to the Sea'. Carlisle, June 1807.

A Treatise on the Progressive Endeavours to improve the manufacture of Cordage. (London, 1808).

'Report on the Drainage of the Marshes from Ancroft Fen to Wainfleet Haven'. Newcastle, April 1810.

'Report on various projected Lines of Navigation from Sheffield'. Newcastle, Sept. 1813.

'Report on the Laneham Drainage'. Retford, 1813.

'Report of George Dance, William Chapman, Daniel Alexander, and James Mountague, respecting the Enlargement and Improvement of the Water-way under London Bridge.' London, Nov. 1814.

Observations on the Necessity of adopting Legislative Measures to dimish the probability of the recurrance of fatal accidents in Collieries. (Newcastle, 1815). Published by the Newcastle Lit. & Phil. Soc.

'Observations on the probable formation of Coal, and on the position & accompanying circumstances of Fossil Trees'. Read at the Royal Society 9 May 1816. Abstract published in *Annals of Philosophy* vol.7 (1816) p.460.

A Treatise containing the Results of Numerous Experiments on the Preservation of Timber. (London, 1817).

'Report on the means of Improving and Enlarging the Harbour of Whitehaven'. Dec. 1821.

'Report on the Formation of a Harbour at Dalden Ness' (i.e. Seaham Harbour) Newcastle, June 1823.

'Report on the Manchester and Dee Ship Canal'. Manchester, June 1825.

'Report on the projected wrought iron Suspension Bridge across the Tyne at North and South Shields'. Newcastle, July 1825.

VII

ENGINEERING ON THE ENGLISH RIVER NAVIGATIONS TO 1760

THIRTY YEARS AGO, under the inspiration of Professor T. S. Willan, I wrote a paper on the engineers of the early river navigations in England. Since then much more has been learnt and I welcome the opportunity of presenting a new synopsis on the subject, but this time with emphasis on the engineering works. The engineers will of course be mentioned, with some corrected or recently discovered dates. Biographical details, however, must await a future publication.

What is needed, and attempted here, is a concise chronological account of the more interesting and important works from Elizabethan times to the mid 18th century; before, that is to say, the beginning of the 'Canal Age'. A complete account even up to this period would exceed the space available, while another chapter (at least) would be required for a proper treatment of river engineering after 1760.

Willan's classic monograph of 1936 is still the best general treatment of early river navigation in England. Histories of individual navigations are admirably set out in the relevant volumes in Charles Hadfield's series *The Canals of the British Isles*, and are summarised by Paget-Tomlinson* who also explains most of the engineering terminology involved. It can be taken for granted that where more specialised references are not given, historical facts have been taken from Hadfield for rivers in the South, the Midlands and North-East, from Hadfield & Biddle for the North-West, and from Boyes & Russell for Eastern England.

Extent of the navigations

The figures given in Table 1 exclude the lengths of tidal rivers navigable by sea-going ships: the Thames to London Bridge, the Avon to Bristol and so on. Important as they are, our concern is with inland waterways navigated by river craft.

On that definition, there were in England from time immemorial rather more than 650 miles of open, or natural, river navigation. York, Nottingham, Peterborough and Maidstone are examples of towns marking the heads of such navigations. In addition, before 1600 a further 90 miles (approximately) had been added by engineering works. This includes the tiny Exeter Canal with its Elizabethan pound locks, but chiefly comprises the Thames from Maidenhead to Oxford and the Lee from Bow to Hertford on which, in both cases, navigation was possible by means of flash locks. Most of the flash locks, but not all, were in mill weirs.

* *The complete book of canal and river navigations.* Waine Research, 1978.

Systematic efforts to improve, or rather to extend, the river navigations were made in the 17th century. During that period the length increased by nearly 200 miles, involving the construction of at least 40 pound locks and numerous weirs, sluices, flash locks and new cuts.

In the next 60 years 370 miles were added to the total length. Or, to put it another way, by 1760 about 700 miles of navigation had been created or improved by engineering works; and this at a time before any of the major English canals were built.

The maximum extension of the river navigations can be dated very roughly to 1815, a date chosen partly because it marks the completion of an important episode of lock building on the Thames. The total length then amounted to almost 1,600 miles. Few of the bigger towns, other than those in the heart of the Midlands, were now inaccessible to navigable river water, fluvial or estuarial.

TABLE 1

Mileage of River Navigations in England

	1600	*1650*	*1700*	*1760*	*1815*
Natural	670	670	650	620	580
Engineered	90	150	300	700	1,000
Total	760	820	950	1,320	1,580

TABLE 2

River Gradients

River	*From – To*	*Length miles*	*Fall feet*	*Average Gradient ins./mile*
Thames	Lechdale - Teddington	129	219	20
Great Ouse . . .	Bedford - St Ives	31	65	25
Aire.	Leeds - Weeland	35	78	27
Nene	Northampton - Peterborough	61	180	35
Lee.	Hertford - Bow	31	111	42
Wey	Guildford - Weybridge	20	72	43
Kennet.	Newbury - Reading	23	128	66
Calder	Sowerby Bridge - Wakefield	25	188	90

Elizabethan navigation works

Exeter Canal No records have yet been found of any pound locks in England earlier than those on the Exeter Canal, built by John Trew of Glamorganshire who was employed as engineer-contractor by the city of Exeter.[1] Construction of the canal began in February 1564. It was a cut 1¾ miles in length from Exeter to a point on the tidal river Exe about two miles above Topsham at the head of the estuary.

A timber weir across the river maintained an adequate depth of water above the cut, and water level in the lower part was retained by a sluice. As this 'opened by the tyde' it presumably had a self-acting pivoted gate or perhaps a pair of mitre gates pointing upstream. The canal, embanked for some of its length, had a width of 16 feet and a depth of three feet suitable for boats or lighters up to 16 tons burden.

There were three turf-sided locks, the lower two being 189 feet in length and 23 feet wide, with a 5½ft. depth of water, and therefore large enough to contain several vessels (a familiar feature of early pound locks in Europe) and to act as passing places. The upper lock was considerably wider, to serve as a basin or dock. The navigation appears to have been opened in the autumn of 1566 and the whole job, including a masonry quay wall 150 feet long on the riverside at Exeter, was completed in 1567. Contemporary documents refer to the lock chambers as 'pools' and the lock gates as 'sluices', so there were seven 'sluices' in all. Some improvements took place in 1581 but thereafter the navigation remained essentially unchanged for nearly a hundred years.

A drawing (Plate 4) dating from *c.* 1630,[2] of 'the Sluices or turnpikes at Exeter water-workes' shows the upper gates to be mitred, with paddles and pivoted balance beams. The stems of the paddles pass through holes in the beams and are levered up a few inches at a time using a crowbar and iron pins. A note records that at the lower end of the lock 'there is but one only whole gate which moveth uppon a great piece of timber, as the other doe accordinge to the diagramme here exhibited'. These details may apply to the 1581 improvements rather than the original structure, but can in either case be taken to represent an example of Elizabethan engineering.

Lee Commissioners were appointed in 1575 to improve the Lee navigation. They gave orders for scouring the river, laying a towpath and regulating the use of the flash locks. There were probably at least a dozen of these between Tottenham and Ware, and some of them seem to have been of the guillotine type, known on the Lee as 'turnpikes'. However, the Commissioners also ordered the construction of a pound lock on a new cut 200 yards in length at Waltham mill, to take the place of a nearby turnpike which was to be transformed to a weir. Built in 1576, with timber walls and floor, the lock definitely had two pairs of mitre gates enclosing a chamber about 60 feet long by 22 feet wide and 10 feet deep. Specifications for the timber work have survived, and the total cost including the cut came to £272.[3]

Serious settlements necessitated reconstruction of the lock on masonry foundations only three years later. This operation, completed in June 1579, was directed and supervised by a 'Mr. Trew': surely John Trew of the Exeter Canal.

Local intrigues led to illegal demolition of the lock in 1592, and the turnpike was reinstated by the mill owner who charged tolls for its use. Despite a life of only sixteen years the historical importance of the Waltham pound lock has long been recognised as the first 'modern' example in England.

Shelford Lock Details of a rather more primitive lock of about the same date, *c.* 1576, have recently come to light.[4] Built by a mill owner and opposed by the river men, it was situated in a new cut alongside the Trent at Shelford and is shown, together with a long weir, on a plan of 1592. The gates, referred to as 'The Lockes' on the plan, are placed about 75 feet apart enclosing a chamber no wider than the cut itself. They are of the guillotine type with posts carried upwards to an overhead beam or lintel, and there are two wheels for lifting or lowering each gate.

Thames It was in 1585 that John Bishop compiled the first complete list of locks and weirs on the Thames.[5] There were in fact no locks from London Bridge up to Maidenhead, a length of 52 miles of open navigation, but in the 63 miles between Oxford and Maidenhead (with a fall of 108 feet) Bishop names 23 locks, of which 19 are known to have been in mill weirs and four in weirs built solely for navigation purposes.

These were all flash locks of the type with retractable paddles (and overfall boards) and rimers bearing on a pivoted swinging beam and a fixed sill (Figure 2.1). Such locks

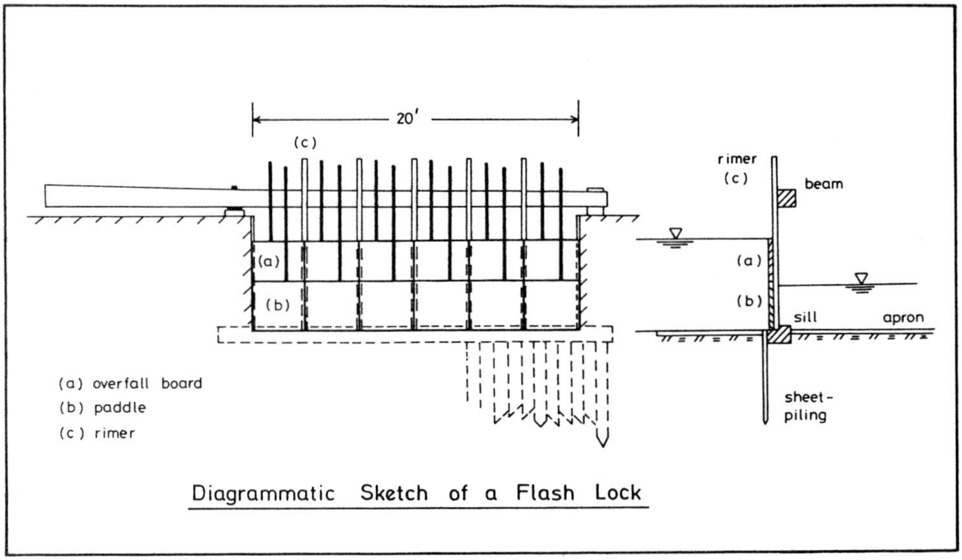

Figure 2.1.

are of medieval origin and, after countless renewals, remained in use until the 18th century. A map of the river near Abingdon, dating from *c.* 1570, shows 'The Locke' with five sets of paddles.[6] Plot, in 1677, gives a clear description of the Thames flash locks,[7] and practically every one in Bishop's list can readily be identified in early 18th-century navigation rolls; while a well known sketch plan of 1786 shows the weir at Whitchurch with its sluice gate and a barge being winched upstream through the opened flash lock.[8]

The weirs themselves must have been considerable structures; built obliquely across the river, upwards of 400 feet in length, and impounding as much as six or seven feet of water with falls of four or five feet as a maximum differential head.

Seventeenth century

Thames: above Oxford Commissioners appointed under an Act of 1605 succeeded in extending the navigation 'for many miles beyond the city of Oxford westward'.

These words date from 1624. More explicitly in 1695 the river is said to have been navigable 'for divers years last past . . . somewhat farther than Lechlade'.[9] No contemporary details of the work are available, but paddle and rimer flash locks were still in use throughout this part of the Thames until the late 18th century (indeed, a few remained until quite recent times), and there is no reason to suppose they differed in any essential way, in type or location, from the original structures. When toll lists and descriptions become available[10] they show barges of 40 to 50 tons burden travelling between Lechlade and Oxford through 20 flash locks in this length of 32 miles, in which the fall amounts to 50 feet. The locks, of the paddle and rimer type about 14 feet wide, were in timber-framed weirs up to 40 feet in total length.

Great Ouse (Part I) The first integrated scheme of pound lock navigation on an English river was carried out by Arnold Spencer (1587–1655) on the Great Ouse between St Ives and St Neots.[11] In 1618 he and a partner acquired the right to undertake the work at their own charge, and on payment of an annual rent to the Crown, in return for any profits. In contemporary terminology Spencer built six 'sluices', but these were certainly pound locks, each with two pairs of mitre gates and a chamber large enough only for a single boat.[12] The chamber walls may have been in timber. The fall from St Neots to St Ives is 28 feet in a length of 15 miles,[13] and the locks were located in cuts or side streams at each of the mills in this part of the river. The work had been completed by 1625.

Spencer hoped to extend the navigation to Bedford. Soon after 1628 he built another lock at Eaton Socon (two miles above the St Neots lock), this time with a double size chamber; he also made two cuts about 100 and 200 yards in length and scoured the river with a horse-drawn scarifier. In this way he enabled barges to reach Great Barford, but financial difficulties prevented any further works.

Thames: Oxford–Burcot In the last 15 miles up to Oxford the Thames has its steepest gradient (24 inches per mile) and there were many shoals; so the larger class of barge used Burcot as the head of navigation. To remedy this state of affairs Commissioners were appointed under an Act of 1624. Considerable sums appear to have been spent during the next five or six years, perhaps on dredging and general improvements, but the main achievement was the construction of three pound locks; at Iffley and at Sandford, replacing flash locks, and in a cut at the head of a re-opened channel of the river (the Swift Ditch) near Abingdon.[14] The locks were completed by 1638. Nuneham flash lock remained, to be used as a stanch for providing, when necessary, an adequate depth of water on the lower sill of Sandford lock. Barges calling at Abingdon continued to use the flash lock there. Interestingly, this was rebuilt in 1649 with excellent masonry abutments. The first pound lock at Abingdon, on the site of the present one, was not built until 1790.

The Oxford-Burcot pound locks, confusingly called 'turnpikes' and as such described by Plot in 1677, had masonry walls and two pairs of mitre gates. The chambers were about 80 feet in length and 19 feet wide, with 4 to 4½ feet of water on the sills. Sandford lock had a fall of 6 feet, which helps to explain why pound locks were needed, quite apart from their greater convenience and economy in water.

Warwickshire Avon This river was first made navigable from Tewkesbury to Stratford during the years 1636–39 under the direction of William Sandys (1607–1669). According to an account by Thomas Habington (who died in 1647), Sandys built 13 'sluices' (at named sites, most if not all at mills) and 'erected also locks and placed many wires [weirs] in the quickest streams' and, seemingly a little later, relieved 'the griveous [*sic*] oppression of waters' on neighbouring land, 'by opening of two new floodgates'.

For the next owner of the navigation various improvements were carried out around 1650 on the river below Evesham, and the 17th-century work reached completion about 1665 with the construction or re-construction of five 'sluices'. These were built by Andrew Yarranton (1616–*c.* 1684) acting as engineer for, and partner in, a syndicate holding a lease on the upper river.[15]

It is difficult to say exactly what was done in each of the three stages of construction, but the final result is known: a navigation 43 miles in length, suitable for 30-ton barges, with seven pound locks and four flash locks or stanches in the upper district (18 miles, 48 ft. fall) and seven pound locks and two stanches in the lower district (25 miles, 36 ft. fall).

Until recently four of the locks (at Pershore, Wyre, Chadbury and Cleeve) retained what was probably their original diamond-shaped chambers: the idea perhaps being to minimise erosion of the earth slopes by the rush of water from the upper gate paddles. The stanches situated below Fladbury and Pershore mills were classic early examples: normally left open so as not to impair the efficiency of the mill wheels, and closed only on the approach of a boat to raise temporarily the upstream water level for navigation before being used as a flash lock. The fall at these stanches, when closed, was two to three feet.[16]

Wey The Wey navigation, designed by Sir Richard Weston (1591–1652), is a remarkable work of river engineering in the 17th century.[17] The Act dates from June 1651. Construction began in August and was carried on with great determination, clearly in accordance with carefully prepared plans. As many as 200 men were employed. Sir Richard found part of the capital, supplied most of the timber, and directed operations until shortly before his death in May 1652.

Work continued under George Weston, who had been instructed by his father, and the navigation was opened in November 1653. It had (and still has) a length of 15 miles, of which nine miles were in cut, with 10 pound locks, four new weirs, 12 bridges, and a wharf at Guildford (Figure 2.2). The total cost exceeded £15,000. All the locks were probably turf-sided and timber-framed, with falls ranging from four feet to eight feet, and together they accounted for rather more than 60 feet out of the total fall of about 72 feet from Guildford to the Thames at Weybridge. Flood gates were placed at the head of the two longest cuts, but they may have been later additions.

The navigation proved to be successful and seems scarcely to have needed any changes or major improvements for many years. In the 18th and early 19th century the cuts were shown as having a top width of about 52 feet[18] and the locks as being large enough to admit barges 84 feet long by 14 ft. 3 ins. wide.[19] These dimensions may well apply to the original construction.

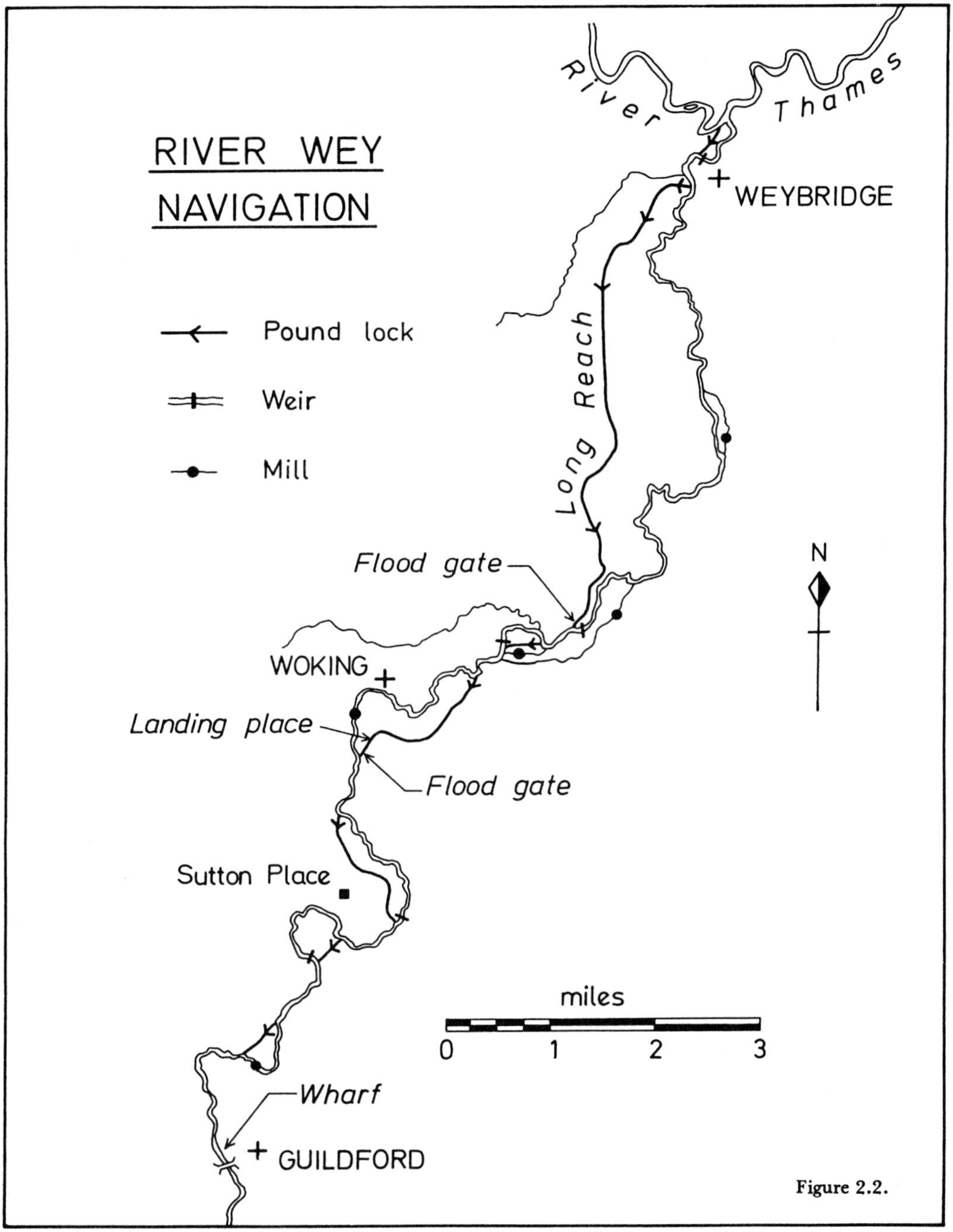
River Thames
RIVER WEY NAVIGATION
WEYBRIDGE
Pound lock
Weir
Mill
Long Reach
Flood gate
N
WOKING
Landing place
Flood gate
Sutton Place
miles
0 1 2 3
Wharf
GUILDFORD

Figure 2.2.

Exeter Canal In 1676–77 Richard Hurd of Cardiff, an engineer 'experienced in such works' was employed by Exeter City Council for a fee of £100 to extend the canal by about a mile to deeper water in the river, and to make a basin at the new entrance with mitre gates (later known as Trenchard's Sluice) of sufficient size to admit small coasting vessels at high tide. Probably at this time Hurd also replaced Trew's weir by a masonry structure nearby: some 250 feet in length with a fall of approximately 6 feet. Total expenditure by the City was upwards of £5,000 including land purchase.

The canal itself still could take only 16-ton lighters. During the years 1698–1701, however, it was enlarged to 50-ft. top width and 10-ft. depth; the three old locks were replaced by a single large turf-sided lock, with masonry abutments, and new flood gates (Kings Arms Sluice) built at the upper end of the canal. From 1699 Daniel Dunnel engineered the works at a salary of £3 per week. Navigation on the reconstructed canal began in November 1701.[20]

Great Ouse (Part II) The navigation from St Ives up to Great Barford had been allowed to fall into a poor state of repair, particularly above St Neots, and Vermuyden's drainage works in the Fens resulted in a somewhat lower river level which made it difficult for boats to cross the shallows at St Ives. Nothing much was done to remedy this state of affairs until the navigation came under the direction of Henry Ashley who, from about 1680, was assisted by his son Henry Ashley junior (1654–*c.* 1730), who also undertook the Lark navigation, opened *c.* 1716, and rebuilt St Ives stanch in 1720.

Improvements carried out by the Ashleys involve the following structures, for which costs and approximate dates are known:[21]

1678	Stanch (and weir) at St Ives	£200
1680	Sluice at Eaton Socon mills	£230
1682	Roxton sluice	£250
1686	Stanch (and weir) below Tempsford	£130
1686	Stanch and overshoot near Barford	£100

As before, 'sluice' in Great Ouse terminology means a pound lock. The stanches were of the paddle and rimer type, with sluice gates and weir adjoining.[22] The lock at Eaton Socon mills replaced Spencer's structure of *c.* 1628 and had a single (not double) size chamber.

The Ashleys *père et fils*, having restored and improved the navigation to Great Barford, now extended it to Bedford by building five pound locks. This was accomplished in 1687–89, the locks being designed to take boats of 30 tons burden. By 1689, in the 31 miles from Bedford to St Ives, the river then had 13 locks and four stanches. The fall in this length is 65 feet.

It is worth recording the name of Joseph Hunt, carpenter of Godmanchester (born *c.* 1650). He may have been employed on some of the new works for the Ashleys, and certainly in the 1690s was often engaged in repairs, making new sluice gates, and so on. He charged 2s. 6d. per day, or 15s. 0d. per week, for himself and 1s. 0d. per day for his son. Labourers working on the navigation received between 1s. 0d. and 1s. 4d. per day.

Early eighteenth century

Aire and Calder In the 17th century boats of 30 tons burden could go up the river Aire on the tide to Knottingley, 24 miles from the junction with the Yorkshire Ouse. Under an Act of 1699 the navigation was extended 23 miles to Leeds and, on the tributary river Calder, 13 miles to Wakefield. Improvements were also made in 12 miles of the river below Knottingley, to Weeland.[23]

The proprietors, named in the Act, consisted of nine members of Leeds Corporation, including the mayor, and nine gentlemen of Wakefield. The Leeds committee had employed the engineer John Hadley to survey the Aire in December 1697. He considered the project to be eminently feasible, requiring seven pound locks between Knottingley and Leeds, one at each of the mills, and deepening of the river in various places as far downstream as Haddlesey where, in a later opinion, a further lock should be built. At about the same time the Wakefield committee commissioned a survey of the Calder from Samuel Shelton. Also, at the request of Parliament, two members of Trinity House studied the possible effect the works might have on the tidal regime of the Ouse and on the trade at various riverside towns. Their sensible and impartial report was written in July 1698.[24]

Hadley gave evidence during the Parliamentary proceedings in 1698 and 1699, and on 9 May 1699 was appointed engineer for the river Aire at a fee of £420. Work began without delay and, despite some damage by a flood in January 1700, the navigation was opened to Leeds in November of that year, the seven locks above Knottingley having been completed as well as a warehouse at Leeds. Not long afterwards two additional locks were built on the Leeds line, one at Methley and the other at the tail of the 1½-mile Cryer Cut newly excavated to bypass a very circuitous stretch of river near Woodlesford.

It is probable, though not yet confirmed, that Hadley died in 1701 (before August of that year). Thereafter the proprietors seem to have relied on James Mitchell, as engineer-contractor, for technical aspects of the works. A carpenter from Bradley, near Huddersfield, he had been involved in building some of the Aire locks. In July 1701, in partnership with James Willans, mason of Dewsbury, he contracted to make the Calder navigable. Four locks were built between Castleford and Wakefield, and this part of the navigation was opened in the summer of 1702.

Meanwhile it had been decided to have two locks (not one) below Knottingley: at Beal and Haddlesey. In April 1702 Mitchell and George Atkinson (1663–1728) of Thorne signed a contract to build these locks and their weirs, and by November 1704 it could be said that the whole navigation 'was near perfected'.

Of the 15 locks, eight were in cuts at the mills, as illustrated in Figure 2.3,[25] but the others required new weirs ('dams' in north country terminology) across the river. As a result of experience, three additional sluice gates had to be provided in Beal dam to avoid flooding. There were flood gates at the head of Cryer Cut; in places the river embankments had to be raised and doubtless a good deal of dredging was needed.

The locks had masonry walls and timber flooring, and were about 60 feet long by 15 feet wide with 3 ft. 6 ins. of water over the sills. At the outset boats appear to have been limited to 15 tons burden but subsequent improvements in the river, without alteration to the locks, enabled boats to carry 28 tons on average; more in

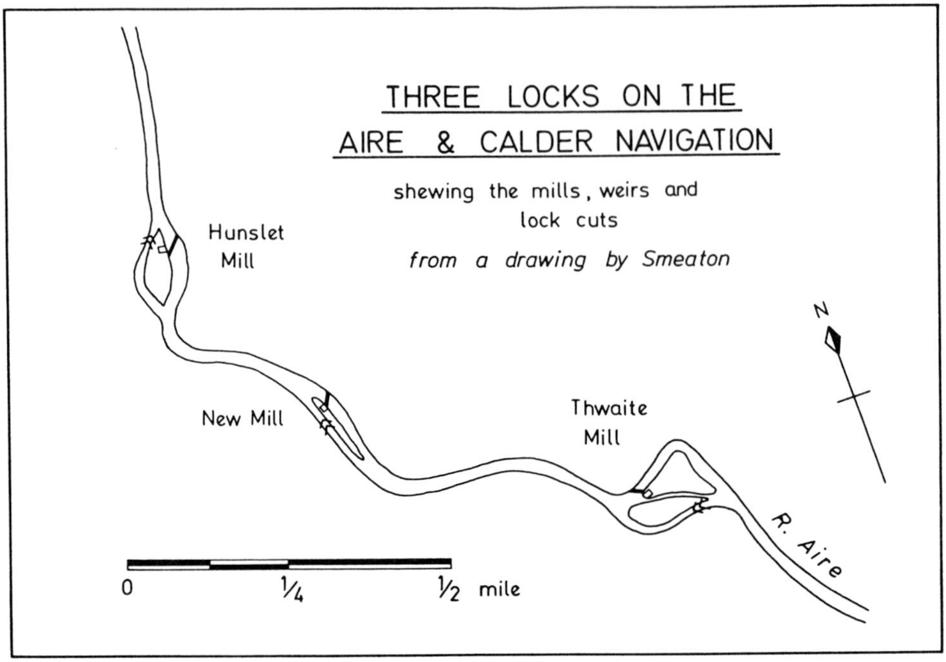

Figure 2.3.

winter, less in a dry summer.[26] The whole undertaking cost over £25,000 but proved to be useful and profitable.

Essex Stour The Corporation of Sudbury obtained an Act in 1706 to make the river Stour navigable from their town to Manningtree at the head of the estuary, a length of 25 miles. Shares in the undertaking were mostly purchased by two London merchants who took responsibility for carrying out the works. The navigation was opened throughout in May 1709, at a total cost of around £9,000. It had 13 pound locks, timber-framed and turf-sided, and 13 stanches of the paddle and rimer type.[27]

Later in the 18th century the locks were walled in timber but retained their characteristic overhead lintels between the gate posts used, instead of land ties, to resist the earth pressure. These and other features of the Stour are well known from the paintings of John Constable, whose father was one of the navigation commissioners from 1781.

Kennet From Newbury to the Thames at Reading the Kennet has a gradient of 66 inches per mile and a correspondingly swift current with many shoals. Under an Act of 1715 work began on building locks at the mills along the river, but it became apparent that success was not going to be achieved in this way. In the hope of finding a solution the proprietors in 1718 appointed John Hore (*c.* 1690–1763) as their engineer. His plan involved cuts totalling 11 miles in length and 20 pound locks. Completed in 1723, the navigation had a length of 18½ miles with a wharf and barge basin at Newbury, wharfs at Aldermaston and Reading, and several new bridges.[28].

From Reading to Sulhamstead there were seven cuts, up to three-quarters of a mile long, bypassing four mills and three bad sections of river where weirs had to be built. Further upstream the navigation was predominantly in cut except for about a mile near Woolhampton and another mile or so near Thatcham. (Figure 2.4). Here some dredging and embanking were needed and a run-off or waste weir.

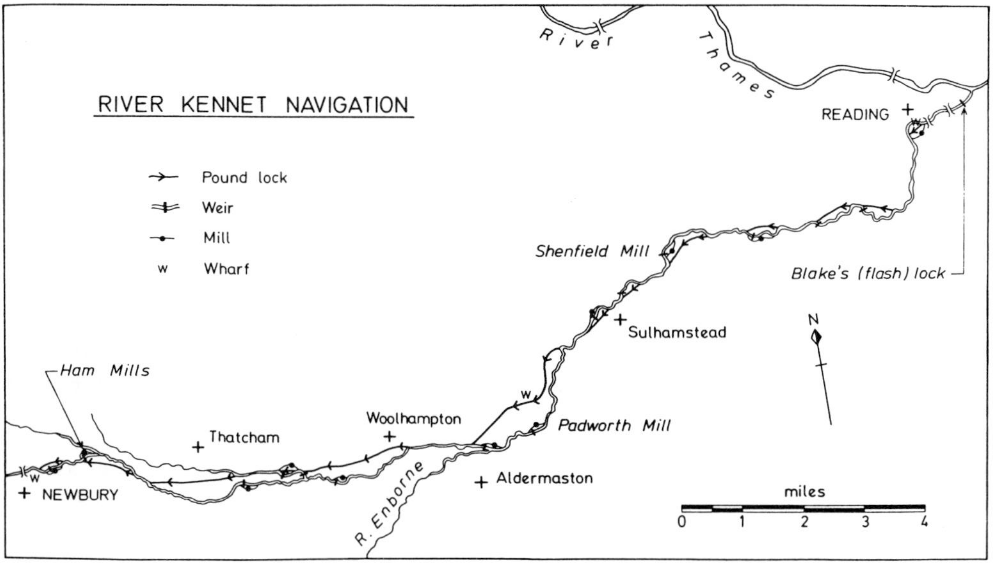

Figure 2.4.

The two locks above and below Aldermaston wharf were in brick, and so (probably) was the lock at Reading. All the rest were timber-framed and turf-sided. Evidence given in 1725 that 100-ton barges passed fully laden from Newbury through Reading to London suggests that from the start the locks had dimensions not very different from 114 feet by 17 feet as in three of the surviving turf-sided locks. Falls ranged typically between five and nine feet.

A letter written by Hore in July 1721 reports progress. Two of the longer cuts are now opened he says, two locks near Reading have recently been framed, the banks near Thatcham are being formed using 'the two large boats' (presumably dredger barges) and work is starting on Padworth cut with its two brick locks. He also refers to the Archimedean screw used in draining the excavations.

According to usual practice at this period, several small local contractors would have been employed, carpenters, bricklayers, blacksmiths for the iron work, and so on. Some of their surnames by chance are known: Potenger of Sulhamstead and

Lawrence of Burghfield, labourers (i.e. engaged on excavation), with Hughes of Shenfield, Grace of Midgham, and Saxon and Slade of Newbury, carpenters.

From an engineering point of view the Kennet is the outstanding early 18th-century navigation. The total cost was upwards of £45,000.[29]

Yorkshire Derwent The engineer George Sorocold surveyed the Derwent in 1694 and again in 1699. The Act for making the river navigable was passed in 1701 but little, if anything, was done until 1720. Joshua Mitchell (who had worked with his father on the Aire & Calder) and a partner then undertook the work in return for a lease on the tolls.[30]

Mitchell built five pound locks, three at mills and two with new weirs; the locks being of a size to take Yorkshire keels 55 feet long by 14 feet beam. The lowest lock, at Sutton mill, was at the tidal limit 15 miles above the confluence with the Ouse. The works, completed in 1724, rendered the river navigable for a further 23 miles up to Malton.

Bristol Avon An Act passed in 1712 enabled the Corporation of Bath to make the Avon navigable from Hanham Mills, near Bristol, to Bath. Nothing was done, however, until 1724 when a group of proprietors took financial responsibility for the undertaking, and in April 1725 appointed John Hore to design and direct the works. The 11½-mile navigation with six masonry-walled pound locks, about 98 feet long by 18 feet wide, was opened in December 1727 at a cost approximating to £12,000. Hore had a Mr. Downs as assistant or resident engineer, and it is on record that the chief mason, Edward Marchant, received the handsome wage of two guineas per week.[31]

With a river of moderate depth and a gradient averaging 30 inches per mile, the engineering problems were less formidable than on the Kennet; moreover the existing mill weirs could be retained. Five of the locks were built in short cuts beside the weirs, with falls of three to six feet. But near Twerton there were several mills, on both sides of the river, and two weirs about a quarter of a mile apart. Here a cut, known as Weston Cut, some 600 yards in length had to be made with a single lock at the lower end accommodating the total fall (about nine feet) of the two weirs combined.

A large-scale map of 1742 by Thomas Thorpe shows this cut and the next three locks downstream at Kelston, Saltford and Swinford, with the weirs and mills.[32] It is interesting to note that the weirs, like those on the Aire (Figure 2.3), are oblique.

Mersey & Irwell The plan by Thomas Steers (*c.* 1672–1750) of the Mersey & Irwell, dating from 1712, is one of the earliest published for any river. Levels are given at nine points along the rivers from Manchester to Warrington (fall 52 feet) and the scheme involves eight pound locks and two cuts.[33]

The locks were built between 1724 and 1736, and also a wharf at Warrington, but whether Steers acted as engineer is not certain. The cuts, to eliminate long loops in the Mersey, followed later.

Don In favourable conditions the river Don was navigable, on the tide, for about 16 miles above Goole to Wilsick House and occasionally for a further eight miles to Doncaster. As part of an effort to improve the navigation Doncaster Corporation sought advice from Joseph Atkinson (1693–1760; son of George Atkinson, the Aire & Calder contractor) and, at the same time, the Cutlers' Company employed William Palmer (d. 1737), engineer of York, to plan an extension up to, or near, Sheffield.[34] Working jointly, with help from Joshua Mitchell, they produced a map of the river in October 1722, to a scale of 2½ inches to a mile. Next year this was published as an engraving (on a reduced scale) with tables of lengths and falls.[35]

Above Doncaster the river was rather steep and swift flowing with dams (weirs) for the mills at places such as Kilnhurst, Rotherham and Sheffield itself. The solution here clearly involved cuts, some of considerable length, with locks, but making use of the existing weirs to avoid further impeding the flood flow. The last four or five miles up to Sheffield were evidently going to be difficult, however, and it was soon decided to terminate the navigation at Tinsley. On this basis an Act passed in 1726, with Palmer giving evidence.

Below Doncaster the river had a very different character, flowing through low-lying country with a gradient of scarcely more than 20 inches per mile from the tail of Doncaster mill dam to Wilsick, and in particular with no weirs. Nevertheless this part of the river had to cope with floods coming down from the higher ground upstream, and it also acted as the outfall for several land drainage schemes in the neighbourhood. Consequently the landowners opposed any measures which might lead to flood or other damage to their property.

At first the plan was to scour the river and make cuts, without locks or weirs, past the worst of the shoals. Then came the idea of building an opening weir with a lock at one especially bad place, and finally it was agreed to have locks and weirs not more than four feet high with very ample sluice gates. Atkinson explained the scheme to Parliament in proceedings leading to the Act of 1727, his evidence being confirmed by Richard Ellison (*c.* 1686–1743) of Thorne who on this and later occasions appeared as expert witness for the Don.

Though it could be assumed that Palmer directed operations for the Sheffield proprietors, the early history of the works is obscure. By 1731, however, about £12,000 had been spent and the navigation now extended seven miles above Doncaster. The two groups of proprietors then amalgamated, formed a management committee and in August 1731 appointed John Smith of Attercliffe as engineer. Described as carpenter or millwright, terms virtually synonymous in a river context, he held this post until 1767.

By 1733 ten locks had been built, eight from Doncaster upwards and two below. A third, at Barnby Dun near Wilsick, followed a little later. The original plan, with a few variations, was now realised except for the last five miles up to Tinsley, and included a 2½-mile cut past Kilnhurst with three pound locks and flood gates at the head. The locks throughout were of the same length as those on the Aire & Calder and one foot wider.

The need now became apparent to bring the river below Wilsick to the standard achieved upstream. Accordingly an Act of 1740 gave powers to make a 2-mile cut, with two locks, between Bramwith and Stainforth and to deepen the channel down to Fishlake. At the same time the plan for completing the navigation to Tinsley was revised. All these works were finished by 1751, resulting in a navigation 33 miles long with nine miles of cut and 17 locks (and two flood gates) suitable for 30-ton vessels.

Smeaton, writing in 1767, said he had seen several of Mr. Smith's locks 'which have stood and answered very well': praise indeed from that master of design and construction.[36]

Yorkshire Ouse In February 1699 the Corporation of York requested the members of Parliament for the city 'to send an Engineer as soone as possibly they can to view the River of Ouze in order to make it more navigable'. They made a good choice, and on 4 May Thomas Surbey arrived from London. (Nothing else is known of this evidently able engineer; his MS report on the Ouse is in York City Archives.) His report, dated 22 May, is accompanied by a chart with soundings and plans of the proposed works, which include a dam and lock at Naburn (six miles below York) and a 1,200-yard cut near Booth Ferry.[37] The Corporation paid Surbey £24 plus expenses, and sought a second opinion from John Hadley. He agreed with the idea of Naburn lock.

Thus encouraged, York sent a petition to Parliament for leave to bring in a Bill. This faltered at the second reading, however, and little more was done until 1725 when William Palmer received instructions to prepare a survey. Part of his map, which includes tidal data, is shown in (Plate 2).[38] The city now succeeded in obtaining an Act, passed in 1727, to levy navigation tolls to meet the cost of improvement works.

The precise nature of these seems still have have been in doubt. The Naburn project though attractive would be expensive; John Perry (*c.* 1669–1733), asked for his opinion, recommended a flash lock above Cawood and several bypass cuts totalling 2½ miles in length, while Palmer favoured a mile-long cut near to the mouth of the Wharfe. In all cases, however, some deepening of the channel would additionally be required.

On grounds of cost the latter expedient was adopted as the sole means of improvement. Palmer directed the operations from 1728, on a retaining fee of £40 p.a. with payments for special works. Timber jetties or training walls were built to contract the stream, and so increase its velocity on the ebb, and the river bed was scoured or 'ploughed'.

Work along these lines continued for several years and doubtless proved effective in removing deposits of tidal silt. But at several places with notorious shoals the bed consists of boulder clay or clayey gravel which would have proved too resistant to respond to this treatment or, indeed, to the 'vigorous dredging' (with, of course, hand-operated equipment) proposed in 1748 by John Smith, who had been brought from the Don to give his advice on the problem. And so at last in 1752 came the decision to build Naburn lock and dam. An account of these works is given later.

Weaver The river Weaver was made navigable from Frodsham up to Winsford, a distance of 18 miles, by the construction of 11 pound locks and weirs.[39] Work began in 1730 under the direction of Thomas Robinson as 'surveyor general'. The locks were finished by the end of 1732 and the whole job completed in 1735, at a cost of roughly £18,000.

An unusual feature is that the locks were built within the river (not in cuts) in locally widened sections to give an adequate weir length. The entire structure of lock, sluice and weir was in timber. The locks, about 60 feet long and 16 feet wide, could accommodate boats up to 50 tons burden.

Lee An engraved map of the Lee navigation in 1740, with levels along the river from Hertford to Bow, shows a fall of 111 feet in this distance of 31 miles.[40] The mill at Ware is situated on the river bank with a 'cistern' (pound) lock nearby, but the 11 mills further downstream are on rather long cuts or side streams. Across the river below the mill-stream intakes, and at some other places, there are four 'turnpikes' and 15 'weirs'. All of these are undoubtedly flash locks. No description of the 'weirs' has been found; probably they were of the paddle and rimer type. But it is certain that 'turnpikes' were guillotine gates with an overhead windlass of the pattern familiar in East Anglia (Plate 3).

In 1741 a new turnpike was built near Broxbourne to the design of William Whittenbury (d. 1757), resident engineer to the navigation, whose map has just been mentioned. The main points in his specification are: the gate to be 14 feet wide and fitted with friction wheels to 'ensure its motion' by the 'Great Wheel' and windlass; and beside the lock gate there are to be three sluices each 6 ft. 8 ins. wide, making a total opening width of 34 feet. When all the gates are closed, a depth of 4½ ft. of water will be impounded above sill level, with a fall of 3 ft. 3 ins. Plank aprons to extend 12 feet upstream and downstream, to protect the bed from erosion.[41]

It is known, in general terms, that a large quantity of water is 'lost' in the operation of flash locks on a busy navigation. But Smeaton and Yeoman, in 1767, actually measured (or, rather, made a careful calculation of) the average discharge in a dry season through a Lee turnpike at Waltham Abbey and found the astonishing figure of two million cubic feet per day.[42] This is equivalent to at least 250 lockfuls in a typical pound lock. No wonder there were frequent conflicts between millers and boatmen, and one is surprised at the long delay in replacing flash locks by pound locks on the Lee and the Thames, an improvement for both rivers which came on a serious scale only around 1770.

Denver Sluice This important work of river engineering, as completed in 1750 across the Great Ouse, is shown in an early 19th-century drawing reproduced in Plate 5.[43] High embankments run alongside the river downstream to King's Lynn; the sluice prevents the tides from overtopping the lower banks upstream on the Great Ouse and other South Level rivers, and the lock provides for navigation at all normal states of the tide. There are in effect two separate structures: (i) The main sluice built in stone-faced brickwork with three arches each giving an 18-ft. clear waterway and equipped with a pair of self-acting pointing gates (the 'sea doors' of Plate 5) and another pair pointing upstream, used in dry seasons to maintain an adequate depth for navigation to Ely and beyond, and for agricultural reasons. The sea doors rise to the level of a large spring tide (1.5 feet above H.W.O.S.T.) which is 17.75 feet above sill level and 19.25 feet above the apron. (ii) A two-arch brick structure on the east side of a 'lock island' containing a supplementary 16-ft. sluice and a navigation lock 52 feet in length between gates, 13 ft. 3 in. clear width, and with 5 ft. 3 in. depth on the sills at normal low water. Like the sluices, the lock has a pair of ebb gates or 'land doors' for dry seasons.

The principal materials used were: locally-made bricks, stone from Ketton, tarras and lime mortar, Baltic fir for bearing piles, sheet piles and planked floors, and oak for the gates. As well as the usual barrows, planks, pumps, etc. there was a winch-operated pile driver and a machine for dovetailing the sheet piles. Work began in June 1746, the lock was opened exactly a year later, the main sluice came into operation in November 1749 and the whole job reached completion by the end of 1750. The costs were approximately: lock £1,100, cofferdams £900, main sluice £3,100.

The work was directed throughout by William Cole, Surveyor (i.e. Engineer) of the South Level from 1731 to 1750, with John Leaford (Divisional Officer 1729-53) as resident engineer. William Robinson of Ely contracted for the lock and cofferdams but the main sluice was built by direct labour. Besides the minute books of Bedford Level Corporation, the authority then responsible for all river and main drainage works in the Fens, surviving original documents include Cole's journals for 1746 and 1749, his meticulous accounts for the entire period of construction, a plan of the scheme in 1747 by Leaford, and a report dated June 1748 by Charles Labelye (1705-1762) who was consulted on the main sluice foundations.[44]

Denver Sluice has a long history going back to the first works by Vermuyden in the 1650s. Here it is relevant to note that the main sluice of 1750 replaced a similar but smaller structure built 1682-84 which suffered a foundation failure in 1713, and had to be demolished; and that the two eastern brick arches were originally built in 1699 by Richard Russell, a predecessor of Cole, for two additional sluices.[45] The lock and supplementary sluice made use of Russell's pier and side walls. A report by Labelye in 1745 ended a strangely long period of indecision as to whether, and how, Denver Sluice should be rebuilt but, although he is usually credited with the design, the 1750 structure differs very materially from his proposals.[46] No changes occurred thereafter until the 1830s when, to derive full benefit from the deepened river bed brought about by the Eau Brink Cut, the present lock and sluice were constructed to designs by Sir John Rennie.[47]

Naburn Lock and Dam Built on the Ouse six miles below York, Naburn lock and dam are the most impressive single works of their kind on an English river before 1760. The contract, dated 18 September 1752, is signed by John Smith of Attercliffe and his son John Smith junior.[48] By February 1753 agreement had been reached with the local landowner on rent payable for the ground to be cut and otherwise to be used by the navigation,[49] and construction started later that year or early in 1754.

John Smith junior (*c.* 1725-1783) acted as site engineer, payments under the contract being made to him,[50] but it can be taken that his father was responsible for design and expert supervision. Work proceeded rather slowly, due to the limited funds available as well as to the engineering difficulties on a river subject to large floods, and was completed in 1757 at a cost not exceeding the contract price of £5,500.

A site plan and a section through the lock chamber are shown in Figures 2.5 and 2.6. The leading dimensions and a brief description may be summarised as follows:[51]

Lock Cut: About 350 yards in length, top width 54 feet, excavated 16 feet below ground level and embanked on the east side.

Lock: 90 feet between gates, 140 feet overall, 20 feet clear width, 23 feet deep from coping to the plank floor. Masonry walls, slightly battered on the front face and backed by counterforts to increase resistance against lateral earth and ground water pressure, founded on timber beams 43 feet long and about 16 inches deep. The wing walls founded on timber piles. Lower sill 22 feet and upper sill 18 feet below coping, these dimensions also being approximately the height of the gates.

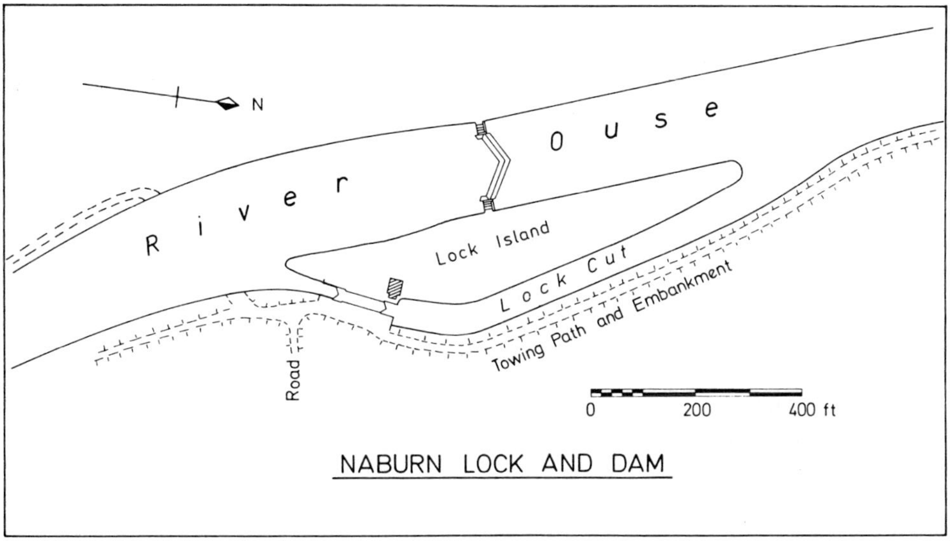

Figure 2.5.

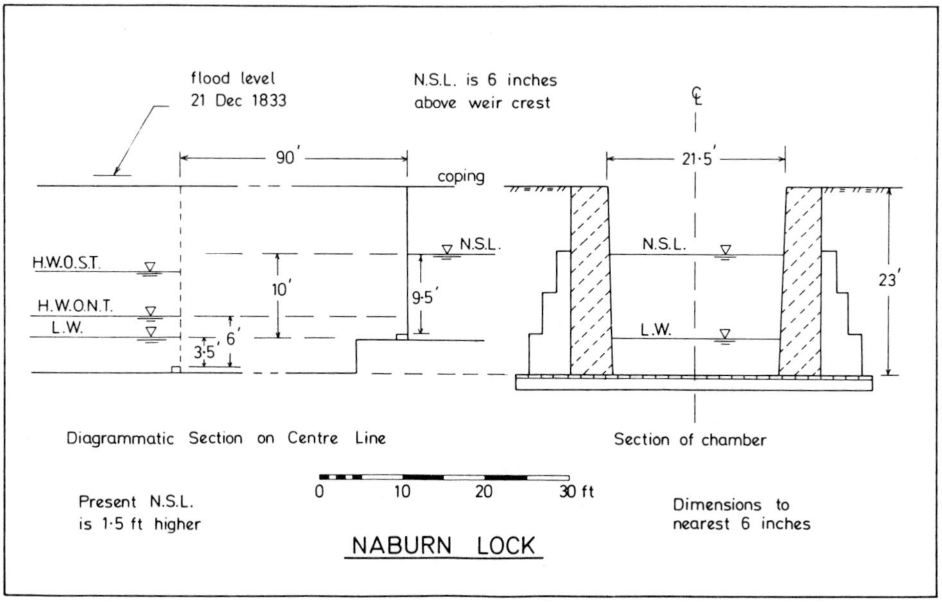

Figure 2.6.

Depth of water: On the upper sill at normal summer level 9½ feet; on the lower sill at low tide 3½ feet (maximum fall therefore 10 feet) and at high water ordinary neap tide 6 feet.
Dam: Built across the river, here 170 feet wide, consisting of a masonry weir V-shaped in plan and 140 feet in length with three sluices 4 feet wide at each end. Weir crest originally 13 feet higher than lower sill level. Normal summer level regulated by the sluices at about 6 inches above the crest and 10 feet above ordinary low tide. Floods in winter completely submerge the weir, rising near to (or in extreme cases above) coping level of the lock. River bed downstream of the weir protected by random stone blocks.

The lock and dam provided a minimum depth of seven feet at all times up to York. Conditions downstream were less satisfactory. A depth of six feet on the lower sill, as at high water neaps, just allowed the passage of fully laden Humber keels carrying 80 tons, but shoals at several places (notably Acaster Selby) for three or four miles below the lock still impeded navigation even for the more usual loading of 60 tons on a draught of five feet.

Improvements in this part of the river had to await the use of a 10-hp steam dredger in the years 1835–37, under the direction of Thomas Rhodes. At the same time Rhodes raised the crest of the weir by 18 inches,[52] thus bringing the maximum fall to its present figure of 11½ feet, and dredged upstream of York. He also rebuilt the lower end of the lock where some of the foundation beams had given way. It was this latter operation which gave him the opportunity of recording the cross section shown in Figure 2.6.

Nene Navigation was always possible on the tidal river from Wisbech to Peterborough, or perhaps a few miles beyond. Commissioners appointed under an Act of 1725 agreed with Thomas Squires and a partner that he would make the river navigable from Peterborough to Oundle, with the right thereafter to levy tolls, and this was done in the years 1726–30. Squires then became sole proprietor and, under a similar agreement, extended the navigation to Thrapston by 1737. No minute books appear to be available for the period of construction, but references in the later 18th century show that pound locks (still called 'sluices' as on the Great Ouse) had been built at each of the 14 mills in this 35-mile length of river, and probably about half a dozen stanches of the guillotine type well known on the Nene until recent times.

The river from Thrapston up to Northampton was surveyed in 1744 by Thomas Yeoman (*c.* 1708–1781) but 10 years passed before an active interest developed in making this western part of the river navigable. Yeoman went over the ground again and produced a rough estimate of £12,000. He also gave evidence on the Bill, which passed in April 1756. Commissioners were appointed and within a year the necessary capital had been raised by subscription.

Yeoman's sketch map indicated on which side of the mills the lock cuts should be made and the positions of the stanches.[53] In November 1757 Ferdinando Stratford (1719–1766) began more detailed surveys. John Case of London and John Smith (accompanied by his father) attended the Commissioners in March and April respectively with plans and proposals for carrying out the work. Yeoman and Stratford examined these and, after considering several other tenders, the contract was awarded to John Smith junior in June 1758. His estimate, with detailed specifications, came to £14,070 for building 20 pound locks and various ancillary works including a total length of 3¾ miles of cut; the locks to be 10½ feet wide at the bottom and

100 feet long between the gates (to take two barges in tandem) with battered stone-faced brickwork walls and timber floors.[54]

Smith acted as engineer and contractor: Yeoman continued in a consultative capacity only, and a clerk was appointed to certify payments for work done. Smith appears to have handled the job efficiently. He made some variations in detail and was asked (and paid) to carry out a few additional items. Yeoman inspected the works in October 1760 and again in January 1762, on the latter occasion signifying his complete approval of the navigation, the formal opening of which had taken place in August 1761. The total expenditure amounted to £15,650.

The cuts by-passed 21 mills and were carefully planned to improve the course of the river. In addition to the 20 locks (one of which, on a long cut, accommodated the fall at two mills) there were two or perhaps three stanches, and a minimum depth of four feet obtained in all parts of the river. A horse towing path, with some 20 crossing bridges, had been provided along the whole length of 26 miles and wharfs were built at Wellingborough and Northampton.

The total fall of 180 feet from Northampton to Peterborough was accounted for approximately as follows: 136 feet in the locks, 10 feet at the stanches (when closed) and 34 feet in the river.[55] The profile in Figure 2.7 shows the division of the fall between locks and river in a typical length of the western navigation.

Our story ends in the years around 1760 with the beginning of the Canal Age and the emergence of a civil engineering profession under the leadership of John Smeaton. In the preceding period the river engineers were as important as those engaged on harbour works, fen drainage and bridge construction, and left as their memorial many hundreds of miles of navigable inland waterways together with much experience of benefit to the next generation.

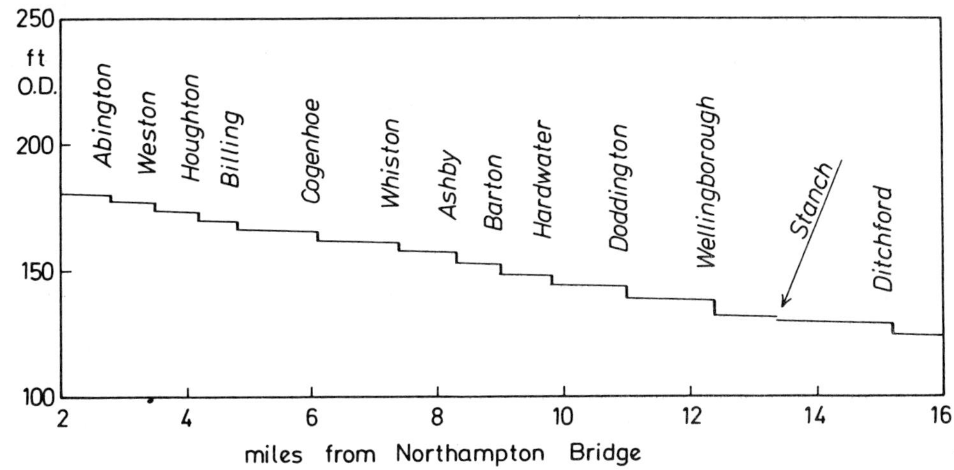

Profile of Part of the Nene Navigation

Figure 2.7.

NOTES AND REFERENCES

1. Memoir of the Canal of Exeter, from 1563 to 1724. By P. C. De la Garde; with a Continuation by James Green. *Min. Proc. Inst. C.E.* 4. (1845). 90–110. STEPHENS, W. B. The Exeter lighter canal. *Journ. Transport History.* 3. (1953). 1–11. CLARK, E. A. G. *The ports of the Exe Estuary.* 1960. 28–40.
2. Bodleian Library: Twyne-Langbaine MS, I, f. 76. The drawings and notes are by Brian Twyne of Oxford University who visited Exeter Canal about 1630 when work was starting on the Thames locks at Iffley and Sandford. PRICHARD, M., & CARPENTER, H. *A Thames Companion.* 1981. 78–80.
3. FAIRCLOUGH, K. R. The Waltham pound lock. *History of Technology.* **4.** (1979). 31–44. Some corrected dates have kindly been supplied by the author in a personal communication.
4. SALISBURY, C. R. Documentary evidence for an Elizabethan pound lock on the river Trent in Nottinghamshire. *Trans. Thoroton Soc.* **86.** (1982). 115–117. Dr. Salisbury reproduces the relevant portion of the 1592 plan.
5. THACKER, F. S. *The Thames Highway; general history.* 1914. 54–55.
6. The original is at Abingdon in the Council Chamber. Photocopies are in BM: Maps 188. a (10) and RGS. S. 242.
7. PLOT, R. *The natural history of Oxford-Shire.* 1677. 232–233.
8. Reproduced in THACKER (*op. cit.*) frontispiece, and opposite p. 66 of HADFIELD, C., & SKEMPTON, A. W. *William Jessop, Engineer.* 1979. The sketch is by John Call, one of the Thames Commissioners.
9. Quotations from preamble to the 1624 and 1695 Acts for Thames navigation.
10. Toll lists for 1731 etc. in *Extracts from the Navigation Rolls of the River Thames and Isis.* 1772. Dimensions of the flash locks and adjacent weirs are given in the printed reports (1791 and 1802) by Robert Mylne on the upper Thames. It is unlikely that the locks had changed in any essentials since the 17th century.
11. WILLAN, T. S. *The navigation of the Great Ouse Between St Ives and Bedford in the seventeenth century.* Bedfordshire Historical Record Soc. Vol. 24, 1946.
12. SUMMERS, Dorothy. *The Great Ouse.* 1973. Plan opposite p. 53.
13. For information on falls of the Ouse and other rivers quoted in this chapter, I am indebted to Miss Elizabeth Shaw of Imperial College who obtained the levels from various river authorities on my behalf.
14. THACKER, *op. cit.* pp. 65–83. See also the same author's *Thames Highway: locks and weirs.* 1920. 125–152.
15. HADFIELD, C., & NORRIS, J. *Waterways to Stratford.* 1962. 15–24.
16. LEWIS, M. J. T., SLATCHER, W. N., & JARVIS, P. N. Flashlocks on English waterways. *Industrial Archaeology.* **6.** (1969). 209–253.
17. MANNING, O., & BRAY, W. *History and antiquities of Surrey*, Vol. 3, 1814. Appendix 3, 54–58.
18. *A Plan of the River Wey. Surveyed by order of the Proprietors 1782.* Engraved map (scale 2 inches to a mile, widths of river and cuts shown at 1 inch to 8 chains). BM: C. 25 d. 2 (189).
19. *Lengths and levels to Bradshaw's maps of canals, navigable rivers, and railways.* 1832. 15.
20. References as in note 1. De la Garde gives a transcipt of the 1699 estimates, the relevant items in which may be summarised as follows:

	£
Compensation and purchase of land	700
Deepening, widening and banking the canal	1,462
Diverting Alphington Brook	150
4 swing bridges and 2 footbridges	200
Works at Exeter Quay	455
2 new cranes for the Quay	100
Kings Arm Sluice	800
New Lock	1,563
Repairing Trenchard's Sluice	55
Dredging the river	100
	£5,585

The estimates quote a sum of £605, additional to the figure given above, for walling the lock chamber but this was not carried out. The engineer engaged in 1698, William Baily, absconded in May 1699 before much work had been done. The appointment and salary of his successor Daniel Dunnel are noted by WILLAN, T. S. *River Navigation in England*. 1936. 81.

21. WILLAN, T. S. *Navigation of the Great Ouse*. (*op. cit.*).
22. The paddle and rimer stanch at Tempsford and its mode of operation are described in PROVIS, W. A. On the locks commonly used for river and canal navigations. *Trans. Inst. C.E.* 1. (1836). 53-54.
23. UNWIN, R. W. The Aire and Calder Navigation. *Bradford Antiquary*. **42**. (1964). 53-85; and **43**. (1967). 151-186.
24. The report, by John Clements and John Bronwell, is transcribed on pp. 82-85 of Unwin's 1964 paper.
25. Smeaton's plan of the three locks on the Aire is in his collection of drawings in the Royal Society, vol. 6, fo. 18.
26. Smeaton, in a report dated 14 September 1767, printed in *Reports of the late John Smeaton*, vol. 1. 1812. 157.
27. CUBITT, W. (in *Min. Proc. Inst. C.E.* **4**. (1845). 111-112) describes the stanches on the Stour as they were before improvements of the navigation in the 1830s.
28. THACKER, F. S. *Kennet Country*. 1932. An 18th-century MS plan of the navigation (scale 3 inches to a mile) shows the cuts, locks (with their falls) and bridges. Inst. Civ. Eng. library. Page Collection: f. 15.
29. *Journ. House of Commons*. 20. 288. 7 March 1726/7.
30. DUCKHAM, B. F. The Fitzwilliams and the navigation of the Yorkshire Derwent. *Northern History*. 2. (1967). 45-61.
31. CLEW, K. R. *The Kennet & Avon Canal*. 1968. 15-18, and extracts from the Minute Books of the Proprietors, kindly supplied by Mrs. Brenda Buchanan.
32. *An Actual Survey of the City of Bath, in the County of Somerset, and of Five Miles Round.* Survey'd by Tho: Thorpe in the Year 1742. Engraved by James Cole (scale 4 inches to a mile). BM: K.37 (9).
33. *A Map of the Rivers Mersey and Irwell from Bank-Key to Manchester with an Account of the rising of the Water and how many Locks it will require to make it Navigable.* Survey'd . . . by Tho: Steers 1712. I. Senex sc. (scale 3 inches to 5 miles). Chetham's Library: Manchester Scrap-Book, item 16. Bodleian: Gough Maps 44 fo. 127.
34. WILLAN, T. S. *The early history of the Don Navigation*. 1965.
35. *A Survey of the River Dunn in order to improve the Navigation from Hull to Doncaster and to continue up to Sheffield.* By William Palmer & Partners taken Anno. 1722. E. Bowen sc. (scale 1 inch to 1½ miles). BM: 1145 (1). Bodleian: (E) c. 17 (451) [30] with accompanying report printed 1723.
36. Smeaton's report on Linton dam (June 1767) in *Reports*, vol. 1, 311.
37. DUCKHAM, B. F. *The Yorkshire Ouse*. 1967.
38. *A Map of the River Ouse . . . Surveyed by Willm. Palmer & Partners in December 1725* (scale 1 inch to 2 miles). BM: K.6 (7). Bodleian: Gough Maps 41K fo. 32.
39. WILLAN. T. S. *The navigation of the River Weaver in the eighteenth century*. 1951.
40. *A Plan of the River Lee from Hartford to Bow Bridge with a Profile of the Fall* (by William Whittenbury, 1740). Scale 1.33 inches to a mile. BM; K.6 (6). Royal Society: Smeaton **6** fo. 87.
41. BOYES, J. *Canals of Eastern England*. 1977. 19.
42. Smeaton's report on the mills at Waltham Abbey, 16 May 1767. *Reports*, vol. 1. 279.
43. County Record Office Cambridge (CROC): R59/31/40/61(i).
44. Bedford Level Corporation records. CROC: R59/31/7 (Cole's journals), 10 (Conservators' Proceedings), 11 (Corporation Order Books), 19 (Account Books). The 1747 plan is R59/31/40/61(ii).
45. Dates, etc. of the late 17th-century structures are derived from R59/31/10, 11 and 19.
46. LABELYE, C. *The result of a view of the Great Level of the Fens*. 1745. His drawings, dated August 1745, are R59/31/40/61(iii) and (iv).

47. Rennie's sluices and lock were built in 1832-35 on the east side of the 1750 sluices with a sill level 6.0 feet lower. His drawings and reports are in the Institution of Civil Engineers library.
48. The original contract is in York City Archives: Acc. 203. I am indebted to Mr. M. F. Barbey for a summary of its contents.
49. Hull University Library: DDPA 7/721 and 722.
50. Accounts Book of the Ouse Navigation Trustees 1724-1834, also in York City Archives. Payments, probably for materials, begin in November 1752. Regular monthly or more frequent payments start in February 1754 and continue to November 1757, with a final settlement in 1758.
51. Details of Naburn lock and dam are chiefly taken from two reports by Thomas Rhodes dated 18 January 1834 and 13 December 1838, both reprinted with a map and section of the lock in *Reports and Plans of the River Ouse Navigation 1834 to 1881* (1881), and his lithographed *Longitudinal Section of the River Ouse from Linton to Selby* (1834) (scale 3 inches to a mile and 1 inch to 20 ft.) in Humberside Record Office: DDMW 7/390. Further details including tide levels and site plans are given in reports by John Coode (18 October 1876) and W. H. Bartholomew (6 May 1880), also reprinted in *Reports and Plans*. A second, larger lock was built alongside the original one in 1888.
52. RENTON, M. On the improvement of navigable rivers, with a description of a self-acting wasteboard at Naburn Lock. *Min. Proc. Inst. C.E.* 1. (1840). 26-27. Renton was Rhodes' assistant.
53. Yeoman's *Plan of the River Nen, from Thrapston to Northampton with the Mills and Locks necessary for the Navigation* (1754) is reproduced in HATLEY, V. A. *Northamptonshire past & present.* 6. (1981). 216.
54. This account of the Nene navigation is based on the Minutes of the Commissioners for the Western Division (1756 ff.) examined at Oundle by courtesy of Mr. G. E. Bowyer, Chief Engineer Welland and Nene River Authority.
55. From levels taken in 1931-32, kindly supplied by Mr. Bowyer.

ADDITIONAL NOTES AND REFERENCES

1. Plate 3 shows Santon stanch on the Little Ouse with its two adjacent sluices and 'overfall'. A note should be added, from Clarke R. H. *Trans. Newcomen Soc. 30* (1955–57) 207–216, that it was built in 1831 by Burrell & Son, engineers of Thetford, as part of a reconstruction (1825–37) of the severn stanches on this navigation originally built under an Act of 1751.
2. Fairclough, D. 'Navigation devices along the River Lea 1600–1767'. *Trans. Newcomen Soc. 64* (1992–93) 21–40. This adds much information to the short account given on p. 37.
3. Jones, P. 'Vermuyden's navigation works on the river Don'. *Journ. Railway & Canal Hist. Soc. 31* (1994) 248–258. From a recently discovered annotated plan of 1633 it is now known that a pound lock (sasse) was built at Turnbridge on the Don and another at Misterton on the Idle, in connection with the drainage of Hatfield Chase by Sir Cornelius Vermuyden (c. 1590–1677). Both had chambers 60ft long by 18ft wide with gates of the guillotine type. The latter dates from 1630, the former probably a year earlier. Moreover, adjacent to Turnbridge sasse was an outfall sluice with 17 openings of 6ft each, seemingly built under the direction of Vermuyden's colleague Hugo Spiering in 1631. Its clear waterway exceeded that of Denver Sluice (see p. 37) but the doors were only 8ft high.

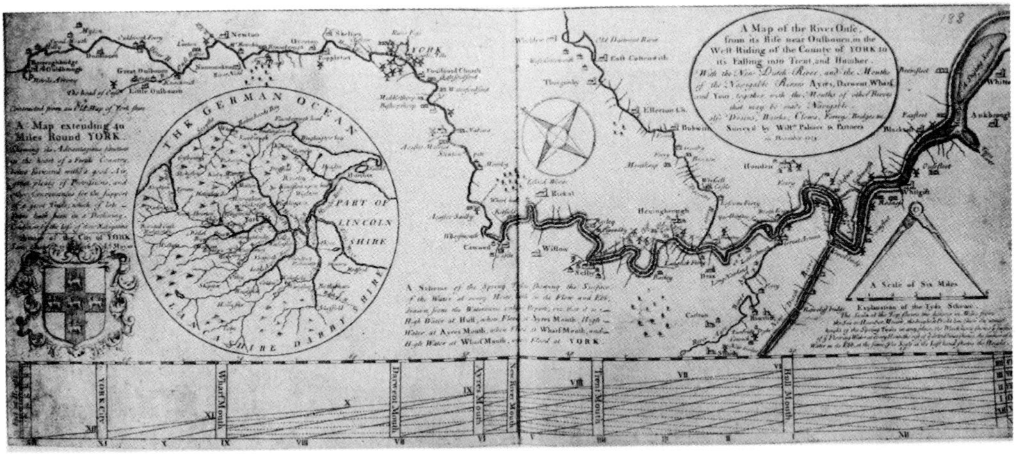

2. Part of William Palmer's map of the Yorkshire Ouse, 1725.

3. Stanch on the Little Ouse at Santon: painting attributed to F. W. Watts (early 19th century).

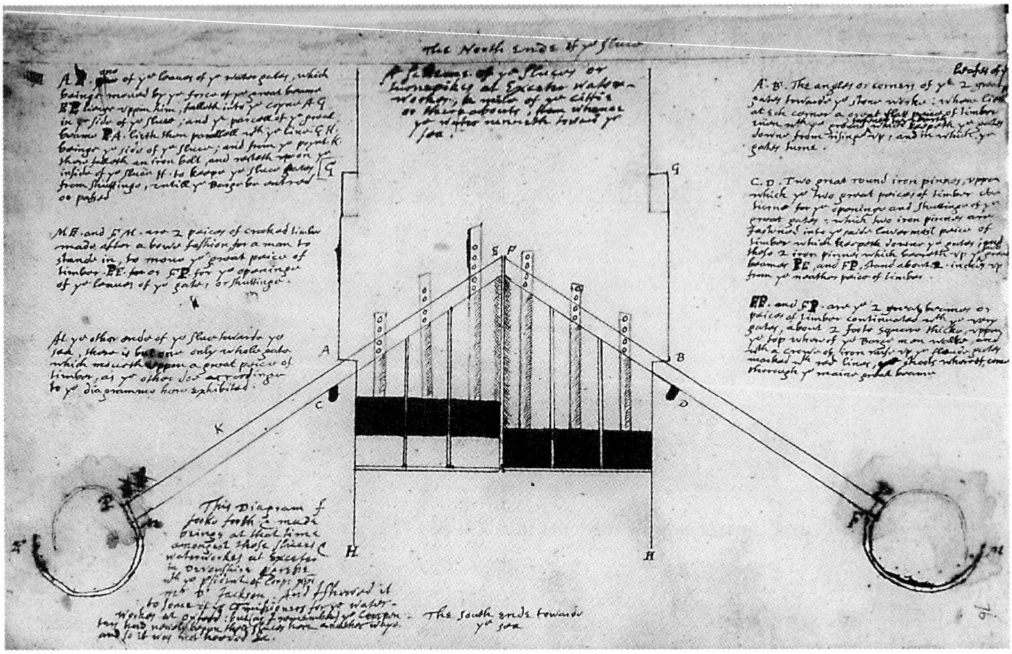

4. (*above*) Exeter Canal lock gates: drawing of c.1630.

5. (*below*) Denver Sluice, as completed 1750: drawing of c.1810.

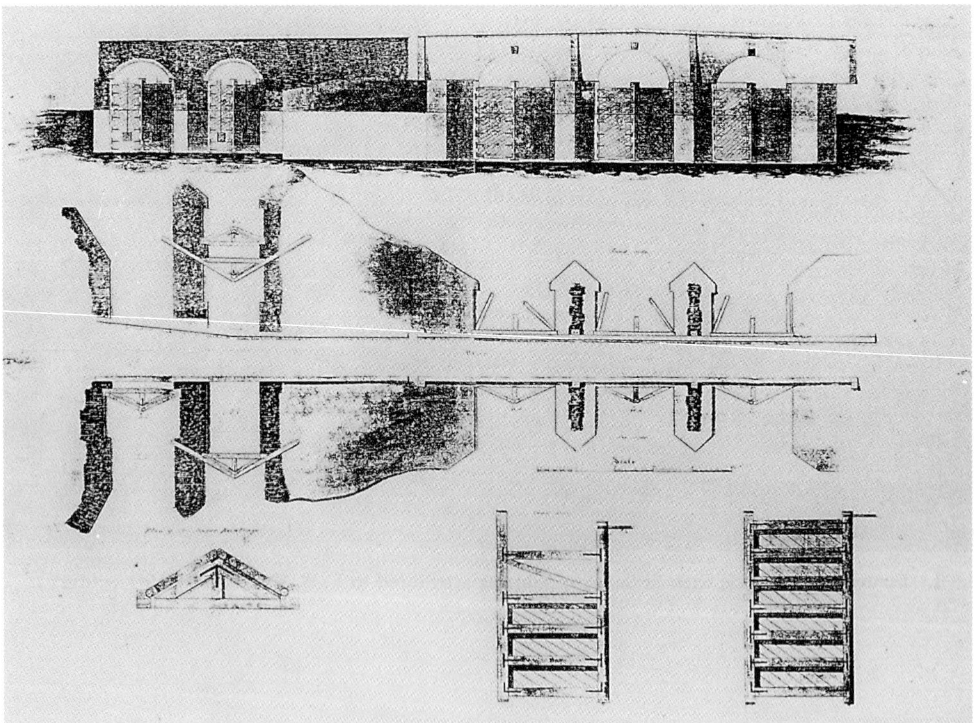

VIII

Engineering on the Thames Navigation, 1770–1845

1. INTRODUCTION

By the mid-eighteenth century the Thames had a long history as the most important commercial inland waterway in England, navigable by barges carrying 150 tons as far as Abingdon, and 60 or 70 tons to Oxford and Lechlade. At least 100,000 tons left the London wharves each year; 60,000 tons passed through Boulter's lock at Maidenhead, about 20,000 tons went upstream of Reading, and similar tonnages came back down the river.

Despite this large amount of traffic navigation was not easy. There were no locks below Maidenhead. Navigation from thence to Abingdon depended upon a series of 18 flash locks mostly, but not all, located in weirs at watermills. Then came three pound locks, built in the 1630s, and from Oxford to Lechlade the barges navigated by means of some 20 flash locks purpose-built in the early years of the seventeenth century. Much of the hauling had to be done by large gangs of men, as a horse towing path existed along only part of the river bank in disconnected lengths. There were troublesome shoals or 'gulls' in many places, and to pass upwards through the flash locks in most of the bigger weirs downstream of Abingdon the barges had to be winched against the current.

It is not surprising therefore that complaints were frequent. The navigation from London Bridge to Staines came under the jurisdiction of the Corporation of London, the 'City', and from thence up to Lechlade (except for a length between Burcot and Oxford) under the Thames Navigation Commission. But these authoritities exerted no more than an administrative function, regulating lock tolls, use of the towpath and so on, and requiring the private owners of the locks, weirs and river banks to maintain their property in good order.

At last, in despair of river improvements, the town of Reading in January 1770 called on James Brindley to plan a canal, by-passing the river, from Sonning to a point known as Monkey Island two miles below Maidenhead. Two months later the City asked him to extend the survey from Monkey Island to Isleworth. Brindley reported in June, but next month the City had further thoughts and ordered a survey of the river itself from Boulter's lock to Isleworth with a view to seeing what could be done to improve this part of the navigation. Brindley's second report, presented in December, was accompanied by a fine plan and longitudinal profile of the river below Boulter's lock to Isleworth surveyed by his assistant, Robert Whitworth.

Meanwhile the Thames Commission employed Joseph Nickalls to draw up a comprehensive scheme for improving the river from Richmond all the way up to Abingdon.

As a result of this activity, petitions were submitted to Parliament by Reading in November 1770, by the Commission in December, and by the City in January 1771. Whitworth and Nickalls gave evidence, and the upshot was that the Commission obtained an Act of Parliament in April 1771, granting powers to build pound locks between Lechlade and Maidenhead (but not further downstream), and to dredge the navigation channel and extend the towpath over the full length. To carry out these works the Commission also obtained power under the Act to raise loans, the interest on which was to be paid by lock tolls.

Work began almost immediately, under Nickalls' direction, and three years later the City responded by starting improvements in their part of the navigation, with Whitworth as engineer.

Pound locks built on the Thames during the period up to 1845 are tabulated in geographical and chronological order at the end of this paper and shown in the sketch map, Fig. 15 (page 173). For administrative reasons the navigation was divided into six districts. These are indicated in the table and map.

The natural gradient of the river in each district varies by only a few inches from the average value of 20 inches per mile throughout the entire length between Lechlade and Teddington.

The flow (in cusecs) as gauged at Teddington is as follows:

average	2,500
'low water' in summer	900–1,300
'high water' in winter	4,000–6,000
typical flood	18,000–25,000
exceptional flood	*c.* 40,000

Typical floods occur on average every ten to twenty years, examples in this century being those of 1915, 1925, 1929, 1947 and 1968. The highest recorded floods in the past 200 years were in March 1774, January 1809, December 1826 and November 1894. In the latter, the river rose to the following heights above normal head water level: at Iffley 3 ft, Caversham 5 ft, Sunbury 7 ft, and at Teddington 6 ft.

2. THE OLD WEIRS

As most of the old weirs remained in use with little modification up to the mid-nineteenth century it is relevant to give a brief description of them.

An example of one of the small weirs across the river below Lechlade is shown in Fig. 1. It consists of a sluice and an adjoining flash lock. The sluice has six 'flood gates' supported by four main posts and three smaller posts or 'rimers', and a fixed cross beam. The flood gates can be lifted or 'drawn' to various heights above a sill beam in the river bed, probably using a handspike, or they and the rimers can be removed altogether. The lock keeper is drawing the 'paddles' of the flash lock. Narrower than the flood gates, the paddles are supported by a sill beam and an upper beam, without rimers. After the paddles are removed, the beam is withdrawn and a pivoted balance beam (on which tne keeper is standing) swung aside. The waiting barge then passes through, and the lock is closed again. In the 6th District the flash locks were about 15 ft wide, so the whole weir in Fig. 1 is not more than 30 or 35 ft wide; nearer Oxford the width could approach 50 ft. Throughout this District the fall at few of the weirs exceeded 2 ft.

The mill weirs between Shillingford and Maidenhead were very different. Fig. 2 shows the layout at Whitchurch. The main body of the weir is a broad, solid dam rising above normal high water level. Apart from water going through the mill race the river flows over the aptly named 'tumbling bay', in effect a fixed overfall weir about 35 ft wide with its crest just below low water level, and through the sluice. The latter, about 30 ft wide, is similar in construction to the sluice in Fig. 1 but with 'overfall boards' above the flood gates to accommodate a fall of 3 or 4 ft. The flash lock, again similar to that depicted in Fig. 1 but with overfalls as well as paddles, and with rimers, is 23 ft wide; seemingly a standard dimension in all the weirs downstream of Abingdon.

The sluices were operated to ensure that the upstream water level did not sink below a specified low water mark in summer or rise beyond a high water mark in normal winter months: usually 12 to 18 inches above low water mark. The resulting gradients in a typical longitudinal profile are indicated in Fig. 3 (page 156). There is little doubt that in floods the whole weir would be submerged.

The total length of these big weirs in some cases exceeded 400 ft and none were less than 200 ft long. In position, they have scarcely changed during the several centuries of their existence. But before 1840 (though at dates I cannot quote) some of the tumbling bays were widened, and by the late nineteenth century practically the entire length, except for the sluice, had been transformed into an overfall weir; the arrangement which we take for granted today. The first weirs to be built *ab initio* in this form cn the Thames were those in the 1st District in the period 1810–15.

ENGINEERING ON THE THAMES NAVIGATION, 1770–1845

Fig. 1. Weir on the upper Thames. Engraving from H. H. Robertson, *Life on the Upper Thames,* 1875. Little change in weirs of this type had been made since the 17th Century.

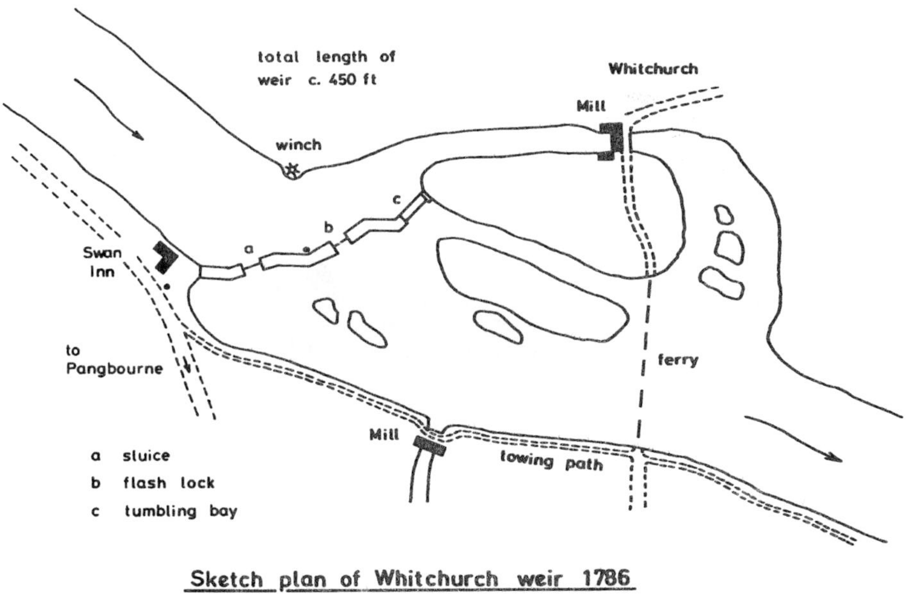

Fig. 2. Sketch plan of Whitchurch weir, 1786. From a drawing by John Call made in April 1786.

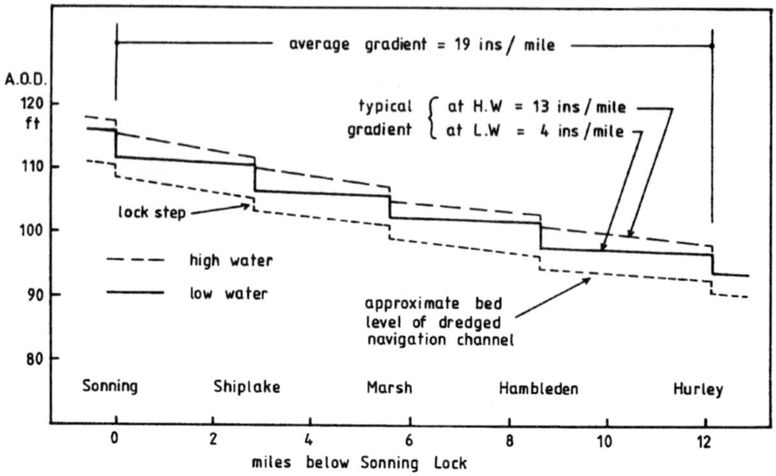

Fig. 3. Typical longitudinal profile of part of the Thames navigation.

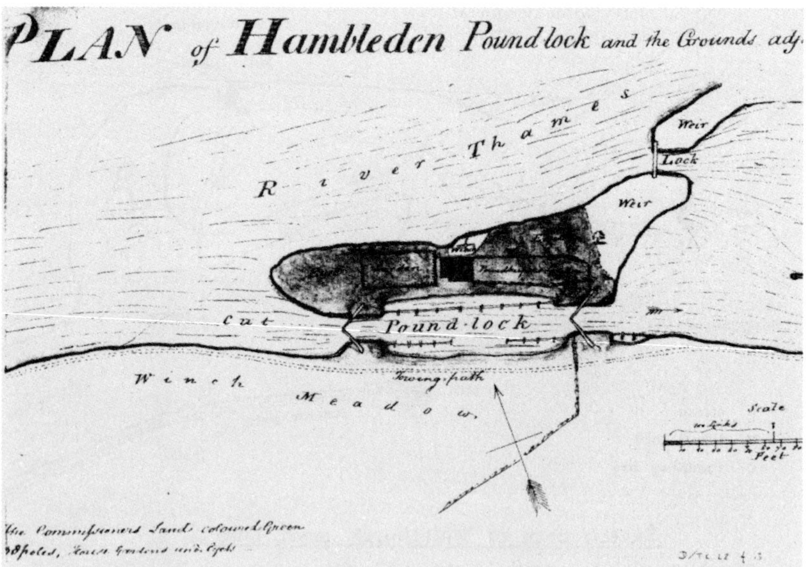

Fig. 4. Plan of Hambleden pound lock. Drawing by Zachary Allnutt, 1815.

ENGINEERING ON THE THAMES NAVIGATION, 1770–1845

Stage I

3. 3RD DISTRICT, 1771–1778

Nickalls was appointed General Surveyor (i.e. Engineer) to the Commission in May 1771 at a salary of £200 p.a. with Thomas Hayward as inspector of works in the 2nd District (15/- per week) and Thomas Withrington as assistant engineer in the 3rd District (1 gn. per week). By the end of June agreement had been reached on the siting and design of the pound locks to be built at Boulter's, Marlow, Temple, Hurley, and Hambleden. Plans for the locks at Shiplake and Sonning followed about a month later and, after some hesitation, at Marsh lock in October. Contracts for each lock were let within a few weeks of the plan being agreed; construction to be in accordance with a model by Nickalls.

The first five contracts, in upstream order, went to Joseph Searle, Thomas Rickett, William Skeel of Bisham (millwright), Edward Edgerly (carpenter) and William Bradshaw of Henley (builder). For the next three locks at Marsh, Shiplake and Sonning the contractors were Bradshaw, Edgerly and John Smith, the latter shortly being joined by John Treacher of Sonning (carpenter). By mid-November Nickalls could report good progress on the first five locks; at Marsh he had recently marked out the line for the lock, and pile engines were being prepared; at Shiplake excavation was nearly finished and the pile engines in a proper state for working, while at Sonning 'the work goes on well' after some initial delays. Just a year later, in December 1772, Boulter's lock was opened; the other seven followed in January–April 1773.

In each case the existing weir was retained and the pound lock built in a new cut on one or other bank of the river, as shown in Fig. 4. The navigation channel had to be deepened by dredging (or 'ballasting' in Thames terminology) for perhaps half a mile below a lock cut and usually for a few hundred yards upstream.

The first eight locks, and indeed all those up to and including Day's lock, near Dorchester, had timber-framed abutments for the upper and lower pairs of gates, and most of them were of the type known on the Thames as 'open-sided'. That is, the sides of the lock chamber were planked up to tail water level but consisted of a sloping earth face above. Guide piles and railings kept the barge in the centre of the lock (Fig. 5). Where a restricted site did not permit open sides, timber walls would be carried up to coping level, with land ties to increase the resistance to lateral earth pressure. An example of these 'close-sided' locks was at Marlow, where the land ties are known to have been 22 ft long.

Fig. 5. Shiplake pound lock. Engraving after S. Owen, 1810. From F. Thacker, *Thames Highway,* Vol. 1.

Open-sided locks, well known already on the Wey, Kennet, and Leae navigations, provided a low-cost type of structure but took longer to fill and empty than locks with closed sides. An intermediate form therefore came to be used with the walls rising just above head water level, the earth slope being entirely in the dry. Also known as 'close-sided', locks of this form (with either timber or masonry walls) became standard on the Thames, whenever site conditions were suitable, from about 1790 and most of the early locks were eventually rebuilt in this fashion. A section is shown in Fig. 6, details of the land ties being taken from a sketch by William Jessop of *c.* 1786.

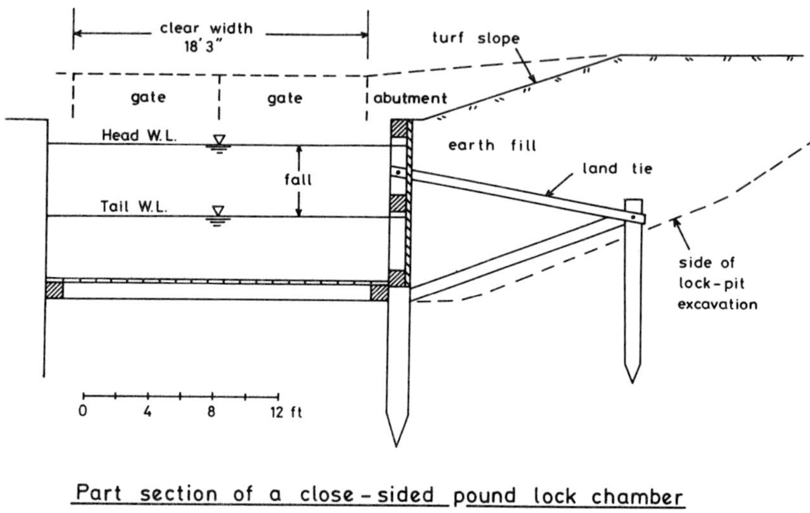

Fig. 6. Part section of a close-sided pound lock chamber.

Piles and planking below water level could be and usually were in fir or 'Memel' timber. But to save as much money as possible, Nickalls used fir above the water line in all or most of the eight locks. It proved to be a false economy, as the upper parts had to be replaced by oak within 10 or 15 years.

But the locks served their immediate purpose of rapidly establishing a much improved navigation up to Reading. They had a length of 133 ft between gates and a clear width of 18 ft 3 in with a depth of about 3½ ft on the sills at low water.

Ballasting operations in the river bed, generally as well as near the lock cuts, were carried out in both the 2nd and 3rd Districts; using the 'bag and spoon' technique with two men in a 6 ton punt equipped with a windlass (Fig 7). The gravel so raised would usually be placed to raise and improve, or in some cases to create, the horse towing path. For routine work one or two ballast boats were used, but in heavy operations as many as six boats would be employed in a comparatively short length of river.

The first ballasting contract went to Henry Paterson of Reading, in July 1771, and it is known that in 1773 Smith and Treacher were at work in the Marlow and Shiplake areas building towpaths, horse bridges and gates and raising about 120 tons of gravel per week at each site from the river bed.

In August 1773 John Clarke of Marlow was appointed inspector of works in the 3rd District, presumably replacing Withrington after the locks had been built. By 1776 the Commission had borrowed, and largely spent, £26,000 for work in the 2nd and 3rd Districts and were about to raise a further £10,000 for the 4th District. But Nickalls left in that year, it being considered that recently he had been spending too much time on other jobs. Thomas Yeoman took his place as General Surveyor for two years on a free-lance or consulting basis.

ENGINEERING ON THE THAMES NAVIGATION, 1770–1845

Fig. 7. Ballasting. Engraving from H. H. Robertson, *Life on the Upper Thames,* 1875. This method of dredging by the 'bag and spoon' technique dates back to the 18th Century or even earlier.

It was to Yeoman's plans that the next two locks, at Caversham and Mapledurham, were built. Opened in May 1778 and July 1777 respectively, in each case construction took almost exactly a year from the date when the plans were approved. They had the same width of 18 ft 3 in as the locks below Reading, but were not quite so long between gates; something like 125 ft seems to have been the ruling dimension between Reading and Abingdon. Smith and Treacher contracted for Caversham lock and may also have built the lock at Mapledurham.

4. 1ST DISTRICT, 1774–1778

In his second report of 1770, on the lower Thames, Brindley remarked that the most effectual way of improving the navigation would be by making weirs and pound locks, penning not more than 5 or 6 ft to avoid flooding the adjacent land, but the expense will be very great for so large a river; and greater than the by-pass canal which he preferred. Nevertheless the river can be made much better than it is, he said, by contracting the channel in the shallow and broad places, most of which were marked in the plan accompanying his report, and £10,000 could usefully be laid out for such works.

The City coupled their application to Parliament for the Maidenhead–Isleworth canal with improvements in the river but, having had their Bill rejected, did nothing until the great flood of March 1774 compelled some action. The flood, one of the largest ever known on the Thames, caused much damage to the banks and towpaths, raised new shoals in several places, and threatened to cut off a loop in the river near Weybridge. A decision was then quickly taken to spend £10,000, as a gift, in the 1st District and to employ Yeoman and Whitworth to make detailed proposals. They reported in June; a week or two later Whitworth was appointed Surveyor (Engineer) at a salary of £250 p.a. and shortly afterwards Charles Truss became Clerk of Works at £100 p.a.

Before the end of July contracts had been made with Thomas Reeve and Edward Cox (separately) for ballasting, at 11d per ton of gravel raised and laid on the towing path, and with Edward Edgerly for all carpentry of sheet piling, towpath bridges, etc. Work began in August 1774. Whitworth submitted regular progress reports and Truss kept meticulous accounts. Already by October 1776 most of the essential new works in the river had been finished, and about £8000 spent. But much remained to be done on the towing path and this, together with a few more river works, Whitworth estimated to cost about another £8000. In addition, future maintenance of the order of £500 p.a. had to be taken into reckoning.

Preparations were therefore made for an Act enabling the City itself to charge towpath tolls. This passed in April 1777, at which time Whitworth published an engraved plan of the river with a summary of the works carried out during the past three years.

By the end of 1778, after an expenditure totalling just over £16,000, the channel improvements had been practically completed, the banks had been stabilised in many critical places, and a horse towing path now ran along the whole length of 30 miles from Putney to Staines.

As for the new works, the first job was to stem the breach near Weybridge. Here the flood had enlarged a small back stream known as Stoner's Gut and the river was rapidly eroding a new channel, with a 16 ft depth of water. As a result Weybridge and the river Wey navigation to Guildford were in danger of being isolated. By the end of October 1774 a dam had been built across the breach, 230 ft in length composed of a double row of sheet piles 15 ft apart and up to 27 ft deep, the space between being filled with earth. The piling was supported by vertical timber piles, walings and iron tie-bars, and further supported on each side by sloping gravel fill.

To prevent a future breach in a rather similar situation across the narrow neck of land at Penton Hook, the river bank was protected by a 'campshot' of sheet piling (held by land ties) 170 ft long and from 13 to 17 ft deep; completed in December 1774. This was the deepest campshot used, but several miles of river bank, in total length, were subsequently protected from erosion in such a manner.

The principal task, however, was to deepen the navigation channel, or to create a new channel, by ballasting. In all, about 90,000 tons of gravel were raised and mostly placed on the towing path. In some places the towpath was built up in the river itself a short distance out from the existing bank, the space between being filled in to ground level. The path everywhere was of course raised well above high water mark, and dozens of timber bridges were built to carry it over streams and creeks.

At Teddington ferry and upstream of Sunbury the river was effectively narrowed by constructing groynes or 'stops'; three at each location. They consisted of a single run of grooved sheet piling driven 5 ft into the river bed with deeper supporting piles at 8 ft centres and a capping beam at low water level. The arrangement at Teddington is shown in Fig. 8, the river here having a width of 250 ft. These

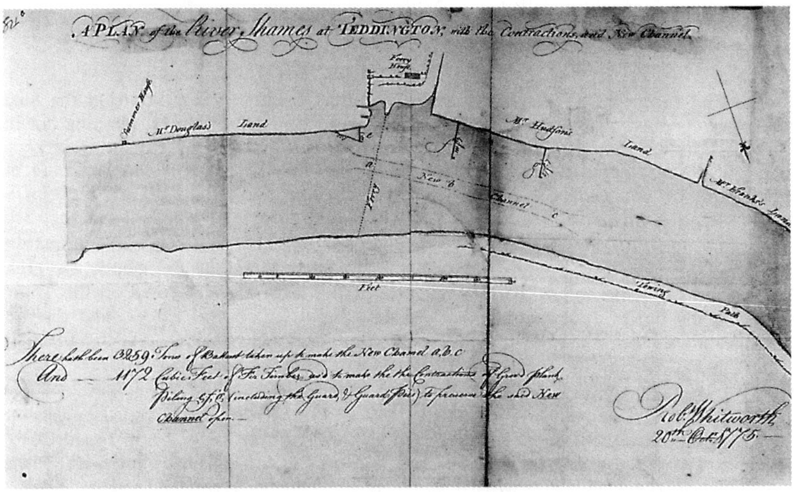

Fig. 8. Plan of the Thames at Teddington showing three groynes or 'stops'. Drawing by Robert Whitworth, 1775.

stops date from 1775, those at Sunbury from 1776. Before the latter were built the channel had been ballasted, but it choked up from want of an adequate stream velocity. The stops at Teddington also proved successful in keeping the channel open, and a few years later similar stops were built with good effect at Laleham.

Another interesting work was the dam built in 1775–76 across the head of a long back-stream about ¾ mile above Sunbury (see Fig. 11, page 167). This deflected the flow down the main channel and raised low water level upstream by several inches: enough to make the difference between a barge being grounded or not. These works and others on similar lines executed somewhat later were much approved by Rennie in 1794.

Stage II

5. 4TH AND 5TH DISTRICTS, 1786–1791

In 1778 John Clarke took over full responsibility for work in the 2nd and 3rd Districts, and from about 1780 to March 1787 he acted as Surveyor to the whole river under the Commission. During this period he had charge of rebuilding five of the original 3rd District pound locks in oak and one, at Hurley, with brick abutments. These works were carried out by contractors. But the next pound lock, at Whitchurch, was the first to be built by direct labour, under the Surveyor's direction and to his plans; with a small management committee appointed by the Commission from among its more active and knowledgeable members.

The Whitchurch committee, formed in April 1786, reported on three possible sites for the pound lock with a sketch plan by John Call, one of the committee: see Fig. 2 (page 155). William Jessop's opinion was sought. He reported in May, confirming the committee's choice of site (between the two islands) and indicating what minor alterations would be required in the weir.

Clarke went to London to order oak and some other supplies; more timber was ordered from a merchant at Pangbourne; William White was appointed foreman at 3/6 per day, and a specialist contractor provided the necessary ironwork.

Work began in August 1786 and proceeded without a hitch. In June 1787 Clarke had two ballast boats dredging the approach channels and was "filling up and making good the earth at the sides of the pound". The lock, opened on 30 June, cost about £1500 including modification to the weir.

Meanwhile, Clarke had been preparing plans and estimates for the next locks at Goring and Cleeve, and supervising repairs at Hurley and Marsh, and was about to start repairs at Shiplake. He also recommended, in February 1787, wider tumbling bays at Shiplake (40 ft), Marsh (50 ft) and Hambleden (60 ft). Clearly, he could not continue to superintend maintenance and improvements in the 2nd and 3rd Districts while undertaking new works many miles further upstream, so in March 1787 John Treacher was appointed engineer in charge of lock construction at Goring and Cleeve.

Within a week of his appointment, Treacher received orders to make four pile engines, to procure timber, tools and sufficient workmen, and to bring a ballast boat from Whitchurch. He soon had about 15 men and 4 carpenters working at each site, and other carpenters making the gates. Built in oak, both locks were opened in December 1787 at a total cost of £2200.

Treacher now (in 1788) became Surveyor to the 4th, 5th and 6th Districts. In March 1791 he was appointed General Surveyor at £130 p.a. with special responsibility for these upper Districts; Clarke to remain as Surveyor to the 2nd and 3rd Districts, with a salary increased to 2 gns. per week.

It was under Treacher's direction, then, that the next two pound locks were built: Benson and Day's. Work on both locks proceeded simultaneously, beginning early in 1788 and reaching completion in December. Up to seven or eight carpenters were employed at each site, under Robert Treacher as foreman, and as many as 20 men. Each lock cost about £1100, but with the ancillary works of lock cuts and towpath the total expenditure came to £1400 at Benson and £1600 at Day's. In the following year a new tumbling bay was provided at Benson, three months work, and the towing path was extended from Day's up to Culham Bridge.

Attention then turned to the river at Sutton Courtenay and Abingdon. At Sutton an extraordinary arrangement had been in existence since the seventeenth century, if not earlier. The mill buildings spanned the river itself, with three side weirs in the river bank upstream discharging into the lasher pool north of the mill. Barges navigated through two flash locks, one just above the mill, the other at the lower end of the large mill pond, and they had to pass under the mill buildings. The two flash locks formed in effect a pound lock of vast dimensions, which took 45 minutes to fill or empty (compared to 5 minutes for a typical lock).

To secure some improvement here, the Commission came to an agreement with the mill owner to remove that part of the buildings over the navigation, to erect a bridge across the river, and replace each flash lock by a pair of pointing gates of the pound lock type. These works were completed by May 1791 under Treacher's direction.

Regarding navigation improvements in the vicinity of Abingdon, three alternatives had to be considered: to rebuild the seventeenth-century pound lock in Swift Ditch and upgrade this old channel, or to make a new cut about 900 yards long in the meadows east of Abingdon (Jessop's suggestion), or to build a pound lock adjacent to the existing weir and flash lock upstream of Abingdon with a good deal of dredging in the river past the town and an enlarged navigation arch in the bridge.

The last alternative was preferred by Treacher and William Vanderstegen, one of the most active Comissioners of his day. But only after a year of discussion and hesitation was it finally agreed in December 1789 to proceed with this plan.

Specifications appeared in March 1790 for the pound lock (to be walled in stone, 120 ft long between gates and 18 ft 2 in wide), for the bridge arch (to be widened to 20 ft and raised 4 ft at the crown) and for dredging, making a towpath, etc. The contractor Edward Edge began work shortly afterwards, and completed the lock in November of that year. But more haste less speed, as the walls had to be rebuilt partially in the summer of 1791. Dredging and towpath construction, by direct labour, and work on the bridge, with John Nock as contractor, proceeded simultaneously with building the lock. The lock cost £990, the whole job £1540.

By 1791 all the flash locks between Maidenhead and Oxford had been replaced by pound locks (with the partial exception of Sutton) apart from one left at Nuneham which, however, was used only as a stanch to impound an extra depth on Sandford tail sill in dry seasons.

6. 6TH DISTRICT, 1789–1792

That part of the Oxford Canal from Banbury to the Thames was opened in January 1790, and the first boat from the Thames and Severn Canal entered the river at Inglesham, a mile above Lechlade, in November 1789. Anticipation of the latter event prompted the Commission to carry out their improvements from Oxford to Lechlade while work was still in progress on the river further downstream. Indeed, a first step had been taken as early as 1787 in making a new navigation channel at Radcot and building an extension of the medieval Radcot Bridge over this new cut.

Work in the 6th District began in earnest, however, after Jessop reported in August 1789. He took accurate levels from Osney mill pond, at Oxford, up to Lechlade, examined the existing flash locks and weirs, and made specific recommendations for new works, the chief of which were six pound locks: at Osney, Godstone, Pinkhill, Rushey, Buscot and St. John's Bridge near Lechlade. Committees were formed to manage the upper and lower parts of the District, and as Treacher was so much occupied further downstream two engineers-in-charge were appointed: Daniel Harris of Oxford and Josiah Clowes the engineer of the Thames and Severn Canal.

Except for a change in location of Rushey lock, Jessop's proposals were followed closely. The locks, on new cuts, were all built with stone walls, a length of 110 ft between gates, and a clear width of 15 ft.

Clowes began work on the locks at St. John's and Buscot in October–November 1789 with John Nock of Daneway, a Thames and Severn mason, as contractor and William Milton of Kempsford executing the cuts. Robert Treacher acted as assistant under Clowes at 1½ gns. per week. Both locks were finished by November 1790. The decision to move the location of Rushey lock about half a mile downstream from Jessop's suggested position seems to have been made in the summer of 1790, but the Minute Book is silent on details. A pound keeper was appointed in January 1791; Robert Mylne in May 1791 refers to the cut as "lately made" and says "the new Pound Lock is well built". The works at St. John's and Buscot also met with his approbation.

Daniel Harris proved less capable. Osney lock, built between November 1789 and November 1790 by Edward Edge with some prisoners from Oxford Gaol as labourers, had to have its side walls rebuilt. Godstone was not much better, surprisingly in view of the fact that the experienced Samuel Weston was contractor. Work started there about August 1790; a year later the lock was built but the cut not yet finished, and it was only in November 1791 that Harris could report completion of both contracts. Plans for Pinkhill lock received approval in May 1791. It was satisfactorily built, with Nock as contractor, and opened before May 1792.

No further pound locks were built in the 6th District until those (Grafton, Radcot, Shifford and Northmoor) dating from the 1890s. The navigation from Lechlade to Oxford remained for a hundred years as shown in Fig. 15 (page 173).

* * * *

Between 1786 and 1792 inclusive the Commission had spent £28,600 on new works: £13,700 in the 6th District and the rest mainly in the 4th and 5th Districts under Treacher's direction. In round figures the tonnage passing upwards through Boulter's lock in 1792 amounted to 68,000 tons; about 36,000 tons reached Reading (half of which went up the Kennet navigation) and some 4000 tons were carried to Oxford and Lechlade.

Stage III

7. ROMNEY LOCK AND WEIR, 1796–1798

The 1771 Act permitted no locks or weirs across the river below Maidenhead. Considerable improvements were made in the navigation channel in the 1st District, as we have seen, and to a more limited extent in the 2nd District, by ballasting and the judicious use of 'stops' to narrow the river in places. But it was clearly recognised, if not by a majority of the Commission and the City Committee, that the navigation would remain imperfect until pound locks were built. The alternatives were by-pass canals, such as the Monkey Island–Isleworth scheme, or a series of long cuts. The latter, advocated by Rennie in 1794, would each have a lock, but no weir; the fall being that of the river itself between head and tail of the cut.

Not surprisingly the idea of long cuts, and the canal schemes, produced strong opposition from the riverside towns, while fears were expressed that weirs would worsen the flood problem in the lower Thames valley. However, pound locks on the river were absolutely necessary. John Clarke, for example, in evidence to a Parliamentary Committee in 1793, said that three were needed in the 2nd District: at Romney (half a mile below Windsor Bridge), near Boveney (two miles above the bridge) and near Staines, and these of course would require weirs across the river.

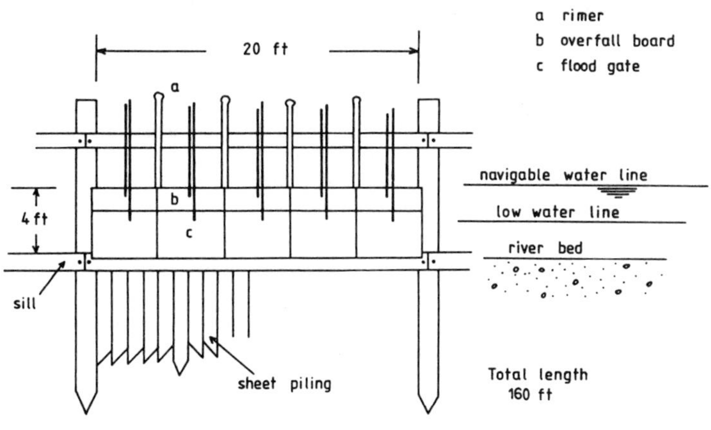

Sketch of one bay of Romney Weir, 1798

Fig. 9. Sketch of one bay of Romney weir, 1798. From a drawing by Zachary Allnutt, 1805.

The answer to this impasse lay in making the weirs either with a very long overfall (tumbling bay) or, in the limit, in such a manner that they offered virtually no obstruction to flow at times of high water and flood.

Weirs of the latter type, in effect, already existed on the river above Oxford, though on a small scale. Eventually, after much internal discussion, the Commission led by Mr. Vanderstegen decided to build one of these 'open' weirs at Romney; with the important proviso that it would be closed only in low water seasons. This concept, ably analysed by Frederick Page, proved to be the key to success in obtaining the necessary Act, in June 1795.

Detailed plans for Romney lock were prepared by Treacher in April 1796. Next month Joseph Kimber of Colnbrook, millwright, signed a contract to build it for £1505; he to find all tools, pumps, pile engines, etc. Clarke supervised the work, reporting completion in December, and ballasting was finished in April 1797. The lock was close-sided in oak. Construction of the weir followed in July–October 1798, under a separate contract with Kimber, at a cost of £590. It had a length of 160 ft, running straight across the river. Fig. 9 (page 163) shows one of the 'open' bays; together they added up to a length of 120 ft. All the flood gates, overfall boards and rimers could be drawn in three hours.

Romney lock and weir proved to be very successful. But for reasons that are not clear no others were built between Maidenhead and Staines until the period around 1820.

8. BLAKE'S LOCK, 1801–1802

We may briefly note Treacher's last work: Blake's pound lock on the Kennet at Reading. This replaced a flash lock used by Thames barges in getting to the Reading wharf, a short distance upstream, and also by barges on the Kennet navigation. Treacher submitted plans and estimates in February 1801. Contracts were made in April with Smith and Treacher (junior) for the lock, at £2110, and with Mordecai Brookes of Hungerford for excavating the 700 yards long head cut, at £680. Opened in the summer of 1802, the lock was close-sided in oak with dimensions of 134 ft by 18 ft 3 in and a fall of about 3 ft.

9. CHANGES IN STAFF, 1800–1810

In August 1800 James Millard was appointed Assistant Surveyor to the 5th and 6th Districts under Treacher. The latter died in January 1802, and his son John Treacher junior became Surveyor to the 4th District. A year later John Clarke died. His place as Surveyor to the 2nd and 3rd Districts was taken by Zachary Allnutt of Henley, son of Henry Allnutt the General Clerk to the Commission from 1771 to 1820. Millard left in 1803, and in February 1804 Treacher took over direction of all three upper Districts at a salary of £165 p.a. For the next seventeen years the Commission had no General Surveyor; Treacher and Allnutt shared the work between them while taking special responsibility for the upper and lower Districts respectively.

In the 1st District, under the City Navigation Committee, Whitworth left in 1784 and Charles Truss, the Clerk of Works, took command with Edward Robinson as assistant. Truss retired in January 1802 at the age of 82. Stephen Leach, having been his assistant since Robinson's death in 1801, now took on the job at £300 p.a. with full responsibility as engineer for the great series of locks and weirs about to be built from Teddington up to Penton Hook, and he continued to serve the Committee for 32 years.

10. CULHAM CUT AND LOCK, 1809–1818

Before his appointment Allnutt had done several surveying jobs for the Commission, one being for a long cut to bypass the unsatisfactory lock at Sutton mill, to which reference has already been made. This was in 1802, but a decision to go ahead with such an expensive and unprecedented work (for the Thames) did not follow until 1809, when Allnutt make a further and more detailed survey.

This 1400 yard cut remains the longest ever made on the Thames. Near its tail end a stone-walled pound lock was built, with the large fall of 7 ft, and a stone road-bridge of 20 ft span. Work began about June 1809 and reached completion just over a year later at a total cost, including land purchase, of nearly £9000.

No other details are available on this highly efficient operation except that it was carried out under Treacher's supervision, and that he received a gratuity of £100 in recognition of his services.

Meanwhile, numerous minor improvements had been made by the Commission, notably in lengthening the old pound locks at Sandford and Iffley from 90 ft to 120 ft, to bring them in line with Abingdon lock. In 1811 a committee of the Commission reported on the entire navigation from Lechlade to Staines, and were able to do so with considerable satisfaction apart from a few places such as Clifton and Cookham, where the need for new cuts was already recognised, and the 2nd District generally, where further pound locks (additional to the one at Romney) were clearly necessary.

Stage IV

11. 1ST DISTRICT, 1810–1815

The first intimation of the solution eventually adopted by the City in bringing pound locks to the 1st District is found in a report by Charles Truss in 1802. He proposed locks at Molesey and Teddington with weirs having a long overfall or tumbling bay, and raising water level upstream by not more than 3 ft. But, following Rennie's proposals of 1794, he advocated long cuts further upstream. This latter method of improvement was again recommended by Rennie and Jessop in 1805, involving four cuts each about 1½ miles in length, and the Navigation Committee wasted a good deal of effort trying to promote such a scheme against opposition from landowners.

Zachary Allnutt's suggestion, also dating from 1805, for locks at Chertsey, Sunbury and Teddington with 'open' weirs (as at Romney) attracted little attention at the time. But in 1809 Rennie came back with proposals for nine locks between Staines and Twickenham: at Penton Hook and Stoner's Gut in short cuts without weirs, and seven others on the river with weirs, two certainly and all probably to be of the open type.

Finally in February 1810 the Committee put forward proposals, drafted by Stephen Leach,

> at once practicable and expedient, the least expensive, and the most likely to pass through Parliament without opposition and yet calculated to remedy the most prominent evils complained of

Quite simply these proposals were for locks and weirs at Chertsey, Shepperton, Sunbury and Teddington; the locks to be 150 ft in length and 20 ft wide, the weirs to have ample capacity for passing floods. In conference with Rennie, Leach estimated construction costs at £40,620. The Act was obtained in June 1810 without difficulty.

Shortly afterwards the Committee with Rennie and Leach visited the sites to decide on the exact positions of the locks. Thereafter Leach had complete charge of the operations. His plans and specifications for Teddington were approved at the end of July. Two weeks later Joseph Kimber and John Dows contracted to build the lock and weir for £11,000 and to excavate the lock cut at 1/- per cubic yard. Separate contracts were made at the same time with merchants supplying timber, and chalk and ragstone for the weir.

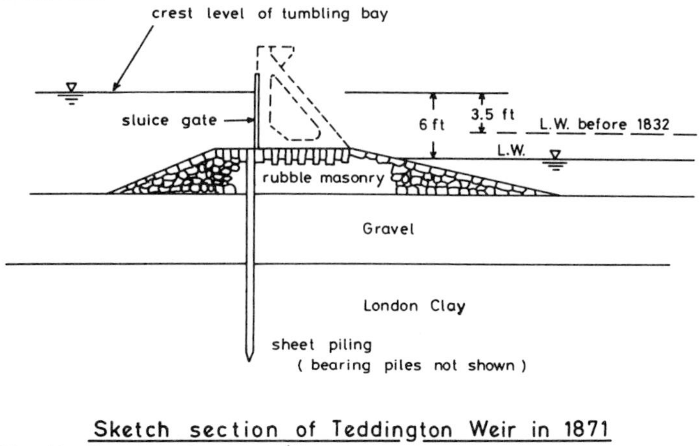

Fig. 10. Sketch section of Teddington weir in 1871. From L. F. Vernon-Harcourt, 1879.

VIII

Work at Teddington began in September 1810. Floods caused delays in November and December, for which reason Leach allowed the contractor an extra £500. The lock reached completion in June 1811, a little behind schedule, and by the end of September a critical state existed. The weir was now two-thirds finished and materials for the remainder were ready. But the cofferdams across the river would have to be removed when water level started rising in the winter, which could mean postponing completion until next July. Leach recommended an immediate advance payment to the contractors and they managed to finish the job in time.

The lock, like the others to be built by the City, was close-sided in timber up to head water level and turf-sided above. It had three pairs of gates, the outer pairs 150 ft apart and an inner pair for use by smaller barges. The weir crossed the river at an oblique angle, with a length of 480 ft. Crest level of the overfall in the tumbling bays was 2 ft 9 in above Trinity Datum or 3½ ft above low tide level at Teddington before removal of the piers of old London Bridge in 1832. After the piers were removed the fall increased to 4½ ft, but subsequent dredging caused a further drop in low water level, requiring the lock to be rebuilt in 1857–58. By 1870 the fall had increased to 6 ft and, following a failure in December of that year, the weir was rebuilt in 1871.

Construction details of the original weir are lacking. In appearance it probably resembled Sunbury weir (Fig. 12) with sluice gates occupying the central portion and tumbling bays on each side. In the 1871 rebuild the sluice bays were 240 ft wide, equal to the total width of the tumbling bays. A diagrammatic section through the sluice is shown in Fig. 10 (page 165).

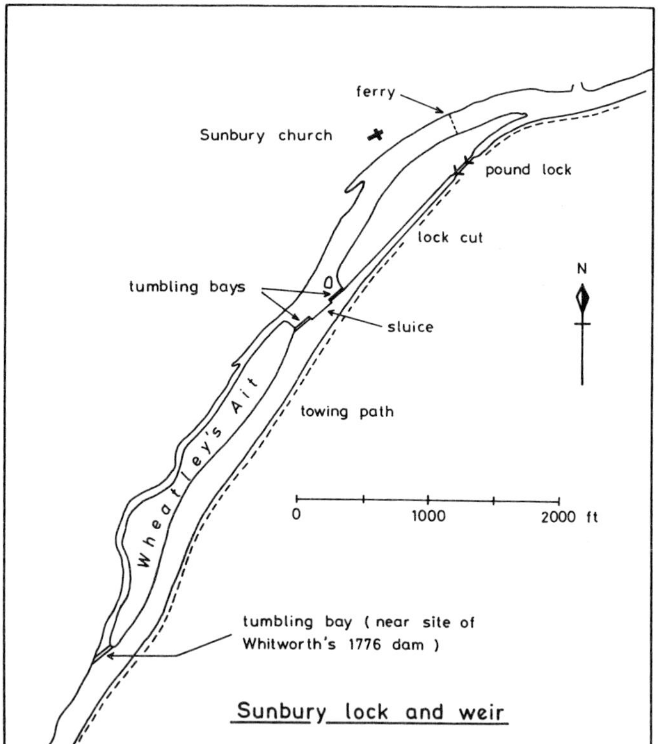

Fig. 11. Sunbury lock and weir.

Fig. 12. Sunbury weir. Engraving from *The Thames from Source to Sea,* 1885.

Construction of Sunbury lock and weir followed in much the same manner. Leach's plans and specifications received approval in March 1811. Kimber and Dows signed the contract three weeks later, the price for excavating the lock cut being 1/6 per cubic yard and for building the lock and weir £10,500. In October the contractor had 150 workmen on site and a 'powerful Steam Engine'. In February 1812 Leach reported completion of the lock and that the cofferdams at each end of the lock were being removed. Kimber died in April, but John Dows carried on and completed the weir by the end of the summer. Two smaller steam engines were employed in the weir works.

A general plan is shown in Fig. 11. The lock cut appears to be long, but it was engineered partly from a side channel of the river. The weir, see Fig. 12, had a fall of about 5 ft. As at Teddington the sluice bays occupy a central position, about 190 ft wide, with tumbling bays, totalling 260 ft, on each side. Fig. 13 (page 168) shows the lock.

Leach prepared the plans and specifications for Shepperton lock and weir in May 1812. The lock was built in a cut about 250 yards in length, beside Stoner's Gut, in the neck of land across the loop in the river near Weybridge. John Mills and William Bough, well known contractors in London's docks, agreed to construct the lock and weir for £8000 and to excavate the cut at 2/- per cubic yard. As before, separate contracts were made with suppliers of chalk (7/- per ton) and ragstone (13/- per ton) and of timber (£4400).

Work seems to have proceeded without any special difficulties and the lock was opened in January 1813. No doubt the weir, only 250 ft in length, was finished in the summer. The total cost of about £16,000 can be compared with Leach's estimate of £14,000 or £16,800 allowing 20 per cent for contingencies.

The site originally chosen for Chertsey lock was changed (to the present position) in April 1811. This required a new Act, obtained in April 1812. Contracts in accordance with specifications by Leach were signed in September: Mills and Bough to build the lock and weir for £9600 and excavate the cut at 2/- per cubic yard, timber to be supplied for £6800 and chalk and ragstone at the usual rates.

The contractor started with a 16 hp engine for pumping the excavation but found he had to add a 5 hp engine to cope with an unexpected large inflow of groundwater. The lock was opened in

VIII

Fig. 13. Sunbury lock. Engraving from W. Tombleson, *Views on the Thames and Medway,* 1834.

September 1813. Work on the weir must have been already well advanced by that time, presumably with a cofferdam across part of the river, as the accounts for the whole job were settled in February 1814. The cost came out at rather more than Leach's estimate of £19,200 even allowing for contingencies.

The weir at Chertsey has a length of 440 ft. The sluice bays, 160 ft wide, have tumbling bays on each side; the fall being about 4 ft.

Encouraged by the success of their works at Teddington and Sunbury, the City lost little time in obtaining an Act for building a similar lock and weir at Molesey, and a couple of years later for a lock at Penton Hook. However, they waited until completion of the first four works in 1814 before starting construction at the new sites.

At Penton Hook only a short length of cut would bypass a long loop in the river, and it was considered that no weir need be built; the lock merely accommodated the natural fall of about 2 ft. Leach produced the plans and specifications in August 1814. Contracts went to John Mills for building the lock (£3250) and excavating the cut (at 1/4 per cubic yard), and orders were placed for timber costing £3650. Opening of the lock to barge traffic took place in March 1815.

Construction of Molesey lock and weir followed the now familiar pattern. Leach's plans and specifications were approved in October 1814. Three weeks later John Mills contracted to do the building work for £7450 and to excavate the cut at 1/6 per cubic yard. Contracts were made at the same time for timber (£6400) and for stone. The lock was opened in August 1815 and completion of the weir followed probably two or three months later. The weir is in two parts: the main weir, about 280 ft long, running obliquely across the river from Ash Island to the lock island, and a tumbling bay at the head of the side channel on the north side of Ash Island. The total cost of Molesey lock and weir is not given, but it is unlikely to differ significantly from Leach's estimate of £17,000.

These six very large locks and the five weirs, costing something like £90,000 to build and completed from start to finish in five years, constitute one of the most outstanding examples of river engineering in English history.

Stephen Leach remained as Engineer to the City Navigation Committee until he died in November 1842. His son, Stephen William Leach, who had been his assistant since 1834, then took up the appointment. This he retained until 1857 when he became Engineer to the Thames Conservancy in which capacity, from 1867 to his death in 1881, he took charge of the navigation for its entire length.

Stage V

In conclusion we look briefly at five locks built by the Commission at new sites, and the rebuilding of three of the older locks, which brought the navigation between Staines and Oxford essentially into the pattern known today.

12. BELL WEIR AND OLD WINDSOR, 1817–1822

The original plans for a pound lock near Egham, at a place called Bell Weir, were drawn up by Zachary Allnutt in 1811. These included a cut 1200 yards long and an 'open' weir of the Romney type. An Act for building the scheme was obtained in 1812, but local opposition to the cut proved too strong and eventually the Commission adopted a less ambitious plan, produced by one of their members, John Rolls, in 1816.

John Clarke and Joseph Longhurst, of Egham Hithe, contracted to build the lock (close-sided in oak) and the much reduced cut for £4600, in February 1817. They probably built the weir also, and the whole job reached completion in December 1818.

The year 1820 saw a determined attack on deepening the worst parts of the navigation channel throughout the 2nd District, with no less than twelve ballast boats at work. Deepening the channel reduces the river gradient, which is beneficial, but results in a greater fall at the weirs and a smaller depth on the tail sills of the locks. In low water seasons this could be critical, and at some locks, including Romney in 1820, a stanch had to be built a short distance downstream to raise water level temporarily for the passage of barges over the tail sill. Eventually, of course, the lock would be rebuilt with a lower sill. At Romney this was done in 1869.

Also in 1820, Allnutt and Treacher jointly prepared plans for locks at Boveney and Old Windsor. The Commission decided to delay construction at Boveney, but raised £10,000 capital for the works and land purchase at Old Windsor. This large sum was necessary as the scheme involved a 1000 yard cut bypassing a sharp bend in the river.

Fig. 14. Old Windsor lock. Engraving from *The Thames from Source to Sea,* 1885.

In June 1821 contracts were signed by George Treacher for building the lock (£2480) and by Ambrose Oliver for excavating the cut (£2660). Other works included a bridge at the lock and another over the cut, and extra pumping in the lock pit, additional to what was envisaged in the contract. The job reached completion in October 1822 at a construction cost of £6780 and a total cost of £9300.

Initially, the lock accommodated the natural fall in the river, about 2½ ft, between head and tail of the cut. However, in 1837 a timber weir was built (direct labour at £620) across the river, which increased upstream depth in low water seasons by 18 inches. Fig. 14 (page 169) gives a good view of this typical close-sided timber-framed lock after some sixty years service. It was rebuilt in 1889.

13. STAFF CHANGES

In September 1821 John Treacher became General Surveyor to the Commission, at a salary of £320 p.a., and Allnutt became Receiver of Tolls. Treacher remained in office until he died in 1836, to be succeeded by his son George Treacher (who had been Assistance Surveyor since 1824). George Treacher held the post of General Surveyor until he retired in 1862, at the age of 71, only four years before the Commission ceased to exist. As we have seen, under the Thames Conservancy, Stephen William Leach held the position of Engineer to the whole navigation.

14. CLIFTON CUT AND LOCK, 1821–1822

Following completion of Culham cut and lock, the Commission considered applying a similar solution to the long standing problem presented by rocky shoals and an awkward bend in the river further downstream near Clifton Hampden. Treacher prepared plans in 1811, but partly as a result of legal complications over land purchase it was not until November 1820 that the Commission applied for a loan to finance the scheme.

John Roe contracted to excavate the 900 yard cut (£2130) and foundation pits for two bridges and for the lock. Richard Clarke carried out the mason's work, while the carpentry for the bridges and lock (piling and gates) was done by direct labour: all under Treacher's direction. The lock had masonry walls and abutments, built in Headington stone brought down river from Oxford. Work was completed in November 1822 at a total cost of £6700.

The weir was not built until 1835, perhaps from an exaggerated fear of the effect on flooding. It raised summer upstream water by about 15 inches, a valuable gain in depth for navigation but not likely to have much influence on flood levels which here rise up to about 5 ft above normal.

15. NEW LOCKS AT MARLOW, BOULTER'S AND SANDFORD

Work was carried out between March 1825 and September 1826 to provide a new lock at Marlow on the present south bank site and to rebuild the weir, possibly in a form similar to that known today with a long overfall. Work started by cutting a 'through' in the old weir for a waterwheel to operate the pumps. Headington stone was brought down to the site, Richard Clarke did the mason's work, Ambrose Oliver made the excavations, John Moss executed the millwright's work, and carpentry was a direct labour job. George Treacher, as Assistant Surveyor, had charge of the works, which were completed to the expressed satisfaction of the Commission at a cost of £4600.

A design for Marlow lock, dated January 1825, shows masonry abutments rising to a level 15 ft above the tail sill and masonry walls of the chamber up to head water level. No doubt the abutments were built as shown but the chamber seems (from the Account Book) to have had timber walls back-filled with chalk. The dimensions were 132 ft between gates, 20 ft clear width, with a step of 2 ft and a 5½ ft fall.

By this time the original Boulter's pound lock was in a bad state of repair and the Commission decided to build a new lock in a better position on the opposite (south) bank, rather than rebuilding, though this involved a lock cut 600 yards long. George Gyngell contracted to dig the cut (£1630), Clarke and Moss again executed the mason's and millwright's work, stone came from Headington (£3000 including carriage) and men employed by the Commission did the carpentry and labouring. As at Marlow a waterwheel temporarily placed in the weir operated pumps, but only for the cut; a 6 hp steam engine pumped the lock pit which was remote from the weir.

Work began in December 1827 and reached completion in April 1829, the total cost being £11,800. With a length of 151 ft, a width of 19 ft and 6 ft fall, with 5 ft of water on the sills, this was the largest masonry-walled pound lock yet built on any English river navigation. In June the Commission voted £100 to George Treacher for services rendered "particularly in the construction of the new Boulter's Pound Lock".

Finally under this heading we note the building of a lock on a new site at Sandford, adjacent to the old one and with similar dimensions except for a greater depth on the sills. It was constructed in 1835–36 at a cost of £3140.

16. COOKHAM CUT AND LOCK, 1829–1830

The need for a lock near Cookham, midway along the 7 miles between Marlow and Boulter's, had never been seriously in doubt since 1770 and it was always clear that a lengthy cut would be desirable to eliminate a bend in the river. As at Clifton, legal complications delayed the scheme. Built in 1829–30, with a 900 yard cut, partly engineered from a back stream, the lock had the same plan dimensions and depth on the sills as at Boulter's but a smaller fall. Indeed the fall originally was only the 2 ft between head and tail of the cut, which increased to about 4 ft when a weir was constructed across the river in 1837. The lock and cut cost £8400, the weir £650.

17. BOVENEY AND BRAY, 1838–1845

Apart from the weirs at Clifton, Old Windsor and Cookham, previously mentioned, the last new works carried out by the Commission were two locks between Maidenhead and Windsor: a distance of 7 miles in which the natural gradient of the river is 24 inches per mile, about the largest along its entire length.

Several plans for a lock near Boveney were considered at widely spaced intervals of time before reaching agreement in 1836. As so often is the case, the simplest plan turned out to be best: a lock cut of moderate length and a low weir running straight across the river. Construction took rather less than a year and was completed in November 1838. The lock, very similar in dimensions and style to that at Old Windsor, cost £2880. The timber weir, 180 ft in length, cost £1050. For the cut (contractor Isaac Gyngell) dimensions are: top width 50 ft, depth 10 ft, head cut 203 yards long, tail cut 83 yards long, total volume of excavation 16,250 cubic yards; cost including turf sides, £650.

Lastly, in 1845 a lock was built near Bray with an 'open' weir of the Romney type. Treacher provided the plans. When closed, at times of low water, the weir generated a fall of 2 ft and increased the depth upstream as far as Boulter's lock, 2 miles above.

The navigation as it was in 1845 is sketched in Fig. 15 (page 173). Apart from building the four pound locks (and final removal of all flash locks) in the 6th District, previously mentioned, the basic pattern is exactly that of the present day.

18. EPILOGUE

Tolls collected by the Commission in several representative years, as given below, reflect the increase in barge traffic following various stages of improvement and the dramatic decline in trade caused by the railways after the 1840s:

1785	£3,500	1841	£14,000
1792	£4,500	1849	£9,500
1805	£8,500	1855	£4,500
1831	£13,000	1866	£1,500

Eventually, neither the Commission nor the City could even provide for proper maintenance.

A new story opens with the Thames Conservancy, under the Crown, in whose hands the river slowly returned to and finally surpassed its former state; to become perhaps the finest, and surely the most beautiful, river navigation in the world.

ENGINEERING ON THE THAMES NAVIGATION, 1770–1845

TABLE
THAMES NAVIGATION TO 1845

Miles		District	Pound locks: date of opening		
			(a) with old weir	(b) with new weir (c) with long cut (d) with new weir built *c.* 1836	
144	LECHLADE		St. John's 1790		
			Buscot 1790		
		6th	Rushey 1791		
			Pinkhill 1792		
			Godstow 1791		
			Osney 1790		
112	OXFORD				
			Iffley 1632		
			Sandford 1632		Sandford 1836
		5th	Abingdon 1790		
				Culham (c) 1810	
					Clifton (c)(d) 1822
			Day's 1789		
93	SHILLINGFORD				
			Benson 1788		
			Cleeve 1787		
		4th	Goring 1787		
			Whitchurch 1787		
			Mapledurham 1777		
			Caversham 1778		
73	READING				
				[Blake's 1802]	
			Sonning 1773		
			Shiplake 1773		
			Marsh 1773		
		3rd	Hambleden 1773		
			Hurley 1773		
			Temple 1773		
			Marlow 1773		Marlow 1826
					Cookham (c)(d) 1830
			Boulter's 1772		Boulter's 1829
50	MAIDENHEAD				
					Bray (b) 1845
					Boveney (b) 1838
		2nd		Romney (b) 1797	
					Old Windsor (c)(d) 1822
					Bell (b) 1818
36	STAINES				
				Penton Hook 1815	
				Chertsey (b) 1813	
		1st		Shepperton (b) 1813	
				Sunbury (b) 1812	
				Molesey (b) 1815	
				Teddington (b) 1811	
19	TEDDINGTON				
0	LONDON BRIDGE				

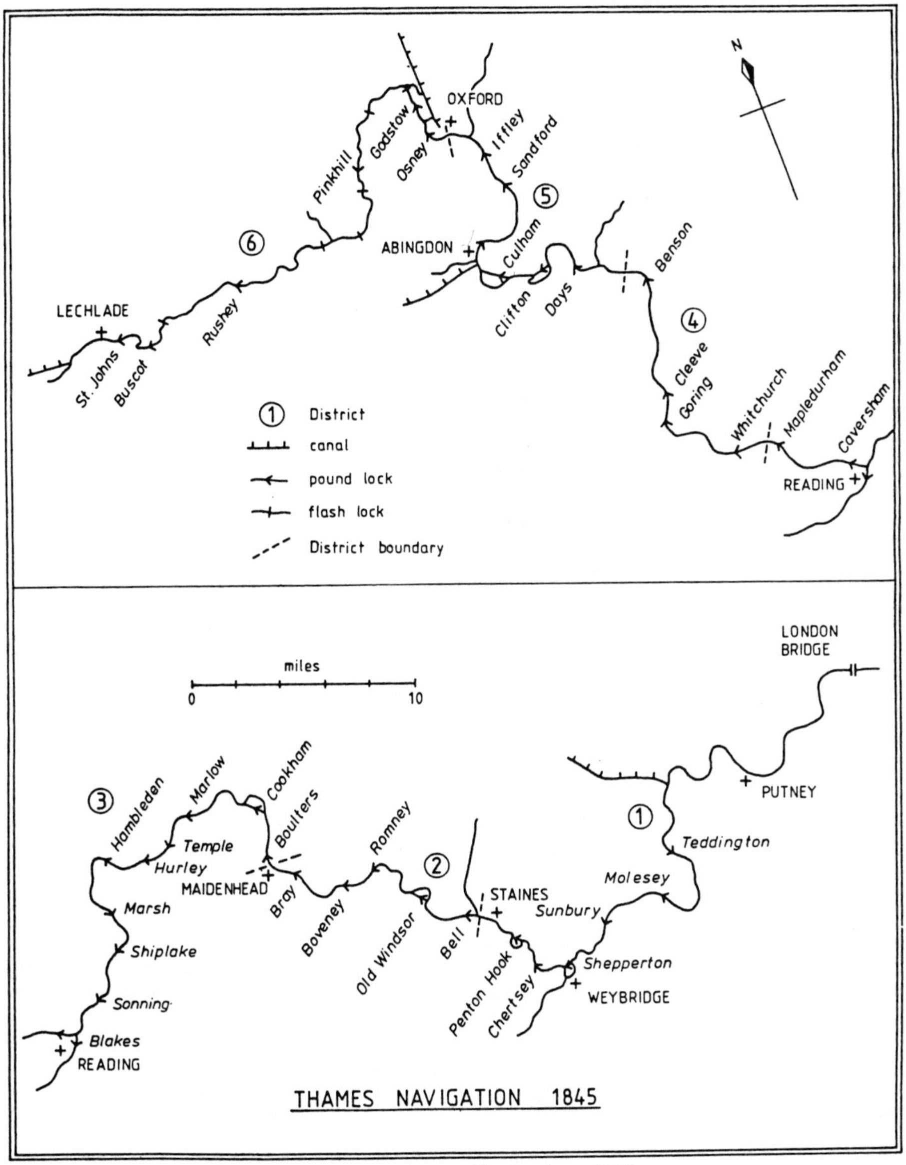

Fig. 15. Thames Navigation, 1845.

ENGINEERING ON THE THAMES NAVIGATION, 1770–1845

SOURCES

All students of Thames history are indebted to Fred Thacker's incomparable *Thames Highway* (2 vols., 1914 and 1920: reprinted 1968), but he left a number of questions on engineering matters unanswered. Principal sources used in the present paper are as follows:

a) Manuscript

1. Minute Books of the City Navigation Committee, 1770–1857, Port of London Authority.
2. Minute Books of the Thames Navigation Commission, 1771–1866, Berkshire Record Office.
3. Engineer's reports, other than those listed below, as entered in the City and Commission Minute Books.
4. Treacher Papers, Reading Central Library. These include Account Books for 1787–1798 and for works in the period 1821–1836 at Clifton, Old Windsor, Marlow, Boulter's and Sandford.
5. Frederick Page, *On the Navigation of Thames River,* 1795, Institution of Civil Engineers.
6. *Plans of the pound locks and lands belonging to the Commission.* Drawn by Zachary Allnutt, 1815 (with some later additions), Berkshire Record Office.

b) Printed Reports and Plans

7. James Brindley and Robert Whitworth, *Plan and Profile of the Thames from Boulter's Lock to Isleworth,* 1770.
8. Joseph Nickalls, *Plan of the Thames showing proposed improvements,* 1771.
9. Robert Whitworth, *Plan of the Thames from Boulter's Lock to Mortlake,* 1777.
10. William Jessop, *Report, Lechlade to Oxford,* 1789.
11. Robert Mylne, *Report, Lechlade to Whitchurch,* 1791.
12. *Parliamentary Committee on Thames Navigation,* 1793.
13. William Vanderstegen, *Present State of the Thames Navigation,* 1794.
14. John Rennie, *Report, Reading to Isleworth,* 1794.
15. Robert Mylne, *Report, Lechlade to Caversham,* 1802.
16. Zachary Allnutt, *Considerations on Improving the Navigation,* 1805.
17. Thames Commission, *Report, Lechlade to Staines,* 1811.
18. George Bradshaw, *Lengths and Levels of Canals, Navigable Rivers and Railways,* 1832.
19. George Rennie, "The progress and present state of our knowledge of hydraulics", *British Assoc. 4th Report,* (1834), pp. 415-512.
20. J. C. Clutterbuck, "The perennial and flood water of the Upper Thames", *Min. Proc. Inst. C. E.,* Vol. **22** (1863), pp. 336-359 and 369.
21. Ordnance Survey, 6 inch maps, 1864-1883.
22. H. W. Taunt, *A New Map of the Thames* (with H. W. and L. W. falls), 1878.
23. L. F. Vernon-Harcourt, "Fixed and Movable weirs", *Min. Proc. Inst. C. E.,* Vol. **40** (1880), pp. 24-42.
24. *General Report of the Conservators of the River Thames,* House of Commons, 1904. (For 1894 flood levels).
25. Charles Hadfield, *The Canals of South and South East England,* 1969.

DISCUSSION

Mr. M. T. Tucker opened the discussion by asking whether elm had been used in the Thames locks and what life might have been expected of it. **Professor Skempton** replied that elm was an option for underwater work, but in practice either beech or more often Memel fir had been used. Above water and for the gates, and probably for the pointing sills, oak alone seems to have been employed after 1786. The life of the lock gates was about 20 to 30 years; they had a great deal of wear. The rest of the lock above water might have lasted 60 or 70 years, with occasional repairs.

The **President** asked what was the origin of the word 'eyot'. Several suggestions were offered from members of the audience, but in a written reply Professor Skempton noted that according to the Oxford Dictionary 'eyot' or 'ait' is a Middle English (i.e. early medieval) term for an islet, especially in a river, ultimately derived from Anglo-Saxon. He thought that in modern times its use was largely if not entirely restricted to the Thames.

Mrs. S. M. Rolt asked why the boatmen sometimes preferred to use the flash locks. Professor Skempton replied that this may have been chiefly because the tolls were less. Regarding a further question on towing, he said that there were two tow ropes; one, which might be very long, was taken round turning posts to the towing path; the other taken to the winch. Once a barge had passed through the flash lock normal towing would be resumed.

Mr. W. J. Whitehead asked why the author had assumed there was a sluice at Caversham as well as a tumbling bay? Professor Skempton replied that this tumbling bay was so much longer than usual (for the eighteenth century), that he may have been wrong and perhaps, in this particular instance, no sluice had existed.

Mr. P. Cooley said that a lot of water would have passed through a flash lock when it was opened, and millers would have preferred to have a pound lock. Professor Skempton agreed. In principle, the pound lock had overwhelming advantages for millers and bargemasters. However, adding to his reply to Mrs. Rolt's question, he said that before the navigation was perfected flashes were still necessary in dry seasons. Thus, at stated times, the flash locks had to be operated and barges could have taken the opportunity of evading pound tolls.

Mr. J. D. Richard referred to the ground called Winch Meadow at Hambleden being a long way from the flash lock. After some discussion and a valuable contribution from **Mrs. Skempton,** a re-examination of the slide (Fig. 4, page 156) showed that she had been correct in pointing out that before making the lock-cut Winch Meadow would have extended up to the original river bank where the winch itself existed, in a position where it could easily have coped with hauling a barge through the flash lock.

Mr. N. D. New said that the steering accuracy required would have been great, and referred to his own experience in using a horse-drawn barge between Abingdon and Lechlade. He had become friendly with one of the lock keepers above the Windrush where paddles were still in use in the weir: they were lifted by hand. In thanking Mr. New for his account of what must have been one of the last occasions on which a horse-drawn boat was taken up the Thames, Professor Skempton referred to a paper by Dr. Michael Lewis and others on the existing remains of flash locks on English waterways, published in *Industrial Archaeology* (1969). By that time, not a single example was in use.

In reply to **Mr. A. L. Little,** who asked how long the river journey would have taken, Professor Skempton said Zachary Allnutt in 1810 reported that in the districts above Staines barges go downstream generally from 25 to 35 miles a day, and upstream from 20 to 30 miles a day. The rates may well have been improved by the mid-nineteenth century, but four days from London to Oxford and three days for the return journey would probably be good going. Replying to a further question he said that perhaps not more than half an hour would be taken in passing through a flash lock, but this had to be compared to about 10 minutes for a pound lock.

Mr. J. W. Butler asked where such terms as ‘tumbling bay’ had come from. Professor Skempton thought they were milling terms but he had not yet found a good book on what might be called the hydraulic engineering aspects and terminology of water mills. Authors tended to concentrate on the machinery.

After some further discussion the President called on **Miss Julia Elton** to propose the vote of thanks. Miss Elton said that on the previous day she had received a book catalogue and had noted a copy of Allnutt's 1805 report; on telephoning, she discovered that Professor Skempton had already purchased the item! He possessed a remarkable ability for finding Minute Books and other sources of information. He had given the meeting a most interesting and informative paper, once again showing the care with which he always prepared his contributions to the Society. It gave her much pleasure to propose that Professor Skempton be given a very hearty vote of thanks. This was passed by acclamation.

ENGINEERING ON THE THAMES NAVIGATION, 1770–1845

CORRESPONDENCE

In a written note Professor Skempton said that after the meeting he had received from Mr. Cooley a plan of Boulter's Lock from an Inclosure survey of 1787 in the County Record Office, Aylesbury. By combining this with a plan of the 1829 Lock, in the Thames Commission records at Reading, the map in Fig. 16 had been produced. It helped to elucidate some of the points made in the paper on one of the classic Thames sites, showing the positions of the old flash lock, weir and winch, Nickalls' timber-sided pound lock of 1772 and the large stone pound lock built by George Treacher.

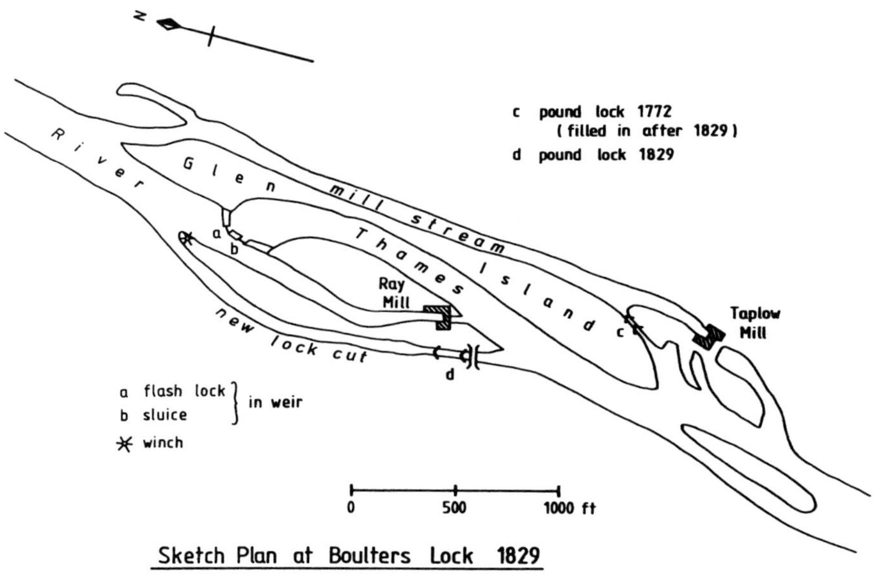

Fig. 16. Sketch plan at Boulter's lock, 1829.

IX

Samuel Wyatt and the Albion Mill

The Albion Mill was built during the years 1783-86 on the Surrey bank of the Thames at the foot of Blackfriars Bridge. Designed to grind corn on a large scale, the mill was outstandingly the most advanced industrial building of its day and the first to be planned, from the start, to incorporate the newly invented Boulton & Watt double-acting rotative engines. The mill was also one of Samuel Wyatt's major works and one to which he devoted a great deal of attention as architect and builder and as the principal promoter of the undertaking. By good fortune many of his letters relating to the mill have survived and can be studied in the Boulton & Watt Collection at the Birmingham Reference Library and in the Tew MSS at the Assay Office, Birmingham. Letters on the mill written by Matthew Boulton, John Rennie and James Watt also exist in these collections. There are structural drawings in the Reference Library; several views of the mill were published as engravings, and in the Public Record Office drawings can be seen of two schemes by Wyatt for rebuilding the mill after its destruction by fire in 1791. All these sources have been used in the present paper.[1]

Letters and financial accounts in the Birmingham collections provided the basis for Westworth's excellent short history of the mill as a business enterprise.[2] Insull has dealt with the same subject at greater length, and includes a dissertation on the sad story of the attempts to rebuild or dispose of the premises after the fire, but he displays a bias against Wyatt.[3] Details of the engines and other technical matters are given by Farey[4] and by Dickinson & Rhys Jenkins,[5] and general accounts can be found in Samuel Smiles, *Life of Rennie*[6] and *Lives of Boulton and Watt.*[7] In all these works, however, little if anything is said about the building, with the sole exception of Farey, who briefly describes the foundations; though by a strange irony he mistakenly attributes their design to Rennie. Such was the unsatisfactory state of the literature, from the point of view of architectural history, when Mosse in 1963 first attempted a proper assessment of Wyatt's achievements at the Albion Mill.[8] He appreciated the importance of the structure; and his account of Samuel Wyatt, while incorporating a number of historical inaccuracies, shows an understanding of the man who had been almost ignored, or unduly criticized, by most previous authors. Finally, in a recent paper, I have given a detailed analysis of the foundations, derived from Wyatt's site surveys and drawings of the mill.[9]

During the preparation of the foundations paper, and in my earlier research with H. R. Johnson on the first iron-framed mills,[10] I came to realize the wealth of material in the Birmingham collections on Samuel Wyatt and the Albion Mill and to see the desirability of presenting a representative selection of extracts from his letters, very few of which have previously been published. It also seemed desirable to make a catalogue of drawings and prints (see Appendix II) and to give a factual description of the mill itself. I have added a brief biography of Wyatt in order to convey some idea of his career and of his works other than the great mill at Blackfriars.

THE ALBION MILL

Wyatt negotiated the lease of the site in March 1783. By July he had produced a set of drawings of the mill, which was to be operated by three Boulton & Watt engines of unprecedented power. Excavation work began in October 1783. The foundations and basement walls were finished shortly before November 1784, and by this date Wyatt had made the final design of the superstructure. Construction then proceeded rapidly. The roof timbers were in position by October 1785, the building was completed early in 1786, and the first engine with its mill-work and associated machinery came into operation in March of that year. The opening of the mill was attended by a crowd of distinguished spectators including Henry Cavendish, Josiah Wedgwood and Sir Joseph Banks, President of the Royal Society. The second engine was working by December 1788; but the third engine had not yet been erected when the mill suffered almost total destruction by fire on 2 March 1791, just five years after its completion. Residential quarters, known as Albion Mill House, were located in the corner of the mill overlooking Blackfriars Bridge. This house escaped the fire owing to the thickness of its party walls, as did the engines, and the west and north front walls also survived. But all the interior of the mill, and virtually the entire stock of grain and flour, disappeared in a holocaust quite beyond the power of the fire engines to control.[11] Arson was suspected, though not proved. Three proposals were made by Wyatt to rebuild the premises, notably in 1795 as a coffee warehouse, and a set of eight beautifully finished drawings for this scheme still exists. Nothing came of these proposals, however, and eventually the west front was converted into a row of private houses.[12] In 1863 all traces of the mill were obliterated by the London, Chatham & Dover railway save for a length of masonry, still showing above the shore-line, which is probably a fragment of the river front wall.

Our knowledge of the structure of the mill is derived chiefly from a large-scale general section drawn by Wyatt, dating from November 1784 (drawing 12 in the catalogue, see Fig. 37a), and from part plans made in February 1787 (drawings 16-19, one of which is reproduced in Fig. 37b). The interior consisted of a multi-bay six-storey timber frame, with 11 in square posts supporting beams 13 in deep and presumably 11 in wide. A comparison of Figs 37a and 37b shows that two extra rows of posts were placed in the eastern bay, after the drawing of November 1784 had been made. The beams carried 8×4 in joists at 18 in centres and floors of $1\frac{1}{2}$ in planking. Typically the beams and joists spanned 12 ft 10 in bays. The walls, in brickwork, rose 64 ft above the footings. The west front wall and the river front wall (the latter with a Portland stone facing in the ground storey) varied in thickness from 45 in at the base to 27 in in the top storey and had large window openings. The east wall was not so thick but had few windows, though two circular openings were provided above the boilers on both sides of the two chimneys (No. 15). Each engine was enclosed by side walls and a massive wall carrying the engine beam. Running into the mill there was a dock 22ft wide, having a depth of 12ft of water at spring tides, by which barges could enter directly from the river, and a carriageway enabled wagons to be loaded or unloaded under cover. The two doors leading to this carriageway from Albion Place can be seen in the west front elevation, Fig. 39a. A light well, about 22 × 11ft in plan, was situated at the centre of the mill.

The central posts, in the upper five storeys, were supported on timber trusses spanning the dock. The other rows of posts rested on masonry pillars or on brick walls in the cellar, and these supports as well as the main walls were carried by deep strip footings, linked together in the east-west direction by inverted barrel vaults; thus forming an immense foundation raft 160ft long by 120ft wide. In this way the weight was distributed as uniformly as possible over the whole area of the ground in order to minimize settlements, and especially to avoid differential settlements due to compression of the underlying alluvium of the Thames. Moreover, the weight of excavated earth was equal to about one-half of the entire building load, so the net pressure imposed on the ground was greatly reduced. So far as I am aware this foundation raft had no precedent, and the timber framing was quite exceptional in its height and the large area of flooring supported without any internal walls.

At the time when Wyatt designed the mill no practical way had yet been devised for providing multi-storey framed buildings with adequate safeguards against fire. During the 1780s several cotton mills were burned down, but the final stimulus towards a solution of this problem may well have been provided by the destruction of so celebrated a building as the Albion Mill. At any rate it was early in 1792 that work began in Derby on the first fire-proof mill, designed by William Strutt.[10]

So far as the external appearance of the Albion Mill is concerned, an engraving (No. 38) published in 1790 gives a general impression, though the architectural detail is poorly represented (Fig. 38). Fortunately, a view of the river front was recorded during the fire by a competent topographical artist, John William Edy, and his fine aquatint of the subject (No. 39) is reproduced in Fig. 40a. This shows, incidentally, how the fire floats were hampered by the low state of the tide when the blaze was at its worst. The drawings for the 1795 rebuilding scheme, already mentioned, include two elevations (Nos 30 & 31) in Wyatt's own hand (see Figs 39a & b). In all essentials these are exact representations of the mill as completed in 1786. Indeed, they may be copies of the original drawings. Lastly we see in Fig. 40b a reproduction of the engraving by Pugin and Rowlandson, published in Ackermann's *Microcosm of London* (No. 41). Dating from 1808, and erroneous in showing the two end bays on fire, it nevertheless provides a good idea of the setting of the mill and of the architecture of the west front, which still existed when Pugin made his drawing.

Wyatt evidently took pride in the design of the elevations, preserving a unity with the four-storey houses on the opposite side of Albion Place and maintaining a proper sense of scale in a building adjacent to Mylne's noble bridge.[13]

SAMUEL WYATT

In 1792 Humphry Repton referred to 'the elegant simplicity of my ingenious friend Mr S. Wyatt's design', and in these words he aptly describes the nature of Wyatt's genius.[14] As architect of the Commissioner's House in Portsmouth

Dockyard and of the very fine Trinity House, London, and indeed of Tatton Park, to which Repton refers, Samuel Wyatt is seen to be the master of a distinguished neo-classical style.[15] His abilities as a structural engineer are amply demonstrated by the Albion Mill. He submitted plans for docks on the Thames and was consulted on the first proposals for a tunnel under the river. He became a member of the Society of Civil Engineers in 1781. In 1792 he received two official appointments, Surveyor to Trinity House Corporation and Clerk of the Works at the Royal Hospital, Chelsea. Two years later he became Director of Works at Ramsgate Harbour, following John Smeaton. All three positions he held until his death in 1807. As a man of business, already running what must have been a considerable building practice and trading as a timber merchant, he undertook the promotion of the Albion Mill enterprise and acted as its general manager, though without much success.

Papworth, who was reliably informed on the Wyatt family, gives the date of Samuel's birth as 8 September 1737 at Blackbrook, Weeford, near Lichfield in Staffordshire.[16] The Birmingham correspondence contains frequent references to his wife and daughter, Louisa, but I have not found where or when the marriage took place. He died at Chelsea on 8 February 1807,[17] and his will, in which everything is left to 'my dear and affectionate wife Ann Wyatt', was proved by his widow ten days later.[18]

Early documentary evidence on Wyatt's career in the Midlands relates to the years 1765 and 1766, when he writes to Matthew Boulton from Kedleston, Derbyshire, about 'Escutcheons for the Dining Room doors and Sconces for the glass frames' and other things for Lord Scarsdale.[19] In 1769 he began building the Pantheon in Oxford Street, London, to the designs of his brother James Wyatt, and for several years he lived with his elder brother John, the surgeon, in Newport Street.[20] About 1774 he took up residence at 63 Berwick Street, a large house near Soho Square, with a timber yard behind it.[21] 'Samuel Wyatt timber merchant' is a constantly recurring entry in the local directories from 1779 until 1793.[22] He and his family were living in Albion Mill House at the time of the fire, and it remained their home for another eight years. He continued to keep an office at Berwick Street until 1803,[23] and it was from this address that his pupils submitted their drawings for exhibition at the Royal Academy.[24]

Some of Wyatt's more important or interesting works are set out in chronological order in the following list, which is based on Papworth[16] and Colvin[25] and on more recent research published in the *Survey of London*, as well as information derived from the Birmingham collections and elsewhere.

1769-72 Builds the Pantheon, Oxford Street, London; architect, James Wyatt[26]

1777-98 Doddington Hall, Cheshire, for Sir Thomas Broughton; architect and builder; plans dated 1776[27]

1780-82 Façade of the Theatre Royal, New Street, Birmingham; designed in 1777[28]

1783-86 Albion Mill, Southwark; architect and builder; engines by Boulton & Watt; millwork by John Rennie

1784-87 Commissioner's House, Portsmouth Dockyard; architect and builder[29]

*c.*1785-92 Tatton Park, Cheshire, for William Egerton; completed 1807-*c.*1812 by Lewis Wyatt following, more or less closely, his uncle's design [14]
1788 'Moveable hospitals' for 'his Majesty's distant possessions': twelve were ordered; each building, about 86×20ft in plan, could be taken down and re-erected in one hour, as demonstrated to George III[30]
1789-90 Heathfield House, Handsworth near Birmingham, for James Watt: at least three sets of plans prepared 1787-89; builder, Mr Glover[31]
1790-91 Dungeness lighthouse[32]
1791-94 No. 15 St James's Square, London, for the Anson family; extensive remodelling of earlier work[33]
1793-96 Trinity House, London; architect and builder;[34] elevation exhibited at RA in 1794
1794-96 Stornoway House, Cleveland Row, London, for Lord Grenville; builder and probably architect[35]
1795 Design for rebuilding Albion Mill as a coffee warehouse
1796 Plans for three docks on the Thames at the Isle of Dogs[36]
1799 Consulted on Ralph Dodd's project for the first tunnel under the Thames, from Gravesend to Tilbury[37]
1800 Patent 2410: 'A Method of Making and Constructing Bridges, Warehouses, and other Buildings, without the Use of Wood'
1802 Design for rebuilding Albion Mill in fire-proof construction, with iron framing of the type described in his patent of 1800
1805 Hackwood Park, Hants; completed by Lewis Wyatt, 1807-13[38]
1806 Flamborough Head lighthouse, for Trinity House Corporation[39]

WYATT'S LETTERS

The following extracts are selected from about one-third of the 125 letters written by Wyatt to Matthew Boulton and to James Watt on the subject of the Albion Mill. Some other letters have been quoted where they are of particular interest, but in general I have chosen to illustrate the history of the mill so far as possible in Wyatt's own words. His signature is invariably *Sam Wyatt* in a firm, well-formed hand which scarcely changes during the twenty years covered by the correspondence.

The letter in which he first mentions the idea of building a steam-powered corn mill in London includes only a few lines on the topic, yet it deserves to be quoted at some length as a characteristic introduction to his style.

London, 18 January 1783: Wyatt to Boulton
I have enclosed a plan of the Victualling Office premises at Weevil [near Gosport]... I have orders to begin the Works immediately therefore the sooner you can conveniently come to Town, to agree with the Board for your premium the better. I have a strong notion that Cast Iron wheels for the greatest part if not all the Mill Work – will be most advisable. I mean to consider you & Mr Watt as principal millwrights – Notwithstanding I am very anxious about this business I am infinitely more so respecting a Mill upon the bank of the Thames, I have a situation in my Eye at the bottom of Water lane Fleet Street, just above Blackfriars Bridge.... I beg my best complmts to your whole family & Mr Watt. Sam Wyatt

P.S. I am going early in the Morning to Portsmo & shall return on Thursday.
turn over
As this was too late for the post last night I send it p. Coach, but I forgot to mention that my journey to Portsm° this Morning is chiefly on Acc^{t} of a New House I am going to build in the Dock yard there for the Commiss^{n} and I have orders from the Navy Board to take a Cursory view of the yard to see what your Engines may be applicable to – I have already told the Comptroller that all kinds of pumping and Mill work may be done – furnaces & forges for working up their Old Iron – Anchors or parts of Anchors at least by tilting forges – Saw Mills – Copper Mills for working up the old Copper Sheathing & Mills for turning Sheaves & makg Blocks.

Sunday Morn. 6 o'Clock

[We may note that the site mentioned in this letter is on the other side of the river from the one finally selected.]

London, 4 February 1783: Wyatt to Boulton
. . . I shall hope for the pleasure of seeing you soon and that you will not only be prepared to treat with the Commiss^{ns} [for an engine at Weevil] but also with your friends here on the subject of a Mill in Town, the Spot is ready; & the most Centrial that you can imagine, it bids fair to be one of the most advantageous concerns I have ever heard of.

[Boulton comes to London, staying with Mr Matthews, his London banker, in Green Lettice Lane.]

Berwick Street, 20 February 1783: Wyatt to Boulton
I should scarce ever forgive myself if we were to lose Hammerton's Ground because I am convinced there is not another situation equal to it – and the Rent is no object compared with the Magnitude of the undertaking of which I am also so well satisfied that I see no one Earthly thing that can prevent its Sucess, therefore I wish you to consider what share of it you chuse to take on your own Acc^{t} or for your friends and I will find Men & Money for the remainder.

Berwick Street, 26 February 1783: Wyatt to Boulton
. . . If you can conveniently see Mr Mylne to Day it would forward the business with Mr Neale.

Berwick Street, [1 March 1783]: Wyatt to Boulton
I have enclosed the proposal I purpose making Mr Neale whom I am to see at 6 o'Clock this Evening – pray let me know if you are coming to this end of the Town or I shall come to you.

London, 5 March 1783: Wyatt to Boulton
Mr Neale yesterday left me the following terms . . . I wrote him for answer that I agreed to the terms except the £25 p.Ann^{m} additional Rent when the Bridge became free. I have this Evening had a letter from him the purpose of which is that he agrees to let me the premises at the foot of Blackfriars Bridge as follows

To pay £100 p.Ann^{m} free of all Taxes untill the 25th Mar 1785 & then to pay

£250 . . . these terms I shall agree to, therefore you may expect to receive a plan of the premises in a few Days.

London, 13 March 1783: Wyatt to Boulton
Enclosed you have a plan of the premises at the foot of Blackfriars bridge which I hope you will consider as soon as you conveniently can . . . Mr Smeaton attended the Victg Board yesterday – he behaved very handsomly . . .

London, 26 March 1783: Wyatt to Boulton
Your favour of the 15th Inst. came in my absence from Home, having been for some time at Portsmouth. I have omitted writing a few Days in hope of giving you more intelligence respecting the fire Mill proposed for the Victualling Board. Mr Smeaton has also been at Portsm° to survey the Tide Mills . . . [re Blackfriars] I believe you will find the plan answer to the dimensions, I have enclosed one made to a Scale of $\frac{1}{8}$ of an Inch to a foot. [Drawing No. 1 in catalogue.]

London, 26 April 1783: Wyatt to Boulton
. . . I am going to Morrow Morning to Mr Cokes in Norfolk & shall return about Wednesday or Thursday next – as soon as I can get the plans of the Mill compleated I will send them down to you.

[Meanwhile, at Soho, Watt & Boulton were proceeding to build an experimental double-acting engine driving a pair of mill stones. Wyatt saw this engine at work, probably on his way up to Lancashire, and shortly afterwards wrote as follows:]

Heaton House, 8 July 1783 Wyatt [to Boulton]
I hope from the Tryal of the Mill you'll be able to form an Idea of the consumption of Coals. I believe the Dressing Mill Sack Tackle & Screening the Corn: may be fairly rated at 1/6 of the power of Grinding.

[From the next letter we see that Wyatt has completed a set of drawings for the mill (Nos 4-8) and with his first partner, William Curtis, he places a formal order for the engine.]

London, 19 July 1783: Wyatt to Boulton
I have herewith sent you a Book of Drawings of the Mill & a Letter to order the Engine . . .

[The enclosed letter, signed by Wyatt and Curtis, is addressed to Messrs Boulton & Watt:]
Having reported to our Friends here, who are to be connected with us in the Mill proposed to be built at Blackfriars Bridge the Satisfaction received on our late Visit to you, respecting the Uniformity of the Power of your Fire Engine when applied to Mill Work: we all concur in desiring you will give immediate Directions for the Construction of a Fire Engine with Powers sufficient to work 8 pair of Mill Stones, Six of which are intended to be kept constantly at Work grinding at the Rate of Six Bushels of Corn p. hour. . . . Mr Wyatt engages to contrive the Building in such a Manner that it shall be ready to receive one Engine as soon as you are able to compleat it.

60

Matthew Boulton replies, 3 August 1783:
I received in due course your two letters & the Book of drawings which I have been prevented from acknowledging sooner by indisposition. I have been very little at Mr Watts lately, from the same cause, & I cannot give you any distinct acct of his advancement in the Corn Mill [engine] drawings but I know he hath made a beginning. Our little mill improves by working & grinds Corn faster than we can purchase it in our Birmhm Market. [Boulton then sets out various calculations for the Blackfriars engine.]

[Watt writes to Wyatt next day, suggesting that ten pairs of stones might be advisable, rather than the eight pairs originally planned, and goes on to say:]
I have contrived some improvements on the mechanism of the engine since I saw you and hope to make this engine the most perfect thing of the kind ever executed.

London, 25 August 1783: Wyatt to Watt
[Wyatt has now decided on 4ft 6in diameter millstones, and agrees that] it seems advisable to encrease the Number of Prs of Stones as there will be an advantage in working a pr for 24 hours & leting them Stand the same time to Cool – so as to work them alternatively every 24 hours . . . I am happy to hear from Mr Matthews that you have agreed to take some shares in the concern.

[In his reply, 21 September 1783, Watt accepts the size of millstones, informs Wyatt of a change in the proposed length of the engine beam and suggests that the cellar of the mill should be 8ft high.]

London, 2 November 1783: Wyatt to Watt
I don't know what apologie to make for my long silence, the chief cause has been owing to a tedious altercation I have had with the Committee for Thames & Canal Navigation & for Viewing encroachments &c on the River Thames – my application to them was for giving up a small projection of ground in the Center of the premises to the River as an Equivalent for advancing 10ft 6In into the River at the N East corner next the Bottle warehouse by which means the Building would form a right angle to the Bridge . . . the Committee joined to that of the Bridge Committee after taking several Views, waited upon Mr Mylnes return from Scotland and his Opinion was so decided in my favour that they have granted me all I asked.

In answer to your last favor, with respect to the Cellar Story there is not the least objection to its being 8 feet high in the clear it will make it much more useful. I have got very forward in sinking the Dock & in a few days shall begin Building the walls – it is 12 feet deep from high water at Spring Tides.

[As Robert Mylne returned to London on 14 October, after a six-week visit to Scotland and the north of England,[40] excavation work for the dock must have started very soon after that date, in order to be well in hand by 2 November. Wyatt's site surveys (Nos 2 & 3) clearly pre-date these negotiations.]

Watt replies on 8 November 1783:
I am very glad that you have settled with the river committee. I have now got the plans of the Engine & mills mostly completed, and shall be glad of your

company here to look over and digest matters, as soon as Mr Boulton comes home, which I hope will be in a week or 10 days.

[In Wyatt's next letter we hear the name 'Albion Mill' for the first time.]

London, 10 December 1783: Wyatt to Boulton

I have seen Messrs Bates & Curtis to Day . . . I dined with the former and as usual our Toasts were 1st Success to the Albion Mill (for so we have named it) & 2nd the Albion Mill Compy . . . I am in some hopes Mr Bates has a friend who will make a fifth partner tis a Mr Freer.

[During the early months of 1784, while work proceeded on the foundations, there was much activity regarding an application for a Charter of Incorporation. The appeal was heard in April but failed, owing to intense opposition from the millers.[2] On 31 May 1784 Wyatt sent a letter to Soho, going over the calculations for the engine and suggesting that the cylinder should be increased from 34 to 40in diameter. Watt gave a very detailed reply confirming that a 34in cylinder was sufficient.]

London, 8 June 1784: Wyatt to Watt

I have your favor of the 2nd Instant in answer to my Letter respecting the size of the Cylinder . . . and I beseech you to consider what I said as arising solely from my anxiety & Zeal for the safety, Honor & prosperity of the Albion Mill undertaking – and you have stated the matter so clear in your letter that we are perfectly satisfied . . .

As I make the foundations to cover the whole Area of the Ground by means of Inverted Arches, I find them very troublesome & have taken a great deal of time, but I have the satisfaction to think that I could not have pursued any mode so effectual for the Safety of the Buildings, we are got about 6 feet above the Ground with the West front wall & I perceive that when the foundation is compleated the Brick work above ground will go on very fast. I wish you to give orders for the Cylinder to be shipped & what other heavy Castings there are to come by Sea, by the time they can arrive I shall provide Room for them. I expect shortly to be in Staffordshr when I shall pay you a visit.

[Wyatt adds a note to Boulton:]

. . . I wish I could have accompanied Miss Boulton [to Soho] but I am waiting to be examined by the House of Lords respecting a petition of Mr Egertons who is applying for a Sum of Money to Build a House [i.e. Tatton Park] . . . The Commemoration of Handel went off to the Astonishment of every body, Mr B— has gained immortal Honr by it.

[By this time there were four shares in the Company, held by Samuel Wyatt, Noah Bates, John Frere (who had taken over from Curtis) and William Matthews with Boulton and Watt.]

Green Lettice Lane, 30 July 1784: The partners to Watt

At a meeting here this Evening after taking a view of the premises at B.friars we find our Mr. Wyatt has made great progress on the West side of the Dock & has made a beginning on the East side of the Foundations for Mill stones & Engine . . . as we mean to confine the Works to the Center Engine entirely at present we are desireous that you should exert yourself to get that Engine compleated with all possible expedition ... We are sorry that we have heard nothing of Mr Rhennie the Millwright which Mr Wyatt informed us you had engaged as he will very soon be wanted.

(*signed*) John Frere, Bates,
Wm Matthews, Sam Wyatt

London, 31 August 1784: Wyatt [to Watt]

... I begin to be Impatient about our Millwright, not having heard any thing of him – I am just beginning the foundation of the Engine House, in which I hope I can not err.

[John Rennie, 'our millwright', now enters the story. As a young man, one year out of Edinburgh University, he had been appointed by Watt to assist in the design and construction of the Albion Mill machinery. Rennie left Scotland in September 1784 and spent two months at Birmingham before coming to Blackfriars. By the beginning of November it became necessary to have, at Soho, an accurate drawing of the mill structure in order to plan the general assembly of the engine, shafting, millstones and the other machinery.]

London, 4 November 1784: Wyatt to Watt

I have sent by the Coach this Evening, in consequence of a Letter I had from Mr Rhennie, a Section of the Mill, the heights of the stories are figured, and as it is drawn pretty correct by a large scale any other dimensions that are wanted may be taken from it.

I have taken the liberty of Drawing the Stones upon the 3rd floor, and I must beg your consideration to that particular point. The Meal troughs & Millers working upon the 1st floor will be greatly incommoded by the Engine, Mill work & Boiler, it will also be Dark & unpleasant, compared with the 2nd floor. There will also be a great convenience in carrying on the business viz by appropriating the Ground floor for Coals &c, the 1st floor for Bran & pollard the 2nd floor for Grinding & also for sacking up & delivering the flour, the 3rd & 4th floors for the Dressing Mills & the 5th & 6th Grainery.

[Wyatt's sectional drawing (No. 12) is shown in Fig. 37*a*. The stones were, in fact, kept on the 2nd floor, but otherwise the use of the various floors followed the scheme outlined in this letter. Comparison with the section of July 1783 (No. 6) shows that Wyatt had made several changes in the design of the superstructure. His new drawing was copied at Soho and details of the engine and all the machinery added (No. 13). At the same time working drawings of the mill work were produced. These are all dated November 1784.

Rennie arrived at the site on 24 November 1784. Next day he wrote to Watt:]

I arrived here last night & was sorrow to learn that Mr Wyatt had gone to Cheshire and would not return for two weeks. I waited on Mrs Wyatt and got his directions & wrote to him directly, to know about the placing of the Smithy, which I supposed you would not have liked me to do without his advice. [Rennie then describes at great length a difficulty in knowing exactly where to place a door in the engine house and makes a sketch which shows, incidentally, that the walls at this side of the building were now up to the level of the first floor beams. Clearly, then, the foundations and substructure must have been finished by October 1784 at the latest.]

London, 22 April 1785: Wyatt to Boulton
. . . I enclose a State of Account of the Albion Mill . . . We are now at Work upon the Engine House which will very soon be up, therefore you will be so good as to consult with Mr. Watt about the proper people to begin upon the Engine. Mrs Wyatt begs to join in Complmts to your self & Miss Boulton.

[The enclosure shows that about £5600 had been spent so far by Wyatt on wages and materials and nearly £900 on rent, purchase and legal charges. Payments of £1200 had been made on each of the four shares.]

London, 19 May 1785: Wyatt to Watt
We are so forward with the Engine House at Albion Mill that I hope you will very soon send us the proper people to fix the Engines . . . we have a fine strong piece of Timber for the Engine beam.

London, 19 May 1785: Wyatt to Boulton
I have wrote to Mr Watt to beg that our Engine may be fixed as soon as possible, we are very foward, the 3rd Tier of Beams are up & the Kiln floor over the Engine House also.

[In due course Malcolm Logan, Watt's engine erector, arrives at Blackfriars.]

London, 18 August 1785: Rennie to Boulton
The [engine] House will be compleatly roofed in the end of next month. Malcom is putting the Engine together so I hope we will soon be ready to do some work. Mr. Wyatt desires to be remembered to you.

London, 3 October 1785: Wyatt to Boulton
. . . I will inform you of the State we are in – the Roof of the Building is compleated & we have just begun the Slating and enclosing the sides, it looks a tremendious piece of Timber Work & every body is struck with it. Malcom is upon the Engine which I hope he will now get forward very fast . . . I am afraid our beginning to work will be protracted by the Mill Work as it proves very tedious, the boiler will be finished in a few Days . . . I had the pleasure of meeting some of your friends at the Mill – Mr Fenton Mr Wilson & anr Gentlem . . . I have a journey to make for about a fortnight into Cheshire.

[Boulton had been worried by some reports that the building was 'reeling' and had to be supported by props. Watt explains (5 November 1785) that] . . . the building has never reeled an iota, but as it has no [internal] walls & consists merely of floor posts set upon one another, the props you mention were put up on all sides to prevent reeling. All the work at the mill seems extremely well done, the only thing wrong seen hitherto is expence which I fear will be very great.

[This is high praise indeed, from Watt, but he was not correct in his rather pessimistic note about expense. The total expenditure on the building, £13 000 in round figures, was certainly high; but the mill had a floor area of approximately 110 000 ft^2. The unit cost therefore worked out at 2*s* 4*d* per square foot, which is practically the same as that of a five-storey mill, of traditional construction, built in 1794-95 for John Marshall at Leeds, and considerably less than Strutt's 'fire-proof' West Mill at Belper.[10]]

Tatton Park, 20 December 1785: Wyatt to Watt
Your favour of the 12th has been forwarded to me here, I left London the same day . . . Messrs Call & Grant propose to take a 1/5 share in Albion Mill.

London, 13 January 1786: Rennie to Boulton
Albion Mill is now in great forwardness. I have no doubt of having her to work next month at furthest. But at present we have so many Carpenters &c that although she was ready it would be impossible to do anything with her.

[On 14 January 1786 Wyatt submitted a 'State of the Account'. This is given in full in Appendix I. It must represent practically the final cost of the building and shows still the same four shares, £2400 having been paid on each. It also shows about £7000 owing to Wyatt. Fortunately Call and Grant (see Wyatt's letter of 20 December 1785) decided in February to loan £10 000 to the company.

The engine had its first run on 15 February and the mill began work on 9 March (letters from Rennie to Watt). Boulton was there on this occasion and stayed for about a month. The engine needed a good deal of attention before it worked properly, and Watt arranged for his best man, the great William Murdock, to devote his skill to the problems.]

London, 12 May 1786: Wyatt to Boulton
I have the pleasure to tell you that the Engine & Mill are both working very well Mr Murdock has really been of great service.

London, 22 May 1786: Wyatt to Boulton
Mr Murdock left on Thursday last . . . The Engine worked from 6 on friday Night with 8Pr of Stones & on Saturday Morning continued to work the same number of Stones five Dressing Mills & both the Sack Tackles . . . In working the Mill 12 Hours we had ground 1000 Bus of wheat. Miss Boulton is at Sunbury Mrs Wyatt called today.

[Nancy Boulton, to whom her father was devoted, is frequently mentioned in the letters. She was probably about the same age as Louisa Wyatt.]

London, 26 May 1786: Wyatt to Watt
I am busy altering the Dock Gates to keep the Water abt 8 feet high into which we propose laying the Suction pipe of the Cold water pump at about 18In from the bottom.

[Rennie was not very happy at the mill, and in a letter to Watt dated 28 June 1786 complains of 'Mr Wyatts strange overbearing disposition' and 'his interference in Matters he is by no means a judge of'. As these remarks have been quoted by several authors, to Wyatt's disadvantage, it is relevant to add that shortly before this time Rennie's vanity had so annoyed Watt, and especially Boulton, that they were seriously discussing the possibility of dismissing him. Boulton adds the comment: 'In point of Natural ingenuity I consider him much inferior to Murdock but can express himself better' (Boulton to Watt, 24 April 1786).

On 4 July 1786 the connecting rod broke.]

Albion Mill, 12 July 1786: Wyatt to Watt
We did not get the connecting Rod compleated so that we could set to work till last Night at 4 o'Clock, we have made an exceeding good Job of it . . . it has half Kil'd me being up almost Day & Night . . . [but] I left the Mill last Night in good spirits.

More mechanical troubles followed and letters were being sent between the mill and Soho every few days. By mid-August, however, this side of the business was in good order.

London, 16 August 1786: Wyatt to Boulton
. . . the Mill has for this week past turned 10 p^r of Stones & Dressing Mills & Sack Tackles & the Engine never went so steady.

[From an engineering point of view the mill was now, and continued to be, a success. An idea of the scale of operations is given by the accounts for 1788 which show that sales ran at a rate of nearly £100 000 per annum[2] – and this was before the second engine began working. Wyatt's letters now become less frequent. He refers to the price of flour, the installation of a new boiler, ordering the second engine, timber for framing the second set of mill-work, installation of the engine and half a dozen other matters, but they need not concern us here. Quarterly accounts more often show a loss than a profit, yet Wyatt remained optimistic.]

Albion Mill, 12 January 1789: Wyatt to Watt
We flatter ourselves that we are now doing very well. As we receive every thing by Land Carriage & deliver in the same manner it creates a vast bustle at the Mill – some Mornings the Country waggons bringing wheat have reached to nearly the Middle of the Bridge from the Mill Gates.

[Indeed, he was devoted to the mill and put its interests, during the five years of its operation, above his other concerns. It might well be argued that he would have been better advised to leave the management more in the hands of the businessmen, but we have to realize that the mill was very much his own creation. Without Wyatt the mill would not have been there, the engines would not have been built and Rennie would not have made his reputation, as he certainly did, from the design of the mill-work.

It may be imagined, then, with what a sense of utter despair Wyatt watched the Albion Mill disappear in flames between 6 and 8 o'clock on the morning of 2 March 1791. And it was he who had to undertake the dismal business of negotiating with the insurance companies,[41] disposing of the engines and boilers and so on.

The Wyatts had moved to Albion Mill House in 1788 and were living there when the fire occurred. Their furniture was hastily carried into the street, a detail depicted by Rowlandson (Fig.40b), amid the jeers and shouts of the mob (letters from Rennie to Watt 2 March and 5 March 1791).]

Soho, 14 March 1791: Boulton to Wyatt
Amidst all the circumstances of our late dreadful Calamity it affords me an unspeakable pleasure to learn that you Mrs Wyatt & family have escaped unhurt & hope you have been preserved to see better days. I should have

instantly come to your aid . . . but I was ill with a sickness & head ache that has lately much incommoded me . . . Please to present my best & sincerest Wishes to Mrs Wyatt for her health & future happiness as well as our worthy Brethren in Misfortune. Excuse this hasty scrawl which I have not time to read & believe me Dear Sir I am your faithful friend.

[A month later Boulton wrote with many inquiries.]

Albion Mill House, 19 April 1791: Wyatt to Boulton
I have your kind favor of the 18th Inst – with respect to myself, owing to the great variety of objects that occupy my mind every Day, I am probably diverted from reflecting so much upon the Chaos of our poor Mill, which alas is no more – yet I cannot help declaring that at Times I am ready to sink with the loss of it – with respect to your Questions I will answer them as well as I am able. [Wyatt then deals at length and in detail with seven questions.

Life at the Mill House, after the fire, could not have been very pleasant for Mrs Wyatt and Louisa. As a postscript to an official letter dated 9 May 1791, about a meeting of the proprietors, we read:]
Mr Boulton's respects to Mrs Wyatt & recommends that she change the Scene & make an Excursion into the Country at this fine Season, as Miss B. will come to Town for a week or fortnight she as well as Mr B beg to have the pleasure of conducting Mrs Wyatt to Soho.

Albion Mill House, 11 May 1791: Wyatt to Boulton
I am requested by Mrs Wyatt to thank you for your kind invitation to Soho, it seems perfectly agreeable to her – Louisa begs her best respects & hopes she is to be of the party. I am at the same time directed to say that we have a Room at Miss Boultons service and Mrs Wyatt will be extreamly glad to accomodate her.

Soho, 27 June 1791: Boulton to Wyatt
. . . we have not fully followed up our enquiries into all the Clues we had found [concerning incendiaries] & therefore recommend to you to cause further enquiry to be made & every other Clue you can find to trace out this Dark affair . . . I have the pleasure to tell you that Mrs Wyatt & Louisa are perfectly well.

Soho, 12 July 1791: Boulton to Wyatt
I expected you would have dined with the Lunar Society at my house yesterday. Mrs Wyatt & Louisa are very well – you know we shall all be happy to see you at Soho.

[Wyatt had probably dashed up to the Midlands, without having time to call at Soho House, for he was very much engaged with drawing up plans for rebuilding the mill as a bonded tobacco warehouse.]

Albion Mill House, 18 August 1791: Wyatt to Watt
[I have sent to the Customs] 6 plans one of each Storey – the ground plan 14 feet high the floor of which will be abt 2 feet above the present Curb of the Dock – which is 6In above the High Tide in Mar last and the 2nd floor will be about 1ft or 18In above the level of the Street above which we have 4 floors of

9ft 6 each (these heights will bring us to the Base of the Roof of our present walls) there was also an Elevation of the River front & the West front to the Street – which were forwarded on Monday last Officially from the Treasury . . . I believe I have seen every Tobacco Merch^{t} thats of consequence . . . in a few days I shall get a memorial from them to the Treasury in our favor.

[The six plans are listed as drawings 24-29 in the catalogue.]

8 February 1792: Wyatt to Watt
Our 2nd Memorial to the Treasury . . . has been forewarded to the Customs and they have referred it to the Tobacco Officer . . . Dance & Jupps plans went to the Board of Customs yesterday.

4 July 1792: Wyatt to Watt
We still hope to hear from the Treasury . . . Mrs Wyatt and Louisa beg to join in best Complmts to yourself & Mrs Watt.

[On 6 December 1793, after various meetings and postponements, Wyatt submitted to the Treasury 'The Memorial of Samuel Wyatt on behalf of himself and other proprietors of the Albion Mill premises'.This refers to the tobacco warehouse scheme and points out that 'it is now more than two years & four months since he had the honour to offer his proposals to your Lordships consideration'. A week later he enclosed a copy of the memorial, in a letter to Watt, and says that Mr Pitt has taken it to read. This business dragged on until the middle of 1794, but finally nothing happened. One source of trouble may have been that the site was upstream of London Bridge, still a medieval hazard to navigation, but in the 1790s the authorities were excessively dilatory in providing docks and warehouses on the Thames, despite urgent representations from merchants and shipowners.[42]

On 11 July 1795 Wyatt wrote to Boulton and Watt telling them that a Bill before Parliament for warehousing coffee had been passed, and the proprietors had agreed on 6 July that he should frame a proposition for rebuilding the mill accordingly. The plans (Nos 30-36) were submitted on 17 August 1795. Among the correspondence on this matter Boulton adds a note on 14 August:]
I beg you will present my kind respects to Mrs Wyatt requesting [her to explain] the best means of makeing the best Brown Bread from English Wheat . . . I remember with pleasure the many sweet morsels I have eaten at A. Mill & wish to eat such at Soho. I have no Maid or Housekeeper that knows how to make Bread . . . I will therefore be very thankful to Mrs Wyatt to write me all necessary particulars.

Albion Mill House, 29 August 1795: Ann Wyatt to Boulton
I fear you think me strangly neglectfull of the request you made of sending a receipt to make good bread. I have made two attempts the last of which was yesterday we think good and the following is the exact method made use of [detailed recipe then follows]. Will you be so good to present Louisa's best respects with mine to Miss Boulton & Comp^{ts} to Mr Mattw.

[To this domestic information Wyatt adds a sad PS:] I am sorry to tell you that our negotiation with the Bd of Excise has not gone on so pleasant as we had hope of.

8 September 1795: Wyatt to Boulton
My brother James has lately been at Durham [he was soon to start designs for rebuilding Soho House] . . . for some time past I have never thought of any thing but coffee. I should be happy to spend a few days at Soho Mrs Wyatt & Louisa join in thanking you for the kind invitation.

Albion Mill House, 16 November 1795: Wyatt to Boulton
. . . the Board of Excise after starting many difficulties & receiving proposals for this last 3 M^{o} have made a report to the Treasury intirely in favor of our premises. My Estimates for reinstating the Building & fitting it up completly for the Service of Coffee Warehouses = £12000 & further for Iron plating the several floors story posts &c £4000.

[But once again Wyatt's hopes had been raised only to be met by delays and indecision. Nearly a year later he writes:]

Albion Mill House, 17 September 1796: Wyatt to Boulton
We are still persevering with the Treasury for Coffee Warehouses and I have been with the Ld Mayor who has an Idea that the Corporation of London sh^{d} erect a Mill upon the site of the late Mill.

[A few months later George Dance visited the site, but by March 1797 it was only too clear that nothing was going to happen to this scheme either.

Lest it may seem that Wyatt was engaged solely on these unsuccessful proposals, we should recall that his finest work, Trinity House, dates from 1793-96; so the picture is not so dark as appears from the letters which, for the purpose of the present paper, have been restricted to Albion Mill affairs.

The correspondence between Wyatt and Boulton continued, but with little about the mill. Indeed, in March 1800 Boulton and Watt withdrew altogether from the partnership, and Wyatt had moved to Chelsea a year before. The last letters refer to a competition for rebuilding the mill in fire-proof construction.]

Chelsea, 2 January 1802: Wyatt to Boulton
You will by most of the Morning prints of Thursday last see an Advertisement for a Plan of a Mill to be worked by Steam & built fire proof, this step the Managers took on Wed^{n} last without the least reference to me, since they hold me so cheap.

[Boulton was in town, and Wyatt writes on 6 and 11 January hoping to meet him], you will find me [at the Mill House] every day if any thing should draw you this way [and he asks for some details of a 36 hp engine.

Wyatt drew up a scheme for an iron-framed structure, using the system of construction described in his patent of 1800, in which the external walls are no longer load-bearing but act simply as walls of sufficient thickness to provide lateral stability. This was a more advanced concept of multi-storey iron structure than anything produced before, or for a long time after, that time. His drawings (Nos 22 & 23) which have survived are those sent to Boulton and Watt to provide information for detailing the engine and are not complete. I have not seen any later documents of Wyatt's concerning the mill.]

APPENDIX I

State of the Account for Building the Albion Mill to Jany 14th 1786 including the Engine, Mill Work &c

				£	s	d
To Sam Wyatts Account of Money paid to Workmen				3582	16	1
To Do. for Materials				8670	11	6¼
To Do. paid to Millwrights by J. Rennie	829	14	9¼			
To Do. what he paid for Materials	49	10	4			
To Do. paid to Smiths	268	8	4½			
				1147	13	5¾
To Do. Payments for Rent, Purchase, Law Charges, Insurance & incidental Expenses				1028	12	9
				14,429	13	10
To sundry Bills on Acct of the Engine & Millwork which remain unpaid Jany 14 1786				2340	5	9
				16,769	19	7

	£	s	d
By John Frere Esq at 8 Installments of £300	2400		
By Joah Bates Esq Do.	2400		
By Messrs Matthews, Boulton & Watt	2400		
By Sam Wyatt Do.	2400		
	9600		
Balance remaining unpaid	7169	19	7
	16,769	19	7

APPENDIX II:
SELECTED DRAWINGS AND PRINTS OF THE ALBION MILL

(Dimensions in inches)

BOULTON & WATT COLLECTION

A Wyatt's site surveys

1 Plan

Scale 1 in to 8 ft. 25×21. Sent to Boulton 26 March 1783.

2 Plan, and

3 Four sections along lines noted in No. 2

Scale 1 in to 20 ft. 18×15. Earlier than September 1783, probably July.
Tracings from Nos 2 & 3 reproduced by Skempton, *loc. cit.*

B Wyatt's first drawings of the mill

4 *West front of the Corn Mill at Blackfriars Bridge*

5 *River front of the Corn Mill at Blackfriars Bridge*

6 *Section from East to West*

7 *Plan of the 2nd floor containing the Mill Work*
8 *Plan of the 3rd Story or Mill Stone floor*
Scale 1 in to 8 ft. Line and wash. 28×20. Sent to Boulton 19 July 1783. The mill as built differs in several respects from these drawings, but all the basic features of the superstructure are present.

C Drawings for the first engine by James Watt
Among this set we may note:
9 *Cylinder for Blackfriars. August 1783*
10 *The Albion Mill Engine. March 1784*
11 *Reverse Drawing of Parallel Motion for Albion Mill. May 1785.*
The earliest known example of Watt's parallel motion mechanism (Dickinson & Jenkins, *op. cit.*, p. 200 & pl. 66). Like many drawings in the collection this is a reversed copy of the original, made by the copying-press, with lettering inserted afterwards.

D Wyatt's general section of the mill
12 *Section of Albion Mill from East to West*
Scale 1 in to 3 ft. Line and wash. 50×34. Sent to Watt 4 November 1784. This is virtually the final design, except that some extra vertical posts were added to the framing. Reproduced here as Fig. 37a.

E Drawings of the mill work, and engine assembly.
Reverse copies, various sizes and scales. All dated November 1784. The first of this set is of interest in the present context:
13 *Reverse No. 1 Albion Mill. November 1784*
Scale 1 in to 4 ft. 30×24. Redrawn from No. 12 with a general assembly of the mill-work, engine and other machinery added and the voussoirs in the inverted arch foundations sketched in. Reproduced by Mosse, *op. cit.* A drawing combining Nos 12 & 13 with original ground level shown, from the site surveys in Nos 2 & 3, is given by Skempton, *loc. cit.*
14 *Reverse Drawing A. 25 Nov. 1784*
This is an extra to the above set, showing details of the engine house.

F Part view of the east wall
15 Scale 1 in to 3 ft. 24×31. Probably by Wyatt, late 1784 or early 1785. Shows four bays of the east wall with a chimney and circular openings in the wall above the boilers.

G Wyatt's plans of the mill in 1787
16 *Plan of the Ground floor. Albion Mill*
17 *Plan of the 1st Floor on which are the Meal troughs*
18 *Plan of the 2nd or Millstone floor* [Fig. 37b]
19 *Plan of the 4th floor on which are the Dressing Mills &c*
Scale 3 in to 20 ft. 24×18. Plans of the eastern part of the mill, showing machinery, walls, framing posts &c.
20 *Plan of Dressing Mills 4th floor*
Scale 1 in to 3 ft. 24×18. Details of machinery.

21 Section of boiler house
Scale 3in to 4ft. 24×18.
These six drawings were sent to Watt on 17 February 1787. They are endorsed in a different hand *Albion Mill as executed – taken May 1787.*

H Drawings for the second engine, ranging in date from February 1787 to July 1787, by James Watt.

J Wyatt's drawings for rebuilding in 1802
22 Plan (signed) *S. Wyatt Feby 26th 1802*
Scale 1in to 12ft. Line and wash. 18×15.
23 Section
Scale 1in to 4ft. Line and wash 30×19. Shows iron framing.

PUBLIC RECORD OFFICE
K Wyatt's proposals for rebuilding the mill as a tobacco warehouse (MR99 & MPD128)
24-29 Plans of the ground floor and five upper storeys. These plans and the elevations (missing in the PRO) are mentioned in Wyatt's letter to Watt, 18 August 1791. The plans show timber framing posts.

L Wyatt's proposals for rebuilding the mill as a coffee warehouse (MPD124 (elevations) & MR98 (plans))
30 *West Front of the proposed Warehouse for Coffee* (signed) *Sam Wyatt August 17th 1795* [Fig.39a]
Line and wash and watercolour. 24×18.
31 *River Front of the proposed Warehouse for Coffee* (signed) *Sam Wyatt August 1795* [Fig.39b]
32 *Plan of the Ground Floor of the proposed Coffee Warehouse* (signed) *Sam Wyatt August 1795*
33-36 Similar plans for the four upper storeys
Scale 1in to 8ft. Line and wash. 22×21. All signed as No.31. The plans show timber framing posts but as an optional extra Wyatt suggested encasing the posts and covering the floors with iron plates (letter to Boulton 16 November 1795). A sixth plan (No.37) shows an enlarged scheme.

GUILDHALL LIBRARY (PR S2/ALB)
38 *The Albion Mill, Blackfriars Bridge. Eastgate sculpt.*
Line engraving. 6½×3¼. Published in *The New London Magazine* vi (July 1790), 329. [Fig.38]. The well-known woodcut in Smiles's *Lives of the Engineers* ii (1862), 138 is based on this engraving.
39 *A View of the Albion Mills on Fire. On Wednesday Morning March 2nd 1791, with part of Black Friars Bridge, Albion Buildings &c. J. W. Edy del. et sculp.*
Coloured aquatint. 14¾×10¾. [Fig.40a].
40 View of the ruins of the Albion Mill
Watercolour drawing. 9¼×6¼. (signed) *SW 1803*. Shows the west front and one bay of the south wall standing.

'MICROCOSM OF LONDON', VOL.2

41 *Plate 35 Fire in London. Pugin & Rowlandson delt. et sculpt. J. Bluck Aquat. London Pub. 1st Sept. 1808 at R. Ackermann's Repository of Arts*
Coloured aquatint. $10\frac{1}{4} \times 8$. [Fig. 40b]

NOTES

1 I am indebted to Mr A. H. Westwood, Assay Master, and to Mr A. Andrews, Archivist of the Boulton & Watt Collection, for their help and courtesy to me while working at Birmingham. Permission to reproduce drawings and prints has been granted by the following authorities: Birmingham Reference Library (Figs 37a & b), Guildhall Library (Figs 38, 40a & b), Public Record Office (Figs 39a & b). I wish also to thank Mr J. M. Robinson for information.
2 O. A. Westworth, 'The Albion steam flour mill', *Economic History* ii (1932), 380-395.
3 A.D. Insull, 'The Albion Mill Story', BA (Hons) thesis, University of Nottingham (1955), 87pp.
4 John Farey, *A Treatise on the Steam Engine, Historical, Practical and Descriptive* (1827), 438, 442-444, 514-515.
5 H.W. Dickinson & Rhys Jenkins, *James Watt and the Steam Engine* (Oxford 1927), 165-166, 200, 214, 220.
6 Samuel Smiles, *Lives of the Engineers* ii (1862), 129-133, 136-141.
7 Samuel Smiles, *Lives of Boulton and Watt* (1865), 353-359.
8 J. Mosse, 'The Albion Mill', History thesis, Archit. Assoc. (1963), 37pp, published, in part, in *Trans. Newcomen Soc.* xl (1971), 47-56.
9 A. W. Skempton, 'The Albion Mill foundations', *Geotechnique*, xxi (1971), 203–210.
10 H.R. Johnson & A.W. Skempton, 'William Strutt's Cotton Mills, 1793-1812', *Trans. Newcomen Soc.* xxx (1956), 179-201; A.W. Skempton & H.R. Johnson, 'The first iron frames', *Arch. Rev.* cxxxi (1962), 175-186.
11 There are numerous contemporary accounts of the fire e.g. *Gent's Mag.* 1791, pt i, 274-275. See also letters from Rennie to Watt 2 March and 5 March 1791 (Boulton & Watt Colln).
12 Edward Walford, *Old and New London* vi (1878), 383: 'The front of the mill remained for many years unrepaired but was subsequently formed into a row of handsome private habitations.' Just when this conversion took place I have not discovered, though it was certainly after 1803.
13 Blackfriars Bridge was designed by Robert Mylne and constructed under his supervision 1760-69.
14 The quotation is given by A. Oswald in his informative articles on Tatton Park, *Country Life* cxxxvi (1964), 162-165, 232-236, 292-296.
15 It is significant that George Richardson in *The New Vitruvius Britannicus* (2 vols, 1802, 1808) chose three of Samuel Wyatt's buildings for illustration. Of the forty-four architects whose works are illustrated, few receive so much attention and only John Carr of York is more frequently represented.
16 Wyatt Papworth (ed.), *Dictionary of Architecture* viii (APS, 1852-92), 78. The Weeford parish registers record Samuel Wyatt's baptism on 27 December 1737. The baptisms of his six brothers, including James, and his two sisters, also appear in these registers, the children of Benjamin and Mary Wyatt.

17 *Gent's Mag*. 1807, pt i, 189: 'Feb. 8. At his house at Chelsea, Samuel Wyatt esq a celebrated architect and clerk of the works at Chelsea Hospital.'
18 PCC, 1807, f. 162.
19 Letters dated 31 July 1765, 6 March 1766, 12 June 1766 (Assay Office).
20 Letter to John Scale, 9 September 1769 (Assay Office); *Survey of London* xxxiv (1966), 345.
21 *Survey of London* xxxi (1963), 232-233.
22 The timber merchant's business was carried on at 63 Berwick Street, 1799-82, at St Catharine Dock, 1782-88, and at 369 Oxford Street, 1789-93. Wyatt held the Office of Works contract for carpentry in the Westminster district from 1780 until his death in 1807 (*ex. inf.* J. Mordaunt Crook).
23 'Samuel Wyatt builder' or 'architect, surveyor & builder', at 63 Berwick Street, are entries in the local directories up to 1803.
24 Royal Academy catalogues, 1785-98.
25 H.M. Colvin, *A Biographical Dictionary of English Architects, 1660-1840* (1954), 734-736.
26 *Survey of London* xxxi (1963), 270-273.
27 D.L. Broughton, *Records of an Old Cheshire Family* (1908), 40, 76-78; G. Richardson, *op. cit.* (1802), pl. 57-60; G. Nares, 'Doddington Hall, Cheshire', *Country Life* cxiii (1953), 344-347.
28 *List of the Proprietors for Building a Playhouse in New Street, Birmingham, with Minutes of their Meetings*, MSS Birmingham Ref. Library.
29 G. Richardson, *op. cit.* (1808), pl. 17-22; *Life of Thomas Telford* (1838), 19; C. Hussey 'Admiralty House, Portsmouth', *Country Life* cxxxv (1964), 774-777, 834-837.
30 Letters to Matthew Boulton 10 December 1787, April 1788 (Assay Office).
31 Letters to James Watt 17 April 1787-29 November 1790 (B&W Colln); drawings, building accounts &c. in Muirhead Colln, Birmingham Ref. Library.
32 *Ex. inf.* J. M. Robinson.
33 *Survey of London* xxix (1960), 148-153.
34 G. Richardson, *op. cit.* (1802), pl.22-23; H. A. Tipping, 'Trinity House on Tower Hill', *Country Life* xlv (1919), 460-467; Trinity House MSS.
35 *Survey of London* xxx (1960), 508-509.
36 PRO plans MPD.40 and MPD.60; House of Commons *Reports* xlvii (1795-96), 155-156, 158-159, 160-163 and Appendix Oo (with plans and drawing).
37 Letters to Boulton 11 February 1799 (Assay Office), 20 December 1799 (B&W Colln).
38 Cockerell Diaries, *ante*, 16.
39 Trinity House MSS.
40 Mylne's diary has been published by A.E. Richardson, *Robert Mylne Architect and Engineer 1733 to 1811* (1955).
41 The building and machinery were insured for £26 000 and the stock for £15 000. The insurance companies chose to pay cash rather than reinstate. For details, with quotations from two of Wyatt's letters, see H.O. Raynes, *A History of British Insurance* (1964), 196-200.
42 Old London Bridge was not replaced by Rennie's fine structure until 1832. For the great but unsatisfied need for docks and warehouses, see Sir Joseph Broodbank, *History of the Port of London* i (1921), 77-91.

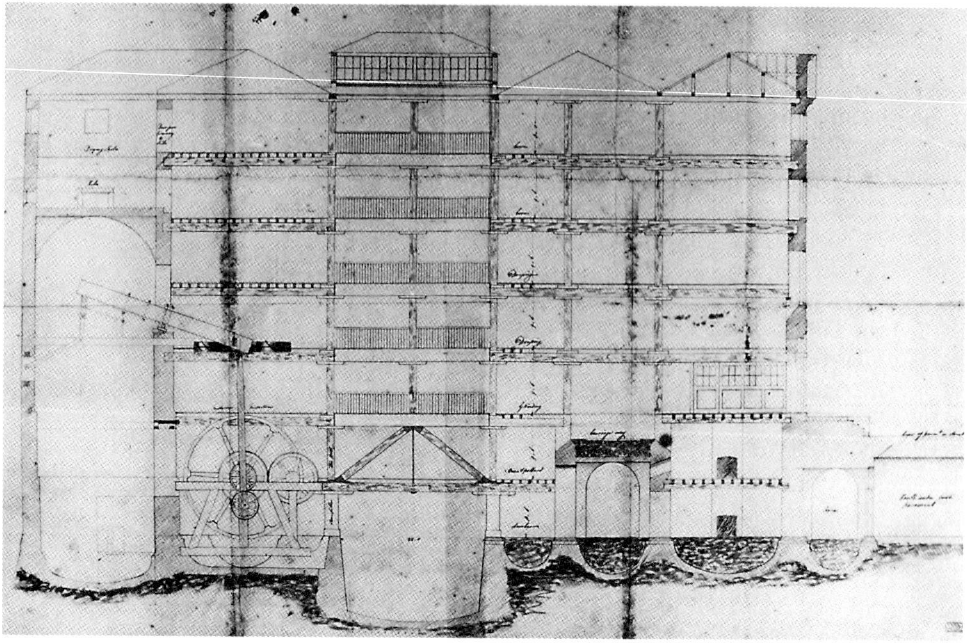

Fig.37a *Albion Mill: cross-section by S. Wyatt (November 1784)*

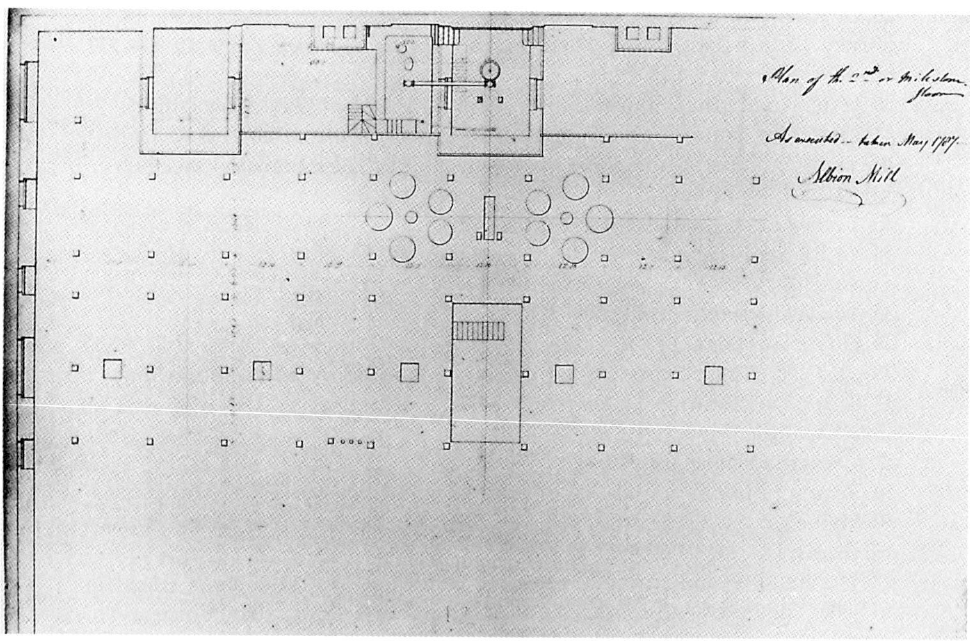

Fig.37b *Albion Mill: part of second floor, plan by S. Wyatt (February 1787)*

Fig. 38 *Albion Mill: engraving published July 1790*

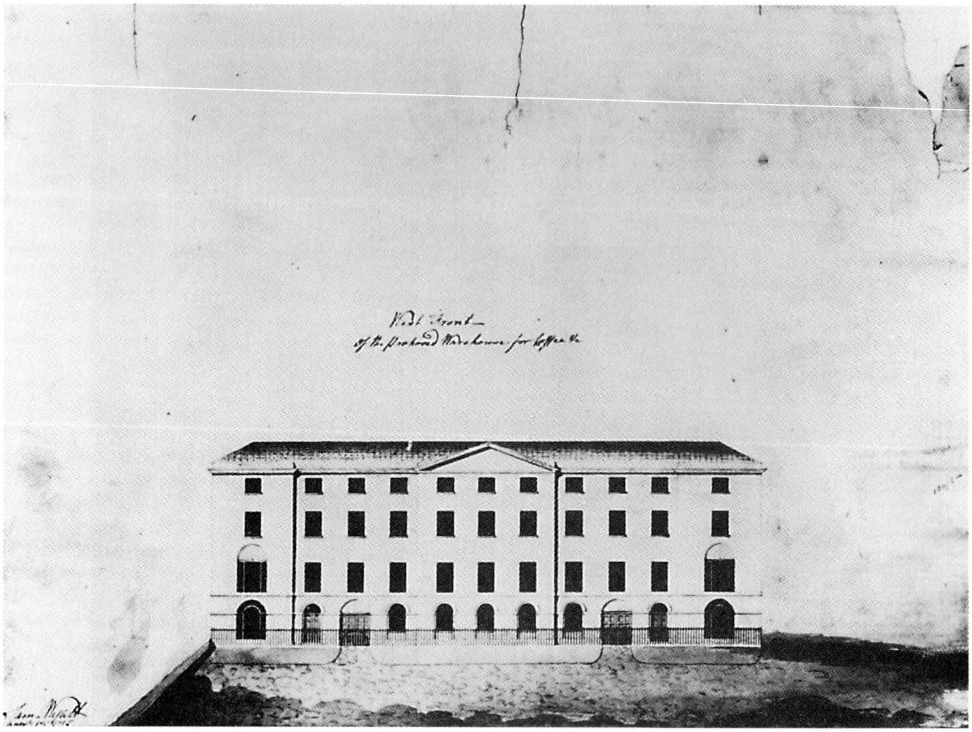

Fig. 39a & b *Albion Mill: elevations of west and river fronts by S. Wyatt (August 1795)*

Fig. 40a *Albion Mill on fire, March 1791: aquatint by J. W. Edy*

Fig. 40b *Albion Mill on fire, March 1791: engraving by Pugin and Rowlandson*

X

William Strutt's Cotton Mills, 1793-1812

BY

H. R. JOHNSON, B.A., A.R.I.B.A.,
and Professor A. W. SKEMPTON, D.Sc., A.M.I.C.E., Member

Introduction

Throughout the greater part of the eighteenth century the textile mills in England were constructed with traditional building technique, using brick or masonry external load-bearing walls, and floors of wooden boards on joists supported by timber beams. The beams in some mills spanned clear from wall to wall, but in other cases the beams were, additionally, each supported by one or two timber columns. The mills were four to six storeys high, usually about 30 ft. wide and roughly 100 ft. in length; these proportions being a function of the mechanical equipment, the need for good day-lighting and the limited power of waterwheels and the early steam engines.[1] The timber-framed mills were economic in first cost, easily built, and readily adaptable to changing requirements. But they were extremely susceptible to fire; especially when filled with such combustible material as cotton. Among those who suffered from this cause were Jedediah Strutt and Richard Arkwright. In 1781 a fire broke out in their mill at Nottingham and

> "raged with such fury, that in two Hours after, the spacious Building was reduced to a mere Shell; all the Machines, Wheels, Spindles &c. employed for spinning and winding Cotton, were entirely consumed . . ."[2]

But this was only one of many similar occurrences, and the need for a fire-resistant construction was very real.

The final stimulus towards a solution of the problem may have been the destruction in March 1791 of the most advanced industrial building in Europe of the period: the Albion Mills in London.[3] At any rate, it was early in 1792 that work began in Derby on the first successful fire-proof mill; designed by William Strutt, F.R.S. (1756–1830), the son of Jedediah. In this six-storey building the load-bearing walls were retained, but the floors consisted of brick arches, paved with brick tiles. The arches sprang from heavy timber beams, protected on the under-side by plaster, and supported by the walls and by two rows of cast-iron columns. Thus, no timber was exposed. The Derby mill was demolished some eighty years ago, but in 1793 Strutt built a four-storey warehouse at Milford, and in 1795 he completed a large six-storey mill at Belper. Both these buildings still exist practically in their original forms and both have the same structural details as the mill at Derby. These were the first multi-storey fire-proof buildings. But they were also the origin of an entirely new form of structure. For in 1797 Charles Bage (1752–1822), who was a friend of Strutt and received much help from him, completed a five-storey flax mill at Shrewsbury, modelled closely upon the Derby

---

[1] W. H. Pierson, "Notes on early industrial architecture in England," *Journ. Soc. Arc. Historians*, Vol. 8, 1949, pp. 1–32.

[2] *Derby Mercury*, 15 November 1781, reporting the fire of 8 November.

[3] *Annual Register*, 3 March 1791.

and Belper mills, but using cast-iron instead of timber for the beams. Thus the earliest iron-framed building was produced; from which the modern steel-framed structures have evolved.[1]

In the subsequent development of iron-framing Strutt also played a part, notably in the North Mill at Belper, erected 1803–04. But his great achievement was in the design first realised in the Derby mill. For this constitutes the real break with traditional construction, and is one of the major turning points in the history of structure. The value of his contribution was recognised by Strutt's contemporaries. Yet within a generation he was forgotten, except by his family, and the present revival of interest in his work is due largely to an important paper published in 1950 by Professor Bannister.[2] In the same year, by a curious coincidence after more than a century of neglect, the Derby mill was described by Mr. W. D. Shepherd[3] who made a correct assessment of its historical significance. Further information, particularly concerning the 1795 mill at Belper, was provided in 1953 by Dr. R. S. Fitton,[4] in a thesis dealing chiefly with the industrial history of the business founded by Jedediah Strutt which, under the direction of his sons William, George and Joseph, became in the first decades of the nineteenth century the largest cotton firm in England.

Stimulated by Professor Bannister's research and, later, deriving much of value from the contributions of Mr. Shepherd and Dr. Fitton, we commenced our investigations in 1953. A great deal of willing help has been given by many individuals and firms, especially the English Sewing Cotton Company, the present owners of the mills at Belper and Milford. A reasonably definitive account can now be made of William Strutt's buildings.

Derby Cotton Mill, 1792–93

In 1802 Britton and Brayley, the noted topographers, wrote:

> "At Derby . . . various works have been established . . . (and) those belonging to Messrs. Strutts' for the manufacture of cotton are particularly ingenious. . . . One of the mills belonging to these gentlemen, is remarkable for its floors being constructed on brick arches, and paved with brick by which means it is rendered absolutely indestructible by fire. This building is six stories high, 115 feet long and 30 feet wide; it was erected in the year 1793, and was the first *fire-proof* mill that was ever built. The second mill constructed on the same plan, was erected by the same proprietors at Belper; but two others have since been built on a similar principle; the first, a flax mill at Shrewsbury: and very lately, an extraordinary large cotton mill at Manchester. From the extreme risk of fires in cotton factories, and the enormous insurance that is demanded in consequence, there can be no doubt but this mode of building will be greatly extended."[5]

---

[1] It might be expected that fire-proof construction would have been used at an early date in warehouses at the docks. But this seems not to be the case; and as late as 1807 Rennie was advising the Navy to build new warehouses at Northfleet "upon the plan of some of the cotton and flax mills lately erected at Derby, Leeds, Manchester and Glasgow." Report of 14 May 1807 quoted by Sir John Rennie, *The Theory, Formation and Construction of Harbours*, London, 1854, p. 59.

Strictly speaking, this construction was not "fire-proof." Under a prolonged attack by fire the exposed bottom flanges of the beams and the exposed columns would lose their strength and collapse. It would now be called "fire-resistant construction."

[2] T. Bannister, "The first iron-framed buildings," *Arch. Review*, Vol. 107, 1950, pp. 231–246. The bibliographic aspect of this paper is particularly valuable.

[3] W. D. Shepherd, "Early Industrial Buildings 1700–1850." Thesis submitted in 1950 for final examination R.I.B.A. We are grateful to Mr. Shepherd for the loan of the sole copy of this Thesis, and for permission to make full use of his work.

[4] R. S. Fitton, "W. G. and J. Strutt, 1758–1830: A Study in Social and Industrial Organisation," Ph.D. Thesis (1953), University of London Library.

[5] J. Britton and E. W. Brayley, *The Beauties of England and Wales*, London 1802, Vol. 3, p. 364. This description of the mill was repeated by S. Glover, *History of the County of Derby*, Derby, 1829, Vol. 2, p. 424 and other writers.

WILLIAM STRUTT'S COTTON MILLS, 1793–1812

This is the *locus classicus* of our subject, and it indicates, at once, the historical importance of the Derby mill. That William Strutt himself was responsible for its design can be amply demonstrated and much of the evidence will be quoted in this Paper, but for the present the testimony of his son must suffice. Writing in 1831, Edward Strutt said of his father,

"He was the first person who attempted the construction of fire-proof buildings on a large scale in this country, and with the most perfect success. The great improvements made of late years in the formation of Castings in iron, have given great facilities to this mode of construction, which is now very extensively in use."[1]

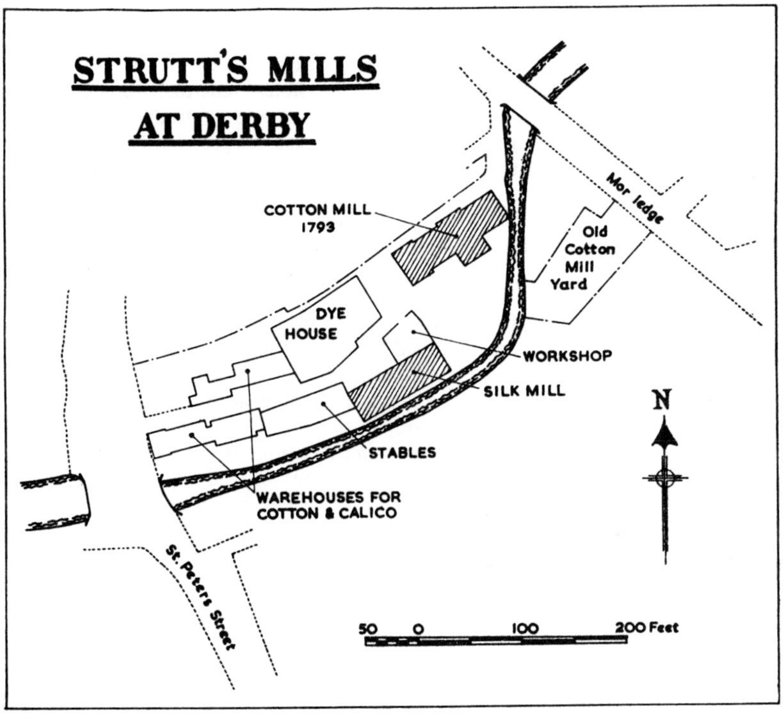

Fig. 1

The "mode" of construction, mentioned by Edward Strutt and by Britton and Brayley, is essentially the brick arch floor. It was this feature which chiefly impressed Strutt's contemporaries; and the use of brick arches instead of wooden floors was unquestionably an improvement of the greatest consequence. Yet today we are at least as interested in the method of supporting these arches, for

[1] "Memoir of William Strutt Esq., F.R.S.," *Derby Mercury*, 12 January 1831. The article is unsigned; but E. Baines, *History of the Cotton Manufacture in Great Britain*, London, 1835, p. 205, attributes the memoir to Edward Strutt (later, in 1856, Lord Belper).

it is here that we see the origin of the modern framed structure. Before discussing this aspect, however, we must establish the location and date of the mill.

In 1792, Boulton & Watt of Soho, were designing a sun and planet engine for a cotton mill for Messrs. J. Strutt & Sons of Derby; and in a letter of 21 June 1792 from Strutts to Soho, a part plan of the mill was enclosed showing the engine-house, the adjacent walls and the end of a neighbouring silk mill.[1] There is sufficient information on this drawing to prove that the mill with the engine-house is the "Cotton Mill" shown on Brayley's map of Derby (1806), at a site 150 yd. south-east of the Market Place, and, in more detail, on an Estate map of Strutts' Derby property[2] made in 1820 (Fig. 1). Moreover the dimensions of the mill scaled from this map, agree with those on a plan made by Boulton & Watt in June 1806 for the installation of gas lighting in "Messrs. Strutts' Mill in Derby." And a section of the mill drawn on the same sheet reveals a six-storey building with arch floors and iron columns.[3]

The identity and location of the mill are therefore defined. As to the dates of construction, the earliest document still surviving is a sketch of the engine-house sent to Strutts by Boulton & Watt on 19 March 1792, probably as a guide in setting out this part of the foundations. But the details of the engine-house cannot have been finalised by that date, for the plan of 21 June, previously mentioned, shows the flues in a somewhat different position. Drawings of the engine were ready by the end of May and, since the legal agreement between Strutts and Boulton & Watt was drawn up in November 1792, this may be the date on which the engine parts were delivered for erection at Derby. As will be seen later, the ceiling of the top storey was probably built soon after May 1793, and it is therefore likely that the mill was completed in the summer of that year; about 18 months after construction had begun.

Turning now to the structure itself, the plan and sections are shown in Fig. 2. They have been reconstructed from the drawings of June 1792 and June 1806, and the map of 1820. At ground level the mill was 115 ft. long and 31 ft. wide, overall, with an internal width of 27 ft. The walls were of brick, diminishing in thickness from 24 in. to 13 in. in the upper storeys where the internal width was 28 ft. 7 in. There were five central bays with arch spans of 8 ft. and three bays on either side of the central section with spans of 9 ft.; and a two-bay wing was built at the west end.

The beams supporting the brick arches rested on two rows of iron columns, dividing the internal width into three spans of 9 ft. The ceiling of the top storey was formed of hollow pots built in arches springing from the tie beams of the roof, and the resulting lightness of construction eliminated the need for columns in this storey. The floor to floor height of the first and second storeys was 10 ft. 3 in. and 10 ft. 7 in. respectively, and 9 ft. 10 in. in the upper storeys. The column height, however, was 8 ft. 2 in. throughout, the rise of the arches being greater in the lower storeys.

Hollow pots were coming into favour at this period, their use (which dates from Roman times) having been revived in Paris in 1785.[4] They were incorporated in the Palais Royal theatre completed in 1790,[5] and in October 1792, in response to a request for information concerning this building, Strutt received a letter from John Walker who had lately been in Paris and had arranged for samples of the pots to be sent to England.[6] Whether Strutt ever received them is uncertain, but in May 1793 he learnt from Matthew Boulton that hollow pots had been successfully used by George Saunders in

---

[1] All the drawings and letters connected with Boulton and Watt, and referred to in this Paper, are in the Reference Library, Birmingham. We are indebted to Miss W. R. N. Payne for her help in the Boulton and Watt Collection.

[2] The original is reproduced by R. S. Fitton, *op. cit.*, p. 375.

[3] The plan and section of June 1806 are reproduced by W. D. Shepherd, *op. cit.*

[4] Bannister, *loc. cit.*

[5] H. Prudent and P. Gaudet, *Les Salles de Spectacle construites par Victor Louis*, Paris, 1903, pp. 33, 42.

[6] This letter is given in "William Strutt: a Memoir," p. 28, a typescript written about 1895 by a member of the Strutt family, in the Derby Public Library. Bannister quotes part of the letter.

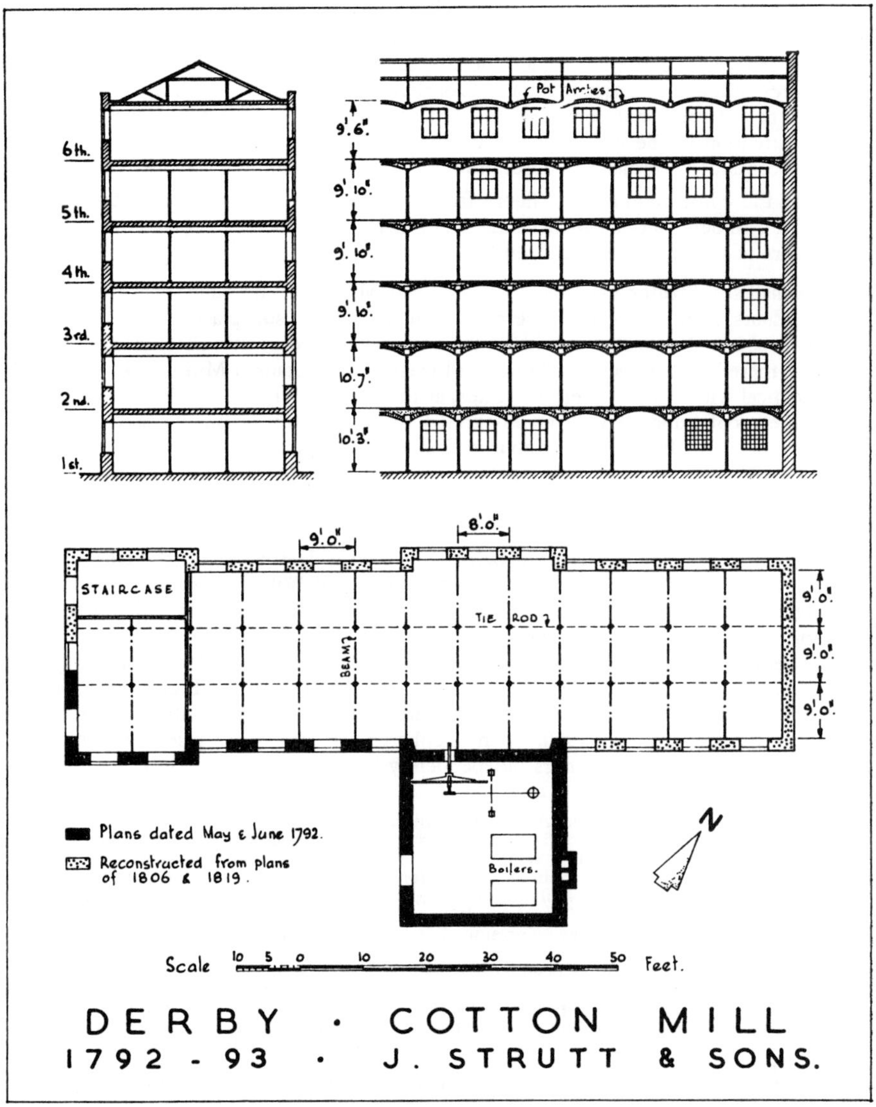
6th.
5th.
4th.
3rd.
2nd.
1st.
9'.6".
9'.10".
9'.10".
9'.10".
10'.7".
10'.3".
Pot Arches
9'.0".
8'.0".
STAIRCASE
TIE ROD
BEAM
9'.0".
9'.0".
9'.0".
Boilers.
N
Plans dated May & June 1792.
Reconstructed from plans of 1806 & 1819.
Scale 10 5 0 10 20 30 40 50 Feet.
DERBY · COTTON MILL
1792 - 93 · J. STRUTT & SONS.

Fig. 2

London,[1] and it may well be that he then finally decided in their favour for his mill at Derby. The pots, like those later used by Strutt at Milford and Belper, were made at the Smalley Common pottery.[2]

Regarding the more important question of the beams, it is clear from the drawing of 1806 that they were about 12 in. wide, although no details are given; and we know that they were made of timber. For, on 13 July 1853, a fire started in the loft, destroying the roof trusses and thus leading to a collapse of the sixth storey ceiling. Practically the whole of the remainder of the building survived (a remarkable testimony to the fire-proof construction), but several lives were lost and the event naturally gained public attention in the newspapers. Fortunately, for our purpose, the report given in the *Daily Mercury* contained a rather full description of the mill including a statement that it

> "differs in the construction of its floors from modern fire-proof buildings, the arches being formed of pot cylinders, about 7 inches long and 3 inches in diameter . . . (and) the girders instead of cast iron are of Baltic fir, cased in iron."[3]

Now, as we shall see, Strutt completed a warehouse at Milford in the same year as the Derby mill. This has arch floors, iron columns and beams of Baltic fir 12 in. sq., plastered on their underside. Moreover, the wooden skewbacks supporting the arches are cased in thin sheet iron, and the beams span 9 ft. It is therefore extremely probable that the structural details at Milford, which are shown in Fig. 5, are identical with those at Derby. Thus, at the Derby mill, William Strutt erected the first multi-storey fire-proof building with an integrated frame structure. But he used timber beams. Therefore this mill cannot be accepted as an iron-framed building, as claimed by Professor Bannister.[4] It represents, nevertheless, a major development in structural history.

MILFORD WAREHOUSE, 1792–93

At Milford, little more than a mile from Belper, the road to Derby crosses the Derwent on a fine two-arch bridge, completed in 1792 probably to the designs of William Strutt.[5] From this bridge can be clearly seen a part of the Old Mill (Plate XL, Fig. 1), the nucleus of which was erected by Jedediah Strutt soon after he and Arkwright had bought the site in 1777.[6] But rather less conspicuously placed, between the Old Mill and the road, is the highly important four-storey structure built as a warehouse, and now known as the Cruciform Building (Plate XL, Fig. 2 and Figs. 3 and 4).

The Milford warehouse seems to have received little attention from contemporary writers and none from modern historians until 1950, when Shephered (*op. cit.*) gave a brief description, with an approximate date of 1795. The Authors independently investigated the building in 1953 and again

---

[1] The letter, dated 8 May 1793, is in the Assay Office, Birmingham. Boulton wrote: "I understand you have some thoughts of adopting the invention of forming Arches by means of hollow pots and thereby saveing the use of Timber in making Floors, and guarding against Fire. Allow me to say that I have seen at Paris floors so constructed, and likewise at Mr. George Saunders' in Oxford St., London, who is an eminent Architect . . . I have therefore no doubt but it might be applyd also with great success and security to a Cotton mill."

[2] John Farey (senior), *General View of the Agriculture of Derbyshire*, London, 1815, Vol. 2, p. 17.

[3] *Derby Mercury*, 20 July 1853. It may be presumed that the reporter saw the pots from the collapsed ceiling of the sixth storey.

[4] From the association of brick arch floors with cast-iron beams and columns in all the early nineteenth century fire-proof mills, and from the statements by Britton and Brayley and Edward Strutt, who make little distinction between the Derby mill and the later iron-framed mills, Bannister, *loc. cit.*, made the assumption that iron beams were used at Derby. He was, however, not aware of the Milford warehouse and the Belper West Mill. H. R. Hitchcock, *Early Victorian Architecture in Britain*, London, 1954, pp. 54 and 492, also accepts the Derby mill as the first building with "a complete interior skeleton of iron."

[5] Britton and Brayley, *op. cit.*, p. 531. William Strutt made plans for the 54 ft. span bridge at Mellor in 1789. (G. Unwin, A. Hulme and G. Taylor, *Samuel Oldknow and the Arkwrights*, Manchester, 1924, p. 222).

[6] The property, known as the "New Mills, near Duffield," consisted principally of two iron forges on the east side of the river, and, on the west side, a paper mill, a warehouse, and a slitting mill. A head of water was provided by weirs "made at vast labour and Expence, and executed with great Judgment." Notice of the Sale, *Birmingham Gazette*, 14 April 1777.

WILLIAM STRUTT'S COTTON MILLS, 1793–1812

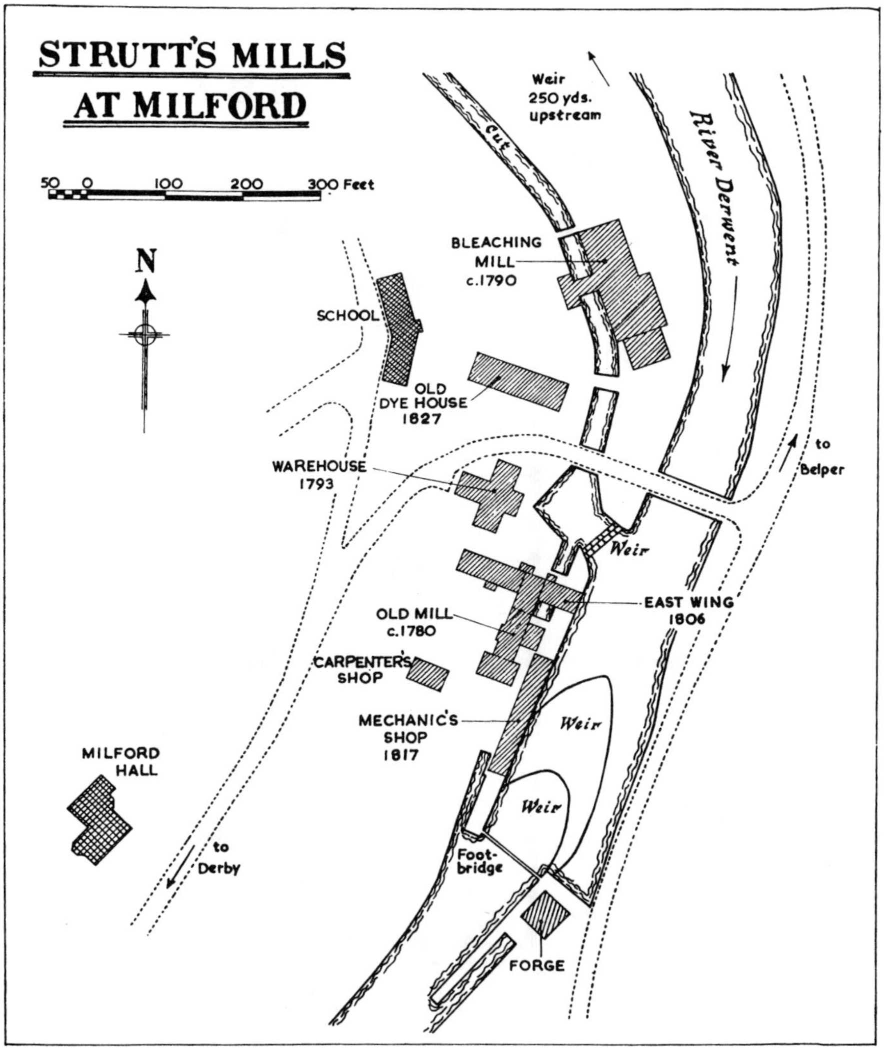

Fig. 3

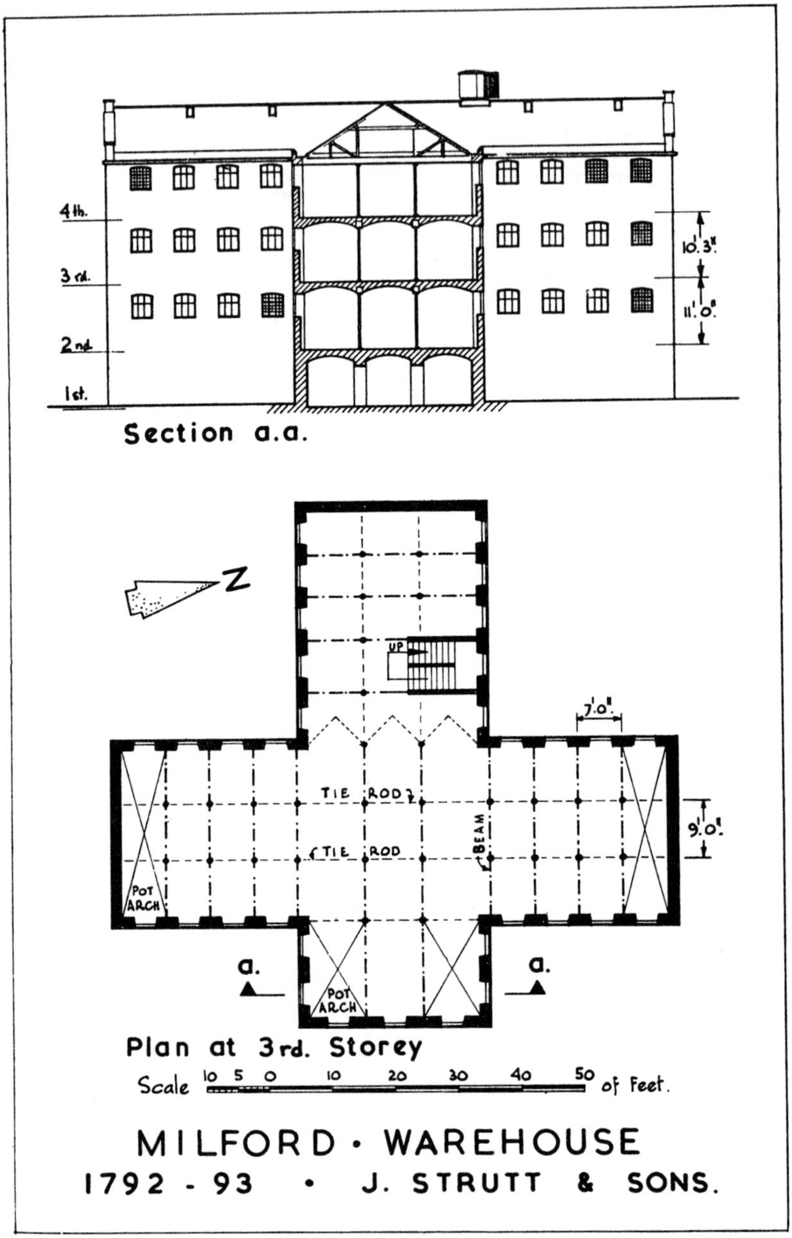
4th.
3rd.
2nd.
1st.
10'.3".
11'.0".
Section a.a.
UP
7'.0".
TIE ROD
TIE ROD
BEAM
9'.0".
POT ARCH
POT ARCH
a.
a.
Z
Plan at 3rd. Storey
Scale 10 5 0 10 20 30 40 50 of Feet.
MILFORD · WAREHOUSE
1792 - 93 · J. STRUTT & SONS.

Fig. 4

in more detail in 1956. By great good fortune it is possible to establish that the roof of this structure was completed by, or shortly before, April 1793. A number of manuscript books exist containing accounts of building work carried out by the Strutts, chiefly at Belper in the '1790's, but a few buildings elsewhere are also included.[1] In one of these books there is a note of "*John Newman's work in sundry places, measured 24th April* 1793." This all relates to roof tiling, and among the items we find:

"South wing of new Warehouse at Millford 31·4 × 36·2
North wing „ „ „ „ „ 37·0 × 31·4
East wing „ „ „ „ „ 22·6 × 31·4
West wing „ „ „ „ „ 45·6 × 31·4"

The lengths agree (with a maximum error of less than 1 ft.) with those scaled from our plan of the structure shown in Fig. 4, and the width we have checked by measurement on the roof itself. For the walls and roof to have been finished by this date, construction must have begun in 1792, though probably in the latter part of the year, and the floor construction would be completed before the end

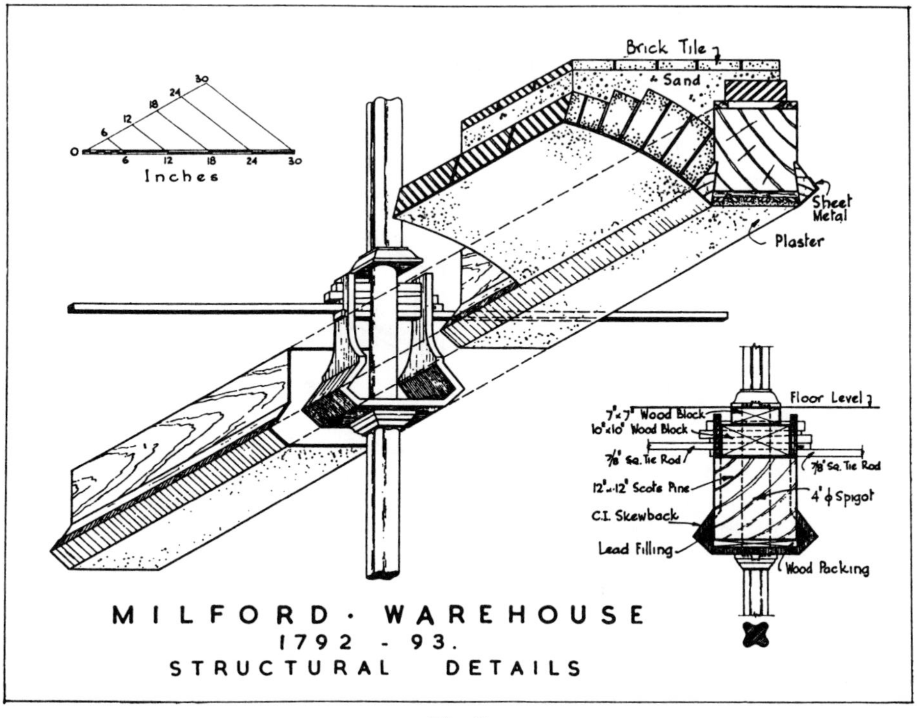

Fig. 5

[1] The Authors are grateful to Dr. Fitton for allowing them to examine the account books while he was working on them. They are the property of the English Sewing Cotton Company.

of 1793. The building was later converted to a mill[1]; and as such is referred to in plans made in 1806 for the installation of gas lighting by Boulton & Watt at Milford. These drawings show a part section of the building with the arches sketched in, and a plan with the positions of the columns marked.

The first storey, or semi-basement, consists of stone piers and brick arch floors. In the other storeys there are brick arches springing from 12 in. sq. Scots Pine (= Baltic fir) beams[2] supported by two rows of cruciform section cast-iron columns, as shown in Plate XLI, Fig. 1. The columns divide the internal width into three equal spans of 9 ft. In the main rooms the central brick arch

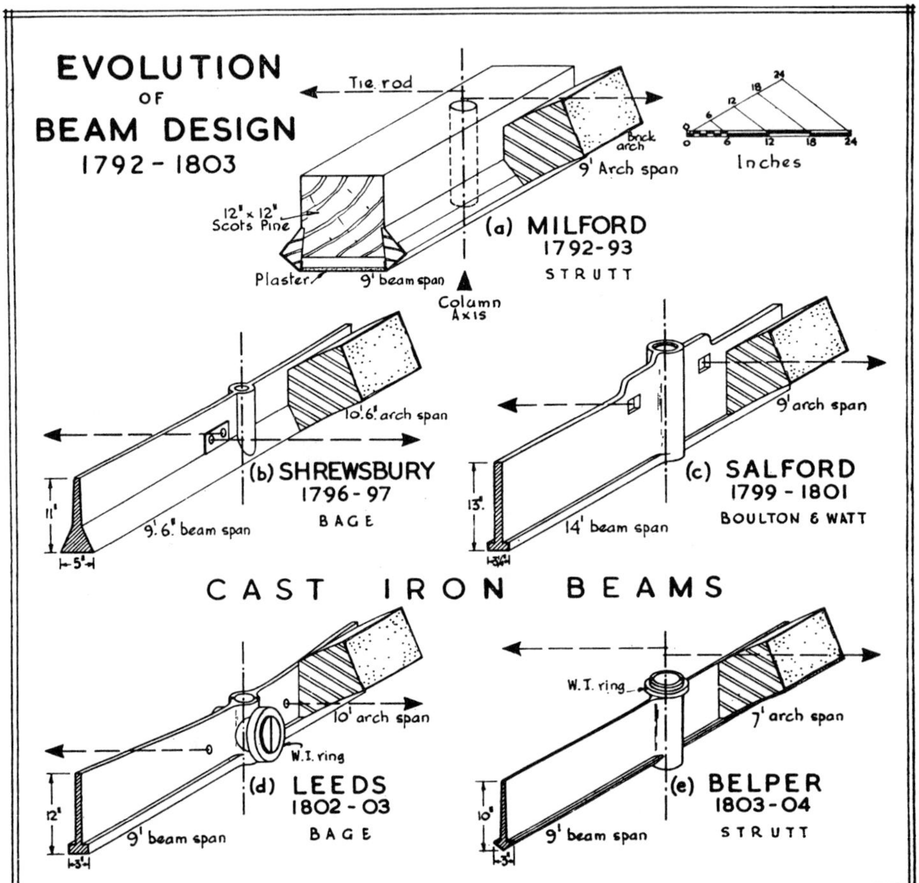

Fig. 6

---

[1] The ventilator on the roof should be noted. It is a characteristic feature of Strutt's care to provide healthy conditions in the mills.

[2] We are indebted to Mr. J. F. Levy, of Imperial College, for examining specimens of the timber from the beams at Milford.

spans 9 ft., and this is flanked by arches of 10 ft. 6 in. span, followed on each side by four arches of 7 ft. span. This span is repeated in the west wing,[1] whilst in the east wing the spans are 9 ft. The brick arches are 9 in. thick at the springing and 4½ in. at the crown. The rise of the arches is typically 13 in. In the outermost arches the 4½ in. brickwork is replaced by hollow pots, presumably to reduce the thrust coming on the walls. The pots are 4 in. diam. and 5 in. high. A hole in the bottom of each provides a key for plaster and flat circular covers are placed over the open ends of the pot to form a base for the sand infilling on which the brick paving is laid (Plate XLI, Figs. 2 and 3).

The chief point of interest, however, is the careful detailing of the beams and of the connection between the columns, the beams and the tie rods (Figs. 5, 6*a* and Plate XLII, Figs. 1 and 2). There is nothing tentative here; Strutt had thought out the structural system very completely.[2] The iron skewback is particularly to be noted. It acts as a bearing plate for the column, as an abutment for the arches and as a fixing plate both for the tie rods and the wedges.

The sequence of construction we envisage as follows. The great timber beams were placed in position as the walls were built; and planks laid across the beams would provide staging for the men working on the inner part of the walls: scaffolding being required externally. After the roof had been completed, holes for the iron spigots would be accurately drilled and the iron skewbacks fitted on the beams, but raised a few inches above their final position. The column could then be erected, the skewback lowered to bear on the column cap, the spigot driven into place and the space between the plate and the beam packed in with wood and lead. Next the tie rods would be inserted and fastened by their cotters, and the whole made firm by the iron wedges. The wooden skewbacks, with their sheet metal covering, would then be fitted on to the sides of the beam and the arches built; the spandrel space filled with sand and the brick tiles laid; the whole operation being repeated in the next storey.

Completed in the same year as the mill at Derby, the Milford Warehouse is the oldest surviving example of Strutt's work. It deserves to be restored and recognised as a monument of unique interest in the evolution of modern design.

Belper West Mill, 1793–95

In 1771 Jedediah Strutt and Richard Arkwright built the first water-powered cotton spinning mill, on Cromford Sough where it joins the Derwent 16 miles upstream of Derby, and 5 years later they were building a second mill there. Between 1776 and 1778 they erected the first of the mills at Belper 8 miles below Cromford. This was at the site of the present South Mill, the position of which is shown in Fig. 7. The partnership was dissolved in 1781, but Strutt retained the works at Belper (and Milford), and in 1784–86 he built another mill at the site of the existing North Mill.[3] All these buildings had timber floors.

Then in 1793, when the Derby mill was nearing completion and work on the Milford warehouse was still in progress, William Strutt began the construction of a six-storey fire-proof mill at Belper to the west of the road running from the bridge through Belper to Derby; and since 1795 this has

[1] The original structure in the west wing has been replaced by iron beams and columns probably in the mid-nineteenth century. The columns in this wing are closely similar to the ones in the Exhibition Dye House at Milford, date 1856.

[2] We are grateful to the English Sewing Cotton Company for providing facilities for investigating these structural details, as also those at Belper North and South Mills.

[3] The dates of the old South and North Mills are taken from *An Old Belper Man's Diary*, the original of which seems now to be lost. But typescript extracts exist in the office of the English Sewing Cotton Company, and we are indebted to Mr. Ryde, Manager of the Belper and Milford mills, for permission to use these extracts. The dates of the old South Mill are confirmed by a statement in *Factories Inquiry Commission, Supplementary Report*, London, 1834, Part 2, Vol. 20, D.I., p. 96.

X

been called the West Mill. An excellent view of the mill was published by Davies in 1811 and this is reproduced in Plate XLII, Fig. 3.[1]

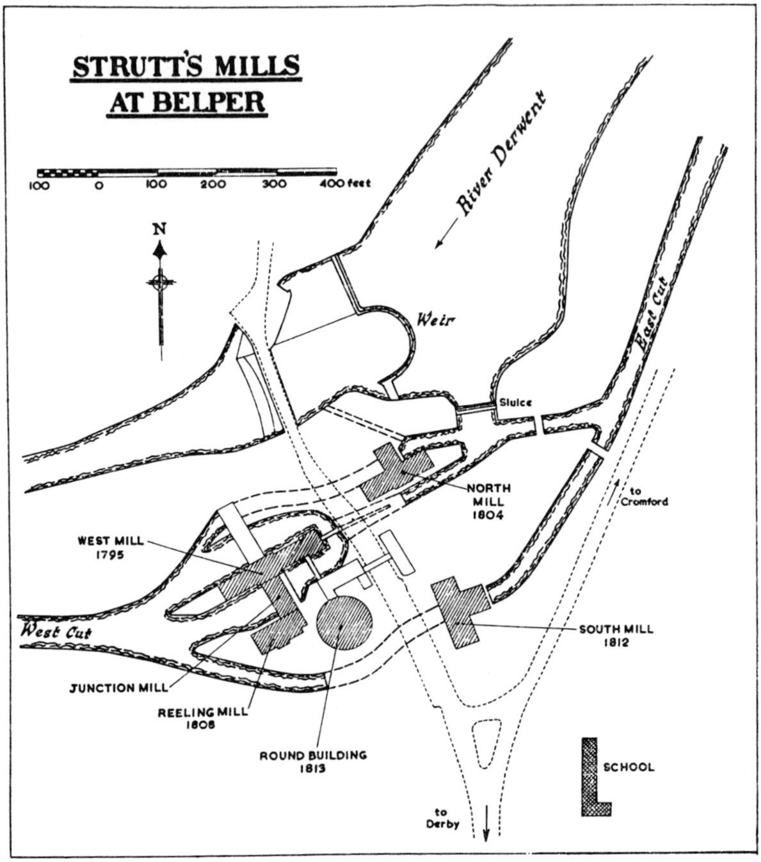

Fig. 7

Britton and Brayley, referring to the three mills at Belper, say:

"The principal of these mills is 200 feet long, 30 feet wide, and six stories high, and like that we have described at Derby, is considered as fire-proof, the floors being constructed with brick arches and paved with brick. The two water-wheels, which work the machinery in this building, are remarkable as well for their magnitude, as for their singularity of construction; one of them being upwards of forty feet long, and eighteen feet in diameter: and the other forty-eight feet long and twelve feet in diameter. As it was impossible to procure timber sufficiently large to form the axles,

[1] D. P. Davies, *A New Historical and Descriptive View of Derbyshire*, Belper, 1811. The engraving is on p. 343.

or shafts, of these wheels in the usual mode of structure, they are made circular and hollow, of a great number of pieces, and hooped like a cask; one of the shafts is between five and six feet in diameter, and the other between eight and nine."[1]

The mill consists, in each storey, of a long room with 22 bays and a four-bay room (the "Counting House") at the east end (see Fig. 8). The overall length is 190 ft. The internal width of the long room is 27 ft., except for the six-bay central portion which has a width of 35 ft. And this greater width is repeated in the east room. The external walls of the first three storeys are built of stone 2 ft. thick, whilst the walls of the upper three storeys are of 18 in. brickwork. The floor of the second

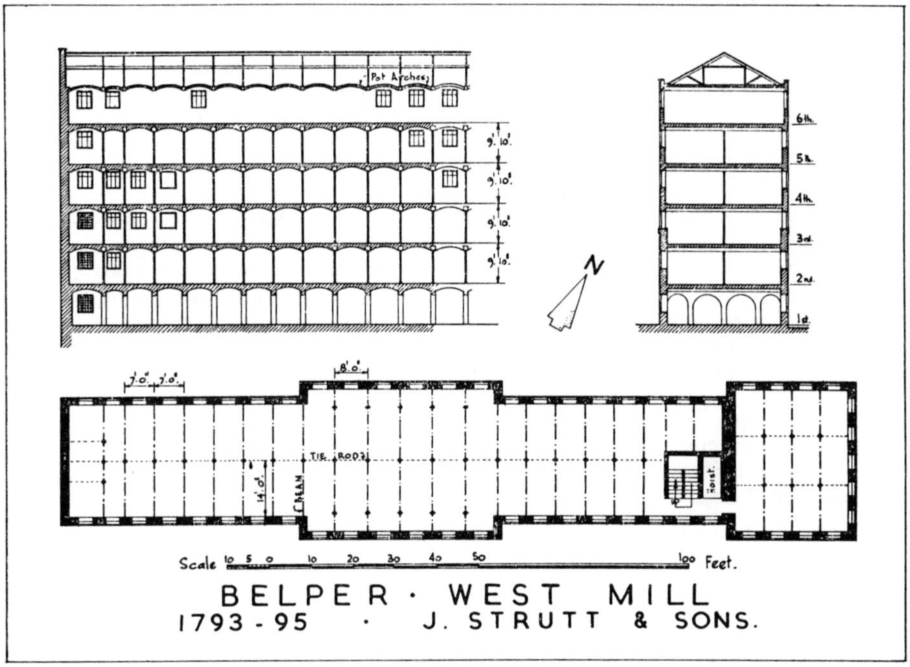

Fig. 8

storey rests on brick piers and arches, the next four storeys are built with brick arch floors, springing from 13 in. sq. timber beams (with skewbacks as shown in Fig. 5) supported on cruciform section cast-iron columns, and the ceiling of the sixth storey is made of hollow pot arches spanning between timber beams which act, also, as the tie beams of the roof frame. In the long room the beams have a span of 13 ft. 6 in. (14 ft. in the top three storeys) and this is maintained in the wide central part by means of columns placed 4 ft. from the walls. In the east room, however, there are two rows of columns and the beam span is thus about 12 ft. The brick floor arches in the wings of the long

[1] Britton and Brayley, *op. cit.*, p. 530. Their description is repeated by Davies (*op. cit.*, p. 346) who adds, however, interesting information concerning two iron waterwheels designed by Strutt and built at the West Mill by T. C. Hewes of Manchester, presumably at some time between 1802 and 1811.

room, and in the east room, have a span of 7 ft. (with a rise of 11 in.), but in the central portion of the long room the arches span 8 ft., and towards each end of the long room there is a narrow bay with a span of 5 ft. The floors are paved with brick. It is therefore clear that the Belper West Mill is closely similar to the mill at Derby.

The interest and historical value of the West Mill, however, is greatly enhanced by the fact that a fairly comprehensive record of its construction and cost has been preserved in the account books at Belper, to which reference has already been made. Some of the more interesting details are as follows.

The earliest dated entries, read

May 3rd 1793 *Rec^d. from Derby an assortment scaffold ropes*
May 12th 1793 5 *Loads of walling stone to New Mill.*

Between January and May 1794 we find fortnightly accounts for a total of 90 "*steps for the Mill stairs,*" and these are evidently for the staircase. Thus by the end of May 1794 the walls[1] must have been nearing completion and, in an account headed "*West Mill. D^tr to Sundries,*" under the date 11 August 1794, there is the note:

To cash paid for slating by John Newman 5935 *sq.* £17. 16. 1.

This is precisely the area noted elsewhere (in an undated entry) for *Slating the roof* 184 *ft.* 9 *in. long by* 33 *ft.* 6 *in. deep,* less an area of 253 sq. ft. for a cistern and several trap doors; the cost being £114.

There is no record of the beams for the third storey floor, but for the upper three floors the beams in the wings of the long room were 12¾ in. sq. and 30 ft. long, and cost £3 each (1s. 9d. per cu. ft.), whilst those in the central positions were 13 in. sq. and 38 ft. long. The roof beams were 13 in. by 9 in. All this timber must have been in position before the roof was slated in August 1794.

A typical entry for the iron columns reads:

July 17th 1794 6 *Cast pillars* 19 *cw.* 1 *qr.* 12 *lb.* @ 12 /– £11. 12. 3.

In a note dated 7 November 1794 we see that *Smiths of Chesterfield*[2] *will do . . . Pillars for the Mill with their appendages* @ £11. 10. 0 per Ton, and this lower price of 11s. 6d. per cwt. appears in the next purchase of iron:

Jan. 29th 1795 13 *Pillars,* 13 *skewbacks, wedges etc. from Chesterfield* 64. 0. 19 @ 11/6 £36. 17. 11.

By March 1795, 280 windows had been glazed and most, if not all, of the third and fourth storey floors had been built, as well as a "*Plaster floor in 6th storey over Counting House.*"

The following entry occurs in the financial year 1795–96:

Pots for the Top Storey, Counting House included 35,609 @ 52/6 £93. 9. 5.

And in *An Old Belper Man's Diary* (*op. cit.*) it is stated that "The West Mill was built 1792–95 and commenced work in 1796." Thus the building was probably finished in the latter part of 1795.[3]

Excavation for the cut, leading water to the wheel, was in progress from May 1794 to January 1795, and the masonry walls of the cut, and the bridge where the cut passes under the road, were being built in 1795–96. Details of the waterwheel are also given, which is evidently one of those mentioned by Britton and Brayley with the hooped shaft, but the second wheel must have been added later;

---

[1] 345,200 bricks were used, at 21s. per thousand.

[2] This firm would be Ebenezer Smith & Co. of Chesterfield who had the second largest iron foundry in Derbyshire. John Farey (senior), *op. cit.*, London, 1811, Vol. 1, p. 396.

[3] The reference to 1792 implies that the site was being prepared in that year for, as we have seen, work on the walls of the mill began in May 1793.

probably between June 1797 and November 1798 when additional building work is recorded together with an account for "Heavy Gear" at the West Mill.

The accounts for the West Mill, apart from the later work just mentioned, were closed in 1797; the chief items being:

Mill building	£4,689
Bridge and cut	£1,142
Waterwheel and water-house ..	£962

The total floor area of the mill (measured within the external walls but including partitions, staircase etc.) amounts to 33,500 sq. ft. Thus the cost of the building was just under 3s. per ft. super. Comparing this figure with the cost of a five-storey timber-framed mill erected 1794–95 at Leeds for John Marshall,[1] we find that the fire-proof construction increased the required capital outlay by roughly 25 per cent.

Finally we may note that the workmen's wages ranged from 1s. 3d. to 2s. 6d. per day; and two of the many entries read:

Jos. Harrison puting Pillars & skewbacks up 24¾ *days @* 2/4 £2.17. 9.
William Moore turning arches & laying floor 38¾ *days @* 2/6 £4. 16. 10.

So far as we know the type of structure built by Strutt at Derby, Milford and Belper West Mill was not repeated elsewhere. But in 1802 Simon Goodrich (1773–1847) submitted a winning design for "Rebuilding the Albion Mills, Fire-Proof," and this was explicitly based upon Strutt's mills.[2] In the report accompanying his drawings Goodrich says:

> "The under side of the beam is to be protected from the action of fire by being covered with a double coating of lath and plaister or with plate iron: the other three sides of the beam are sufficiently protected, being buried in brickwork. The beams might be of cast iron, but the preference is given to wood, since being fully secured against fire, it is cheaper and more to be depended upon for strength in this position, than cast-iron."

The Albion Mills, however, were not rebuilt.

Early Iron Beams

It is probable that about this time other engineers also had doubts concerning the suitability of cast-iron for beams but, in fact, once this material had been used for the beams in the mill at Shrewsbury it was adopted in all subsequent fire-proof buildings until, in the mid-nineteenth century, wrought iron took the place of cast-iron. And before describing Strutt's next building of consequence, which was erected in 1803–04, we must outline the developments in iron beams during the preceding 7 years.[3] The Shrewsbury flax mill was built in Castle Foregate for Thomas and Benjamin Benyon in partnership with John Marshall, and its design was carried out by Charles Bage, who also had a small share in the partnership. Construction began in the summer of 1796 and William Hazeldine supplied the beams and columns.[4] The Boulton & Watt engine was installed by September 1797.

---

[1] We are indebted to Mr. W. G. Rimmer of Leeds University for access to the Marshall records (in Leeds University Library) and for his help in examining this material.

[2] The valuable collection of Goodrich's papers, *Journal* and drawings are in the Science Museum, London. See also, E. A. Forward, "Simon Goodrich and his work as an engineer," *Trans. Newcomen Soc.*, III, 1922, pp. 1–13.

[3] Except where noted, all the information in this section of the Paper has been obtained from field investigations by the authors and from the following sources: John Marshall MSS. (Leeds University Library): Boulton and Watt Collection (Birmingham Public Library): Bage-Strutt Correspondence (Shrewsbury Public Library). Full accounts of the mills at Shrewsbury, Salford, Leeds and Glasgow will be published elsewhere.

[4] T. Minshull, *The Shrewsbury Visitor's Pocket Companion or Salopian Guide*, Shrewsbury, 1803, p. 47.

From a letter written by Bage to Strutt early in 1796 we know that he received much help from the designer of the Derby mill, and most of the features of Strutt's fire-resisting mills were incorporated at Shrewsbury, including the brick arch floors and the cruciform iron columns.[1] Even the beams, although of cast-iron, were evidently adapted from Strutt's beam with its skewbacks (Fig. 6*b*). Moreover, the Shrewsbury beams, like the timber beams, were made continuous over the supports, although their section is qute unsuitable to resist the negative bending moments in a continuous beam at the columns.

If Bage's mill was directly inspired by Strutt, so this mill formed the starting point for the second iron-framed building: the seven-storey cotton mill at Salford, near Manchester, erected 1799–1801 for Messrs. Philips and Lee. Working drawings in the Boulton & Watt collection show cast-iron beams of 12 ft. 10 in. span with a section similar to that at Shrewsbury. Late in 1799 it was decided, however, to increase the span to 14 ft. and the beams were made correspondingly larger. Shortly afterwards the section was altered to the more economical flanged (or inverted T) section shown in Fig. 6*c*.[2] This was an important advance, but the beams were still continuous over the columns; and made in two equal lengths bolted together at the centre of the building, as at Shrewsbury. The Salford beams were thus not altogether satisfactory, but the more logical form of a simply supported beam, which eliminates the negative bending moments, was adopted in a flax mill completed by Benyons and Bage in Meadow Lane, Leeds, in July 1803.[3] Charles Bage was again responsible for its design. Only the foundations now exist, but details of the beams can be reconstructed (Fig. 6*d*) from a sketch made by Goodrich when he visited the mill[4] and from dimensions of a beam tested by Bage.[5] It will be seen that the Leeds beam has a section of better proportions than the Salford beam, and it has a pronounced turtle-back profile, which, of course, corresponds to the bending moment diagram in a simply-supported beam.

Bage also developed a theory for the strength of flanged iron beams which he outlined in a letter to Strutt, dated 29 August 1803.[6] It leads to a formula almost identical with that later published by Eaton Hodgkinson,[7] and is in good agreement with tests on actual beams. Moreover, in the same letter, Bage describes tests on a full scale iron roof truss, the span of which is the same as would be required for the Leeds mill. Thus by 1803 the cast-iron beam had attained a rational form and was subject to theoretical analysis. William Strutt was in close touch with this development, and he designed the next iron-framed building, the North Mill at Belper, incorporating these improvements together with a more satisfactory type of connection between the beam and column. The North Mill will be described in the next section of the Paper, but it is worth noting here that between June 1803 and December 1804 Bage built a second mill for Benyons, in Shrewsbury, at Castle Fields, presumably with the same structural details as those for the Leeds mill; and that in a large cotton

---

[1] Bage acknowledges his personal debt to Strutt, for technical advice, in a number of letters. But in one dated 28 October 1811, after describing a disastrous fire in a timber floor building at Shrewsbury, he speaks for posterity in saying, "How much we are obliged to you for teaching us how to make buildings Fire proof."

[2] The working drawings of the final design have not survived. But the flange can be clearly seen in a drawing dated January 1801, and Fairbairn, *op. cit.*, p. 3 gives dimensions. The Salford mill was destroyed by bombing in December 1940.

[3] *Factories Inquiry Commission*, 1834, *op. cit.*, C.1, p. 37.

[4] Goodrich *Journal*. Entry dated 4 August 1804. S. B. Hamilton, "The use of cast-iron in buildings," *Trans. Newcomen Soc.*, XXI, pp. 139–155, describes an addition to the mill, built about 1816. Mr. E. A. Forward first realised the connection between Goodrich and Benyon's mill at Leeds, but it has since been widely assumed that Goodrich saw the building described by Dr. Hamilton. The beams in this later building, however, differ in many respects from those sketched by Goodrich.

[5] *Life of Thomas Telford*, London, 1838, p. 682.

[6] A. W. Skempton, "The origin of iron beams," *Actes VIII Int. Conf. History of Science*, vol. 3, pp. 1029–39, Florence, 1956, gives a full account of Bage's theory and its comparison with test results.

[7] E. Hodgkinson, "Theoretical and experimental researches to ascertain the strength and best form of iron beams," *Mem. Manchester Lit. & Phil. Soc.*, 2nd series, Vol. 5, 1831, pp. 407–544.

mill erected 1804–06 in Glasgow for Henry Houldsworth, the beams had a type of connection similar to Strutt's at the North Mill.

Belper North Mill, 1803–04

Under the date 12 January 1803, the *Derby Mercury* carried the following news item:

> "About three o'clock this morning a most tremendous fire broke out in one of the large Cotton Mills belonging to Messrs. Strutt, at Belper, which raged with incredible fury, and in a few hours destroyed it, with all the valuable machinery, water wheel etc. The cause of this dreadful accident is not yet known. The loss will be immense as no part of the property is insured."[1]

From *The Old Belper Man's Diary* (*op. cit.*) we read that the fire destroyed the North Mill (i.e. the timber floor building of 1786) which was then rebuilt fire-proof and started work again at the end of 1804. This new mill still exists in its original form as a five-storey structure with an attic, almost exactly as depicted in the engraving published in 1811 by Davies (see Plate XLII, Fig. 3), on the opposite side of the road from the West Mill quite close to the weir (Fig. 7). Sections of the mill, showing the machinery as well as the structure, were drawn in 1812 by John Farey (junior) to illustrate his article on "Manufacture of Cotton" in Rees' *Cyclopaedia*, and are reproduced in Plate XLIII. The mill is iron-framed throughout; including all the members of the roof.[2] The brick arches typically span 7 ft. although in two central bays of the main block the span is 8 ft. The beams span 9 ft. and the storey height is 10 ft. 2 in.

Concerning the waterwheel, and the structure above it, Farey gives the following account:

> "The whole motion of the machinery is taken from the great waterwheel, situated underneath the wing, in the lowest room of the mill; and as it is of so great a size, namely, 18 feet diameter and 23 feet long, that no cast-iron girder could be thrown across it strong enough to support the arches for the wing above it, a strong stone arch is built over the wheel; and to resist the thrust of this arch, two strong iron bolts are extended across it, and render it as strong as possible, so that the iron columns may be raised upon it as safely as they could upon foundation piers, like the others. But as a precaution against overloading the walls which, as they include the waterwheel, would ruin every thing if they settled, the arches of the wing immediately over the wheel are built, instead of solid brick, with small pots like garden pots, so that they are light, but sufficiently strong to bear any thing which is ever required to be loaded upon them."[3]

As for details of the iron-framing, these were the subject of careful sketches made by Goodrich in 1804 during a visit to Belper when the mill was actually under construction. The relevant page of his *Diary* is reproduced in Fig. 9 and, by the removal of part of the floor, we have been able to confirm the accuracy of these drawings[4] Each end of the beam consists of a half socket (as at Glasgow) which fits around the column spigot and seats in an annular recess in the column capital. This method of seating would alone prevent two adjacent beams from being pulled apart, but in addition a wrought-iron ring was provided around the top of the sockets. The large tolerances necessary in such work would provide sufficient flexibility to avoid continuity over the supports,

---

[1] *Derby Mercury*, 13 January 1803.

[2] The attic was used as a school-room on Sundays.

[3] Rees' *Cyclopaedia*, London, 1819, Vol. 22. The article is not signed, but there is unambiguous internal evidence for its date and authorship. The illustration is in the 2nd volume of Plates. John Farey (senior) in 1815 (*op. cit.*, Vol. 2, p. 16) mentions that his son had made measured drawings of the Strutt mills.

[4] Goodrich does not state that he was in Belper when he made this drawing, but on the same sheet of his *Journal* he describes (with accurate dimensions) the Round Building there (Fig. 7). This curious structure originated from an idea of Sir Samuel Bentham (who was Goodrich's chief at the Admiralty) and its construction began in 1803 (*An Old Belper Man's Diary* and *William Strutt: A Memoir*, *op. cit.*, p. 29). The sheet with the North Mill details is not dated, but the adjacent pages were both written in August 1804. It is reproduced by permission of the Director, Science Museum.

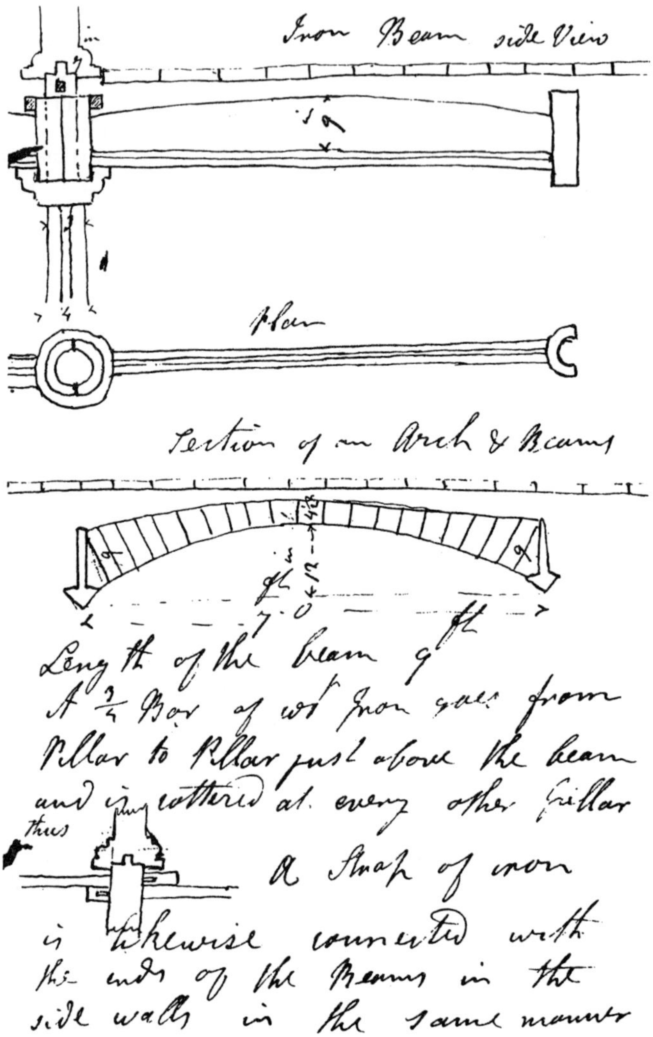

Fig. 9. Beams in Belper North Mill
(Goodrich, *Journal*, August 1804)

and permit small differential settlements to occur without causing secondary stresses in the beams. This system, with minor variants, has been found in most of the iron-framed buildings we have examined, built in the 20 years following 1804.[1] The tie bars are fastened in the upper end of the spigot, as at Milford, and are wholly concealed within the arches, and Goodrich notes that ties are also used to join the beams in the walls of the building. Farey was much impressed by the concealed ties and remarks that:

"The ceilings of the room, therefore, consist of regular arches, which have a very good appearance, and make the most firm and solid floors above that can be imagined."

This "very good appearance" is due also to the care with which Strutt has provided an arris to the lower surface of the beam flange, to continue the line of the arch soffits. Once more, then, he has achieved a system most carefully thought out in all its details; and the simple but pleasing architecture of the building itself, as well as the logical structure, proclaims a designer of taste and skill.

Moreover, we know from the correspondence between Bage and Strutt that the latter was carrying out tests on arches, early in 1803, and that he was now, in his turn, benefitting from Bage's work on beams. In a letter dated 19 May 1803 Bage says that he hopes to hear

"the result of your experiments on Arches, the subject being somewhat complex. I think it became a question, when I had the pleasure of seeing you,[2] how far the failure of one pillar would affect the rest."

And later in the same letter he writes:

"If on a balance of advantage and disadvantage you should at length prefer Iron beams, I should be glad to submit my reasoning on the strength and shape of them to your examination."

Strutt did, of course, decide to use iron beams, and in a letter of 29 August 1803 Bage writes:

"My conscience has long upbraided me for having delayed to send you some theorems for ascertaining the strength of cast-iron beams. But with the care of building two Mills,[3] the ordinary business of the present Mill . . . I really find but little time."

The "theorems" are enclosed in the letter, and have been previously mentioned in the present Paper. It adds immeasurably to the interest and importance of these early iron-framed mills when we realise that their designers undertook some of the first tests and theoretical discussions on iron beams and roof-trusses, and voussoir arches.

Milford Old Mill, East Wing, 1805–06

The east wing of the Old Mill at Milford extends from the clock tower to the river; five bays long and five storeys high (Plate XL, Fig. 1; and Fig. 3). The keystone of a doorway in the north wall is dated 1805,[4] and in the Boulton & Watt gas plan of June 1806 the east wing is shown as a "new building now erecting." The clock is also dated 1806. The iron beams, with a span 9 ft., and the structural details, are very similar to those in Belper North Mill. The columns have the usual cruciform section, and the brick arch floors span 9 ft. with a rise of 10 in.

A slight digression may be permitted here for a glance at the history of the Old Mill. As stated earlier, the site at Milford was purchased in 1777, and presumably no time was lost in constructing

---

[1] It should be noted, however, that the rather more cumbersome design of Bage (Fig. 6*d*) was used in the eight-storey Sedgwick mill of McConnel & Kennedy in Manchester (1818–20); and is illustrated by Fairbairn as late as 1854 (*op. cit.*, p. 137).

[2] Bage visited Strutt on his way from Shrewsbury to Leeds in February or March 1803.

[3] The mills to which Bage refers are those at Leeds and Shrewsbury, Castle Fields.

[4] Set in the wall there is also a stone recording the flood level of 10 February 1795: but this stone has evidently been transferred from its original position.

X

the cotton mill there. This five-storey structure, 67 ft. long and 30 ft. wide, occupies the position shown in Fig. 3 and six of its nine bays can be seen in the photograph in Plate XL, Fig. 1. The wheel house is on the river side of the mill. Timber floors, which may be original, exist in the top three storeys, comprising $\frac{3}{4}$ in. boards on 6 in. by 3 in. joists at 18 in. centres, supported on 13 in. sq. Scots pine beams, 7 ft. apart. The storey height is 8 ft. 9 in. The mill was extended for a length of 50 ft. to the north, possibly in 1783,[1] and a few years later a wing 85 ft. long was added to the west (Fig. 3); timber floors still being used.

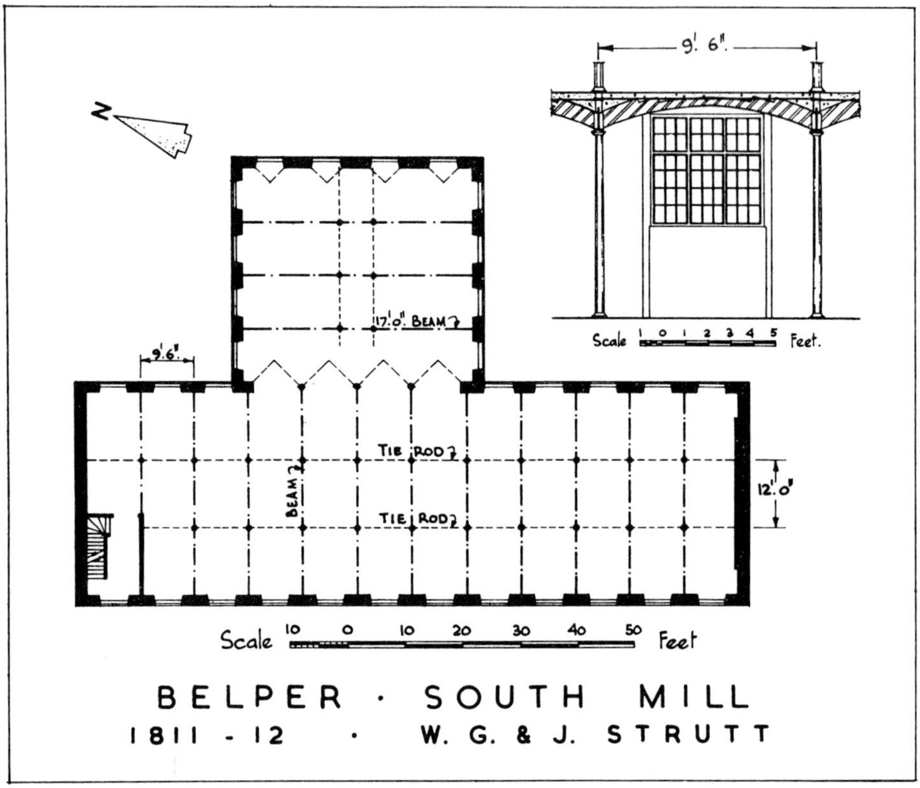

Fig. 10

At some date between 1802 and 1806 the two lower floors of the main block were converted into fire-proof construction with brick arch floors and a central row of iron columns.[2] A similar row of columns has been inserted in each of the three upper storeys, to strengthen the floors, and this

[1] The *Derby Mercury*, 22 May 1783, carried an advertisement for masons and bricklayers required by Mr. Strutt at New Mills near Duffield.

[2] The mill had not been made "fireproof" in 1802 when seen by Britton and Brayley, *op. cit.*, p. 531; but the arch floors are shown in Boulton & Watt's drawing of June 1806 for gas installation.

modification may be of the same period. There are also columns in the west wing, but they are definitely later. The east wing, as we have seen, is iron-framed and was built 1805–06. The south wing, which is also clearly shown in Plate XL, Fig. 1, four bays by three in plan, appears to be structurally identical with the east wing. The date of construction must be later than 1806, for it is not on the Boulton & Watt plan of that year; but it is shown in the 1820 map of the Estate of W. G. & J. Strutt.[1]

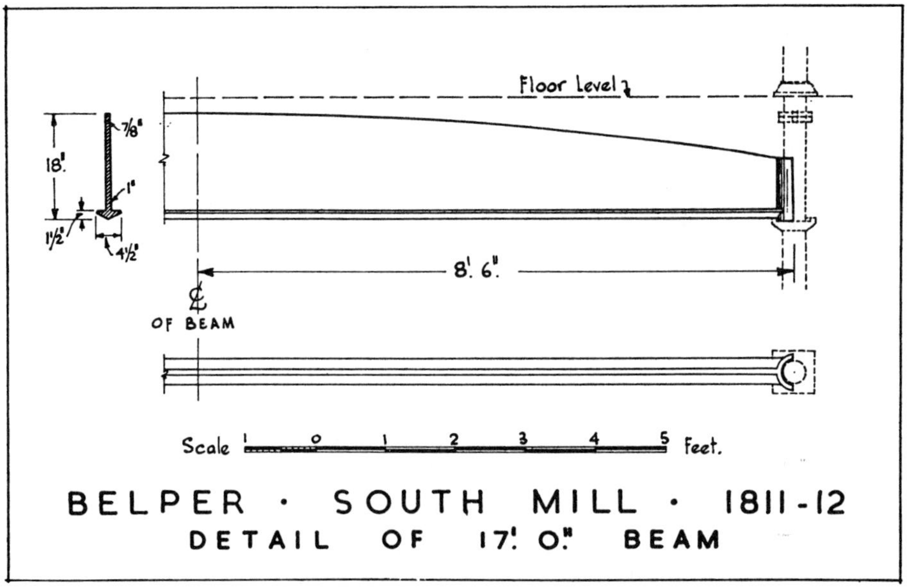

Fig. 11

Belper Reeling Mill, 1807–08

A short distance south of the West Mill at Belper (Fig. 7) a six-storey structure known as the Reeling Mill was built 1807–08.[2] The internal width is 36 ft., but the beams still span 9 ft. (there being three rows of columns instead of two as at the North Mill and Milford East Wing). The arches span 9 ft. 6 in. with a rise of 12 in. The beams and columns are very similar to those in the North Mill. At a slightly later date the West Mill was joined to the Reeling Mill by the Junction Mill. The floors of this building are above the waterwheels and are supported by trussed timber beams spanning 32 ft. between the walls without internal support.

Belper South Mill, 1811–12

The location of this fine example of an early iron-framed mill is shown in Fig. 7. It has five storeys, with an attic, a length of 118 ft., a width of 40 ft. and a four-bay wing projecting to the east, the basement of which housed the waterwheels. The brick arches of the floors span 9 ft. 6 in. as at the

[1] The Estate map is reproduced by R. S. Fitton, *op. cit.*, p. 377.
[2] *Old Belper Man's Diary*, *op. cit.*

Reeling Mill; but the span of the iron beams is 12 ft. in the main body of the building, and 17 ft. in the east wing where the floors bridge over the wheels (Fig. 10). This is a bold use of iron beams, for the period, and there can be little doubt that Strutt gave considerable thought to their design. Indeed, these 17 ft. turtle-back beams, of which a section is shown in Fig. 11, and the South Mill itself, fittingly represent the culmination of Strutt's work in the 20 years following his design in 1792 of the first fire-proof mill. It may be noted that the tie rods end in an eye which fits over the column spigot and that no wrought iron ring exists (in contrast with the North Mill, Fig. 9). Greater freedom of movement of the beams is thus achieved.

Evidence for the date of the South Mill is contained in a diary of notes and jottings kept by various members of the Strutt family between 1810 and 1826. The diary came into the possession of John Hunter, Esq., J.P. and extracts were printed in the *Belper News and Derbyshire Telephone*[1] for 26 May 1911. The relevant passages read as follows:

> "The South Mill was built in 1811–12 and the first wages were paid for the week ending Decr. 19th 1812.
>
> The first yarn was spun on Jany. 1st 1813, from 96 spindles, numbered 20's twist.
>
> Ann Robinson was the frame tenter."

Attention was first drawn to the South Mill by Professor Bannister (*loc. cit.*) but he attributed its construction to the year 1797, and Dr. Pevsner also gave this date.[2] The accuracy of the Strutt diary, however, can hardly be questioned; and Davies in his book of 1811 (*op. cit.*, p. 346) says:

> "These Mills (at Belper) are four in number, the first of which was erected in 1776 by Jedediah Strutt Esq. (but) was lately pulled down for re-erection. The principal of those now standing is 200 ft. long 30 ft. wide and six stories high. . . ." [3]

Here, then, we have a contemporary account which proves that in 1811 the old mill had recently been demolished and the new building not yet completed. Moreover, in the engraving published by Davies (see Plate XLII, Fig. 3) the South Mill would appear conspicuously had it then been in existence. The artist, in fact, must have made his drawing in 1810 or early in 1811, between the demolition of the old mill and the erection of the new mill. The South Mill has a well-lit and neat interior of remarkably modern appearance (Plate XLIV). It could hardly be less evocative of Blake's "dark satanic mills."

Summary of Strutt's Mills

The leading dimensions of the various fire-proof and iron-framed mills built by Strutt, with their dates and chief materials of construction, are given in the accompanying Table. The progressive structural development will be clearly seen.

Acknowledgements

We wish to record our indebtedness to the English Sewing Cotton Co. for providing facilities for investigating the mills at Belper and Milford. Mr. J. Woolley and Mr. A. Selvey, engineers of the mills, and Mr. F. W. White, master mason, have given much willing and valuable help. Dr. R. S. Fitton has been most generous in allowing access to the Strutt account books, and Mr. W. G. Rimmer

[1] Under the heading "Old Belper—Its People and Environments." An unreferenced typescript of part of this article exists at the offices of the English Sewing Cotton Co. Mr. F. T. Harrison identified the article as one appearing in the *Belper News* and Miss Carlotta Hacker discovered the original in the newspaper files which are kept in Chesterfield. We are not aware of the present whereabouts of the diary.

[2] N. Pevsner, *The Buildings of England: Derbyshire*, London, 1953, pp. 32 and 59.

[3] I.e. the West Mill.

has also very kindly assisted us in the examination of documents which he is studying. We have benefitted greatly from the unpublished research of Mr. W. D. Shepherd on the Derby mill and have received valuable advice from many librarians, but particularly from Miss W. R. N. Payne of Birmingham and Mr. R. E. James of Shrewsbury. Miss C. L. Hacker has contributed much by her work in the British Museum, at Chesterfield and at Derby.

Summary of the Fire-Proof and Iron-Framed Mills Built in Derbyshire
By William Strutt, F.R.S. (1756–1830)

Date	Building	Number of storeys	Overall dimensions	Floors		Columns		Beams	
				Material	Spans	Material	Storey height	Material	Spans
			ft.		ft.		ft. in.		ft.
	"Fire-proof" Mills								
1792–93	Derby, Cotton Mill	6	115 × 31	Brick arch	8 and 9	Iron	9 10	Timber	9
1792–93	Milford, Warehouse	4	90 × 31	"	7 and 9	"	10 3	"	9
1793–95	Belper, West Mill	6	190 × 31	"	7 and 8	"	9 10	"	13½
	Iron-Framed Mills								
1803–04	Belper, North Mill	5 + a	123 × 31	"	7 and 8	"	10 2	Iron	9
1805–06	Milford, East Wing	5	47 × 31	"	9	"	8 9	"	9
1807–08	Belper, Reeling Mill	6	90 × 40	"	9½	"	9 10	"	9
1811–12	Belper, South Mill	5 + a	118 × 40	"	9½	"	10 0	"	12 and 17

Discussion

The President said the Paper was most valuable because it was the joint work of two authors, an architect and an engineer; each supplemented the other. It contained history which was being published for the first time, and it corrected statements previously made that were now proved to be inaccurate.

Mr. K. R. Gilbert pointed out that William Felkin in his book *A History of Machine Wrought Hosiery and Lace Manufacture*, 1867, stated that William Strutt had devised a system for thoroughly warming and ventilating a large building. This suggests that Lee might have got his ideas on central heating from Strutt.

William Strutt's scientific ability was inherited from his father, Jedediah, who was Arkwright's partner and had made an important invention in machine knitting. Lord Rayleigh, however, who was named John William Strutt, belonged to an entirely independent family.

Charles F. T. Young in his *Fires, Fire Engines, and Fire Brigades*, 1866, referred to "fire-proof" buildings in inverted commas and was contemptuous of their value. He argued that with naked iron columns there was a danger that the heat of a fire would be conducted from one storey to another, thus spreading the fire; and that heated iron columns would crack and break when exposed to a stream of cold water. He considered that iron columns should not be used unless coated with a fire-proof material.

It would therefore be interesting to know if any fires had actually occurred in any of the Strutt mills and, if so, whether the structure had proved advantageous or not.

The President thought the fire danger to the cast-iron column was due to the core not being central when it was cast so that the hollow column was often thicker on one side than the other. Also, the application of water on one side of the column would produce uneven stress which could cause it to crack.

Mr. Rex Wailes thought it could be of interest to make a further study of central heating; they might well find that William Murdock was the pioneer in this country of heating by passing hot water or steam through pipes. Murdock's bi-centenary recently had been ignored by Birmingham and the country generally. No life of Murdock had ever been written and if anyone was so disposed, and had ready access to the papers in the Boulton and Watt Collection in Birmingham, it would be a wonderful job of work to do.

Mr. J. Foster Petree was reminded by one of the Authors' illustrations of a Paper on "Shrinking-on," by Rhys Jenkins,[1] and asked whether there was any evidence that the iron rings used to connect discontinuous beams were shrunk on. Rhys Jenkins, in his Paper, said that he could find no evidence of shrinking as a process earlier than the early 1800's. Various members had questioned that, and Mr. W. J. Tennant instanced built-up wooden masts; but he personally could find no evidence for iron mast hoops earlier than about 1795. There was some indication that the hoops were sometimes warmed before being put on, but they were not heated like a tyre on a wheel. He would like to raise another point, possibly outside the subject of the Paper. In one of his books, Samuel Smiles referred to a cast-iron church at Mossley Hill, near Liverpool. Could the Authors give any information about it?

Mr. Wailes said, following Mr. Petree's reference to Rhys Jenkins's Paper,[2] that he would like to put it on record that Rhys Jenkins was the man who looked at anything, including wind and watermills, for evidence of wrought iron rings being shrunk on to wooden shafts, and he believed he had found no such evidence whatsoever. The rings were usually driven on.

Mr. Johnson, replying to Mr. Petree, said the tie-rings were in some instances attached to the tie-rods but in other instances were separate. They were not shrunk on; they just fitted over the spigots of the column sockets. He had seen buildings demolished in recent years where it was quite easy to lift the ring off the spigot or off the column. They were fixed, or secured, temporarily by cotters or wedges driven down into the spaces between the sockets on the beams. There was a space of ½ in. between the column and the socket on top, that allowed for manoeuvrability in fixing the beam to the column and setting it and then the rings were just slipped over and the wedges driven into the gap in the column socket. There was no evidence of shrinking.

The second point he was not competent to deal with.

Dr. S. B. Hamilton said John Cragg (1767-1854) of Liverpool built churches in and around that city with cast-iron columns, arches and roofs; the walls were of slates with a finish which could be limestone, or sandstone dust, or whatever one wished. The wall construction of these buildings was the subject of British Patent No. 3761 of 1813. The architectural designs were by Thomas Rickman (1776–1841), an architect who had made a special study of Gothic ornament; but Rickman repented of it after a time because he did not think the aesthetic effect was satisfactory; the cast members were too slender and their faces too smooth.

He thought the Authors had made a most valuable contribution to our knowledge of the use of cast iron in building, and their researches had filled gaps and corrected errors in previous work by Sir William Fairbairn, Professor Turpin Bannister and himself.

Mr. G. B. L. Wilson reverted to steam heating and mentioned Mr. A. F. Dufton's Paper read before the Newcomen Society in 1941 on heating,[3] and asked whether it was not hot air heating at Belper. If steam were used, would it not have condensed in the pipes?

---

[1] *Trans.*, XV, 1934–35, pp. 73–75.
[2] *Trans.*, XV, 1934–55, pp. 73–75.
[3] A. F. Dufton, "Early applications of engineering to the warming of buildings," *Trans.*, XXI, 1940–41, p. 103.

WILLIAM STRUTT'S COTTON MILLS, 1793–1812

The PRESIDENT recalled the Authors' reference to plans by Boulton & Watt in 1806 for the gas lighting of the east wing of Milford Old Mill. He thought William Murdock's name should be included in the Paper because he had a very close connection with the plans and with the gas-lighting of the mills.

As the first water-powered cotton-spinning mill was erected at Cromford, the President asked if any of those large watermills still existed in that neighbourhood, or whether they had to rely only on drawings and account books.

The Paper was of great interest to him because it contained references to so many engineers and to water power, textile machinery, fire-proofing and fire fighting, which subjects were all in his Department of the Science Museum. The Society had only published one Paper dealing with fire prevention, the late Rhys Jenkins had dealt with English fire extinguishing engines, 1625–1725.[1]

In the eighteenth century it seemed that fires were mostly in warehouses and in store houses at docks in England. They were probably recorded because they were spectacular fires. In the nineteenth century the most disastrous fires appear to have been at warehouses and mills in America; because while fire protection rules were getting stricter here, in America they were building larger and quicker and with less control by legislation. But in England theatre fires were beginning to become very serious. In the early twentieth century the worst fires seemed to be in department stores, grain elevators and warehouses in America, which was perhaps natural.

On the question of wood and iron, one imagined that wood should never be used and iron was always preferred, but unprotected iron columns were dangerous as an iron beam, which was in contact with flame, could spread the fire and therefore all ironwork should be encased.

With regard to wood, timber of large scantlings, particularly oak and other hard non-resinous kinds, soon charred and the hard charcoal formed on the outside protected the core. So long as the charcoal remained in position, it prevented the oxygen in the air assisting combustion, and therefore partly-consumed timber could remain in place and carry quite a considerable weight.

"Pot floors" were widely used today but the pots did not resemble the ones they had been shown; instead they were hollow open earthenware, lying in rows in the floor, with reinforced concrete between the rows.

Much has been written on fires and their prevention. *Engineering* on 23 November reported a conference on fires in industry. One speaker stated there were at least 50 causes of fire in industry, and many of the causes were now due to heating equipment, lighting and electrical faults, but the main cause of fires was still excessive friction in bearings of all kinds, and sparks caused by foreign objects getting into machinery, which had been known to ignite textiles while being processed. The President suggested that this statement today was probably just as true in the old mills, which they had been shown on the screen.

Dr. C. ST. C. B. DAVISON said friction in bearings had been mentioned by our President as a cause of fire. Actually at the beginning of last century bearings for line-shafts in factories were just about 25 per cent. efficient. In other words, most of the power going into the line-shaft was being transformed into heat in the bearings. That must have been a very dangerous situation in many places where they were employed.

He would like to know if Professor Skempton had found any information about mathematical design in the Strutt mills, because about that time, he believed, they began to apply more generally mathematical solutions for mechanical problems instead of empiric methods. He would welcome any such information that may have been discovered among the papers that had been examined at the Strutt buildings.

---

[1] *Trans.*, XI, 1930–31, pp. 15–25.

WILLIAM STRUTT'S COTTON MILLS, 1793–1812

Mr. W. H. Bailey said that he would have liked the Authors to go further and tell more of the mechanical engineering work in Strutt's mills. There still existed a workshop which was built in 1817 with machine tools of apparently the same date. A pillar machine which was a slotter on one side and a drilling machine on the other had gone to the Science Museum, and the Sheffield Trades Historical Society had its eye on one or two more machines and on the iron roof framing. It was a very interesting place and he was very sorry that it had to come down. The Strutts were very keen on casting the dates on their ironwork and 1822 appeared on a waterwheel and millwork of which there were photographs. Very wide waterwheels had been used but they were replaced by water turbines about 1907.

Mr. B. Spiller said that he had brought along a few lantern slides of buildings rather earlier than the Derby mill, which contained a suggestion of cast-iron columns. Some of the slides were of designs, line drawings, for breweries, one being for Whitbread's Brewery, London, and the evidence suggested that they were built somewhere about 1784. Other slides showed designs by Sir John Soane, dated 1790, and the columns drawn were so slender as to suggest cast iron; they could not have been of wood.

Professor Skempton, replying to Mr. Gilbert, said that as to central heating in the Salford Mill there was a letter from George Lee to the younger Watt dated January 1802, saying: "Please to acquaint your father and Mr. Southern that we have admitted Steam into the Pillars of the building and the Elongation was as expected. . . ." From that he concluded that the idea of steam central heating was one of the elder Watt's ideas, or at least originated in the firm.

As to the column at Salford, Mr. Johnson and he last summer had seen it on the scrap heap in the mill yard. The fragment was only about 4 ft. long, but could certainly be taken as one of the columns. In fact an old foreman there, who remembered the building, had told them to look around because he thought some of the scrap still remained. A mill at Glasgow, built 1804–06 for Henry Houldsworth and fitted by Boulton & Watt with central heating, also must have been successful, although there was some expansion trouble.

Professor Skempton said he had not gone into the form of central heating which William Strutt devised, but he thought it was some sort of hot air system. Strutt was also interested in the ventilation of buildings and some of his original ventilators were still to be seen.

It did not appear that any fires took place in the fire-proof buildings while William Strutt was still alive, although the new North Mill was built in 1803 because the old timber-frame building had burnt down. There were plenty of fires in the timber buildings. Strutt's Derby mill of 1793 had a fire in 1853 which started in the timber roof, but did not get further down into the building, which might suggest that Strutt's fire-proof construction did stand up quite well under fairly severe conditions.

As to the naked iron columns, modern practice suggested it was dangerous, but Dr. Hamilton could say more about that. Throughout much of the nineteenth century columns were usually left as bare iron and there were several failures, but on the whole they seemed to have been moderately safe where used with a proper floor construction.

Professor Skempton said that in the Account Books for the West Mill there were items concerning the cost and construction of the waterwheel. It had a shaft of enormous diameter built like a barrel because no single timber big enough could be found. Two iron waterwheels were also built at Belper, and he believed one of them included the first example of the bicycle wheel type of structure with thin spokes in tension. Its invention was always attributed to Strutt, and he suspected it occurred in waterwheels of 1804–05. He had also read that Strutt was the first to put the governor on a waterwheel.

Professor Skempton then dealt with some of the points mentioned by the President in his remarks.

As far as gas was concerned he thought Sir Arthur Elton should be consulted, but the drawings were definitely Murdock's.

Regarding waterwheels, it seemed that none was now in use at Belper or Milford. They had been replaced by turbines.

The question of warehouses at the docks: it was most curious that fire-proof construction was not applied to dock warehouses until long after the mills. In 1807 John Rennie was pleading with the Admiralty to adopt that construction, and all the examples Rennie quoted were textile mills. This seemed to indicate quite clearly that, by 1807, no fire-proof buildings had been used in the dockyards.

With regard to mathematics, the indications were strongly positive. Bage wrote a letter to Strutt in 1803 outlining a theory for the strength of iron beams, which was almost identical with that published in 1831 by Eaton Hodgkinson. Bage also made tests on beams and on full-scale iron-framed roof trusses. As to the mathematics of columns, Bage and Strutt were in correspondence on that point, and had a reasonably satisfactory simple theory as early as 1796.

Commenting on the slides shown by Mr. Spiller, Professor Skempton said he had seen a photograph of Soane's drawing and he had been struck by the very slender columns. He could not imagine them as anything but iron. There was, therefore, a strong suggestion that multi-storey buildings were being at least thought of, if not built, with timber beams and iron columns before 1793. At one time he thought that Strutt was the first to use iron columns, because he could not see much point in using them unless in the interests of fire prevention, but this view was probably wrong. There was a mill in Derbyshire, built in 1785 (although he was not sure of the evidence for the date), and that mill, which still existed, although not in its original form, is said to have had iron columns with fully exposed timber beams and floors. About that time iron columns may have been coming in for multi-storey buildings. If that was correct, it would suggest a clear four-stage development from the timber frame to the iron frame.

Lastly, Professor Skempton regretted that, so far, no portrait of William Strutt had been discovered.

PLATE XL

Fig. 1. MILFORD, OLD MILL

Fig. 2. MILFORD, WAREHOUSE

PLATE XLI

Fig. 1. MILFORD WAREHOUSE, THIRD STOREY

Fig. 2

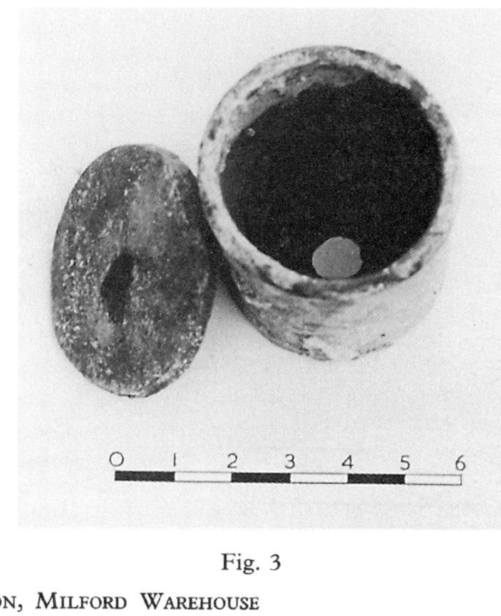

Fig. 3

HOLLOW POT CONSTRUCTION, MILFORD WAREHOUSE

PLATE XLII

Fig. 1

Fig. 2

STRUCTURAL DETAILS, MILFORD WAREHOUSE

Fig. 3. VIEW OF BELPER NORTH AND WEST MILLS
(by H. Moore of Derby; Davies 1811)

PLATE XLIII

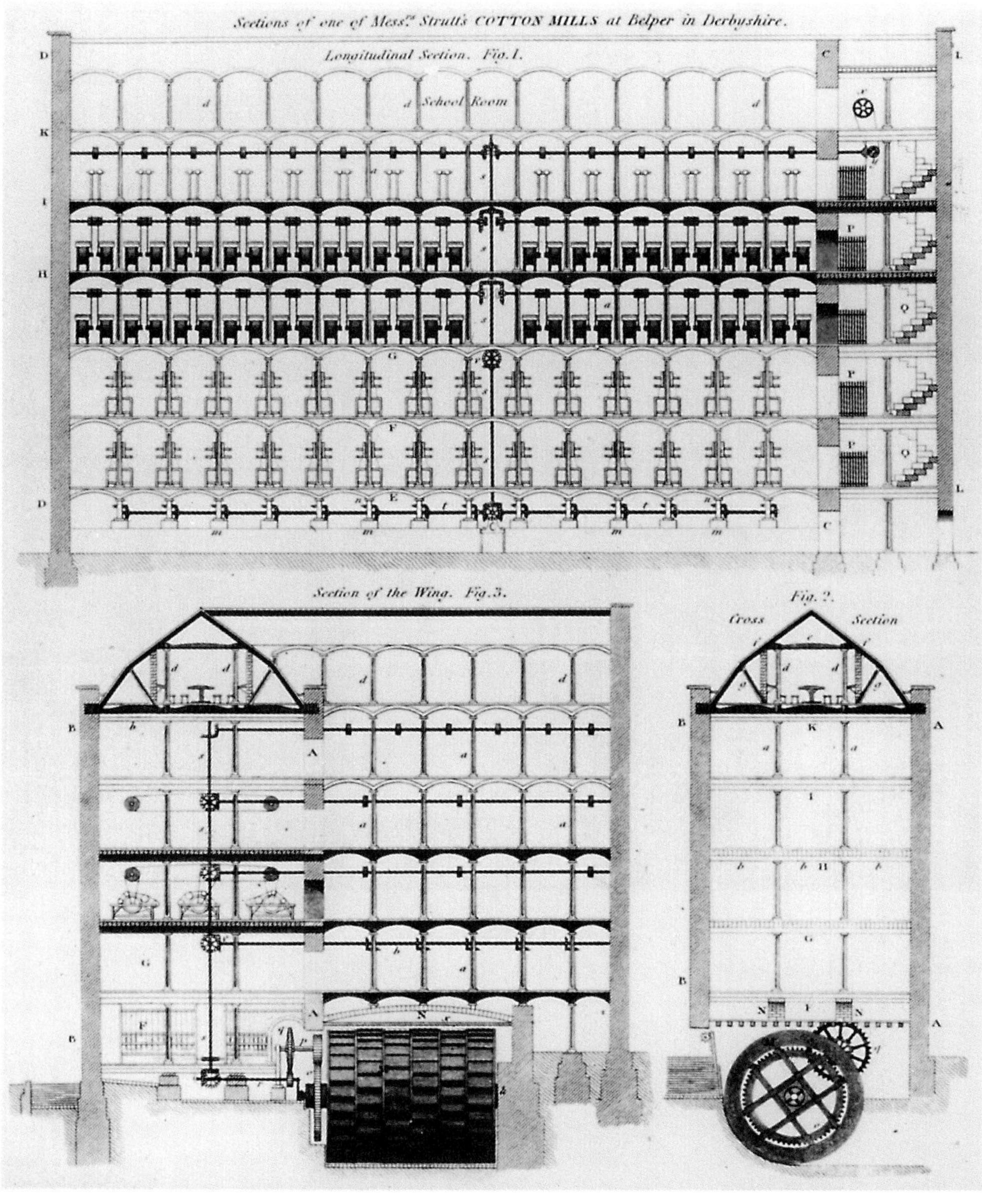

SECTIONS OF BELPER NORTH MILL
(Farey 1812)

X

PLATE XLIV

BELPER SOUTH MILL

XI

Engineering in the Port of London, 1789–1808

INTRODUCTION

During the second half of the 18th century the tonnage of ships entering the Port of London from abroad increased by nearly three times and their number doubled; as did the tonnage of colliers and other coastal vessels:[1]

Year	From foreign ports		Colliers and coasters	
	ships	tonnage	ships	tonnage
1700	1,300	157,000	5,600	278,000
1750	1,700	234,000	6,400	512,000
1793	3,500	655,000	9,600	964,000
1830	5,200	952,000	18,300	2,330,000

Yet few improvements were made in port facilities and, in particular, virtually no commercial dock accommodation existed before the opening of the first of the West India docks in 1802.

The Navy at Deptford and several ship-building yards on the Thames had dry docks and some small wet-docks. At Rotherhithe there was the Howland Dock (opened in 1699) of 10 acres extent, originally built for the fitting-out of ships: later re-named Greenland Dock and used by the whaling trade. And at Blackwall the ship-builder John Perry in 1790 opened his Brunswick Dock of 8 acres, also used for fitting-out and by whalers. But all ships from the East and West Indies and America, and from the Baltic and other European ports, had to moor in the river and off-load into lighters; the cargoes being landed at various quays. These included, for goods attracting customs duty, the Legal Quays between London Bridge and the Tower and the so-called Sufferance Quays, mostly on the opposite bank. Colliers also off-loaded into lighters, which usually went further upstream, and only the smallest vessels unloaded directly at riverside quays. Reasonably efficient up to the mid 18th century, this system became intolerable by the 1790s from excessive delays and serious losses by plunder; with as many as 880 ships frequently at the moorings, where accommodation could properly be provided for about 600, and at some places large ships were lying 12 or even 16 abreast in the tiers.[7]

The initiative for reform came from London merchants headed by William Vaughan, who wrote and published in 1793 a tract in which he gave reasons for concluding that major improvement of the port depended upon the construction of enclosed wet docks equipped with warehouses; the latter, where necessary, being under the control of customs officers.[2] He envisaged such docks at Rotherhithe, the Isle of Dogs and St. Katharine's, but thought a site at Wapping to be most suitable.

Following his lead, plans were produced in 1794 by John Powsey[3] and, after more detailed surveys, by Daniel Alexander. The latter scheme was for two linked docks having a total water area of 35 acres and a 3 acre entrance basin at Wapping, and either a second entrance at Shadwell (as proposed by Powsey) or a ship canal nearly 3 miles long from the eastern dock to the Thames at Blackwall.[4] The merchants backing this plan raised £800,000 by subscription and petitioned Parliament for a Bill in February 1796. Meanwhile the Corporation of London, as the statutory body responsible for the port, considered plans for improving the Legal Quays,[5] building warehouses on the south bank, constructing a canal across the Isle of Dogs and providing large docks (without warehouses) on the Isle to remove some of the shipping from the river.[6]

These plans, and several others, then came before a parliamentary committee which reported in May 1796.[7] Daniel Alexander and John Rennie gave evidence and estimates for the Wapping proposals based on the former's 1794 plan;[8] Ralph Walker, on his own initiative, submitted an alternative dock plan and also looked into the proposed ship canal across the Isle of Dogs;[9] George Dance, Surveyor to the City, dealt with the Legal Quays and warehouses,[10] while William Jessop, consulting engineer to the City, sent in plans and estimates for the dock at the Isle of Dogs and an additional dock for colliers at Rotherhithe.[11] The committee came out in favour of wet docks and associated warehouses, but refrained from specific recommendations at this stage.

In November 1796 Alexander produced a more elaborate plan for the merchants' docks, now increased in total area to 44 acres with entrances at Wapping and Shadwell, but without a canal to Blackwall.[12] In February 1797 the City petitioned for a Bill to build the docks and canal in the Isle of Dogs. By June 1797 an influential group of West India merchants realised this site might suit their purpose better than Wapping, and the City naturally welcomed the idea. Dance and Jessop with John Foulds assisting, for the City, and Ralph Walker, now acting as engineer for the West India merchants, together worked out a scheme and reported in August.[13] They proposed two linked docks totalling 31 acres, a 5 acre entrance basin at Blackwall, and a lock at Limehouse, together with a canal three-quarters of a mile in length, 23 feet deep and 176 feet wide at the water surface. The docks were to have ten stacks of warehouses and the Blackwall entrance gates a depth of 24 feet at high water springs.

An engraved plan of the Isle of Dogs canal and docks appeared in November 1797 with a printed report,[14] the scheme was formally approved by the City and the West India merchants, and a petition for a Bill went to Parliament in February 1798. After further discussions the docks plan was modified in some respects, including an increase of the water area to 43 acres (still with a 5 acre entrance basin), Jessop and Walker produced revised estimates in November,[15] and a new engraving was published.[16]

Meanwhile, Vaughan's group of merchants, now without their West India colleagues, adhered to the Wapping scheme of November 1796. This so-called 'Merchants Plan', supported by evidence from Alexander and Rennie,[17] and the 'City Plan' for the West India docks and ship canal, were examined during the months of February, March and April 1799 by parliamentary committees.[18],[19] Counter-proposals for rebuilding London Bridge, to allow shipping to pass upstream as far as Blackfriars Bridge, were also considered. But this idea, though leading to some interesting designs (notably Telford's 600 foot span iron bridge), was never a practical proposition; and the Act for the City Plan was passed in July 1799, followed in June 1800 by the Act for the London Docks at Wapping.

The East India Company had little concern in these negotiations. They owned several warehouses in east London, including those in Cutler Street, and their ships entering the port, though large in size, numbered no more than about 60 a year, and moored at Blackwall well away from the congested areas. However, the advantage of wet docks for loading and unloading became so apparent that early in 1803 proposals and estimates for docks at Blackwall were commissioned from Rennie and Walker.[20] Their plan envisaged two docks totalling 18 acres and a 3 acre entrance basin, making some use of the existing Brunswick Dock. On this basis a petition went to Parliament in April and the Act was passed in July 1803.

All three dock schemes were modified soon after obtaining the Acts, and the as-built dimensions are summarised in Appendix II. By 1806 the ship canal, the West India and East India Docks had been completed, and also a major part of London Docks, with vast warehouses at the London and West India. The largest fully laden ships ranged from about 500 tons in the London Docks to 800 tons in the East India. Quayside accommodation was provided for at least 200 ships, and as many more could lie at moorings within the docks. The total area of impounded water amounted to 100 acres, or 110 acres including the shallower Greenland Dock, a capacity quite without precedent elsewhere.[21] A map of the Port at this stage is shown in Fig. 1.

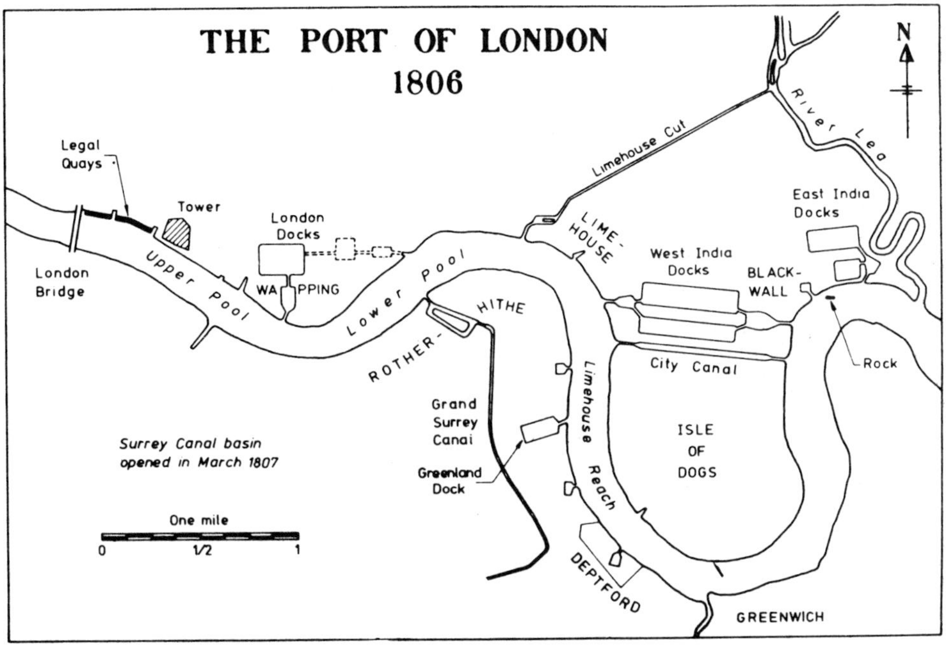

Fig. 1. Map of the Port of London in 1806.

Meanwhile, progress was being made in removing an outcrop of rock in the river bed at Blackwall and providing new East India moorings, and the Grand Surrey Canal Co. were building an 8 acre dock at Rotherhithe. These works, completed in 1808, round off the classic phase of improvements which form the subject of the present contribution. It is hoped in a subsequent paper to deal with the next period which ends in 1833. By that time London Docks and the Commercial Docks had been completed, the Grand Surrey Docks enlarged, St. Katharine's Docks were fully operative (with splendid warehouses) and, with two smaller docks also functioning, the total area of impounded water for commercial shipping amounted to 184 acres plus 19 acres in entrance basins, together with 45 acres of timber ponds; and about 4,300 ships entered the docks annually.[22]

The history of London's docks is given in general terms by Sir Joseph Broodbank[23] and John Pudney.[24] Some technical details can by found in Elmes,[25] Telford,[26] Rennie[27] and Shepherd-Baron,[28] and a study of the West India Docks has recently become available.[29] But a need has long been felt for an account of the engineering works in the Port of London, especially in the early formative years, together with a record of the engineers and contractors responsible for design, supervision and execution of the works.

1. BRUNSWICK DOCK (1789–1790)

As a prelude to the main story, a brief description will be given of the Brunswick Dock, built as a private venture by John Perry in his yard at Blackwall. Work began early in 1789 with John Powsey as engineer and Thomas Wilkins as assistant. There were, in fact, two linked docks. The western dock, used for fitting-out and equipped with a mast-house 120 feet high, had an area of 5 acres. Its entrance consisted of a single pair of pointing gates 44½ feet wide in the clear with 16 feet of water over the sills at ordinary high-water springs;[30] capable of admitting 1,250 ton East Indiamen, unladen. The smaller (eastern) dock of 3 acres, used chiefly by whalers, had an entrance lock 140 feet long between gates, a width of 32 feet and sills 13½ feet below H.W.[30] An outline plan is shown in Fig. 10,[31] and William Daniell engraved an excellent view of the dock in 1803.[32]

The dock walls were built with timber piles and sheeting[33] probably supported by land ties. The opening of the dock took place on 20 November 1790 and the first ship was masted in October 1791.[34] Apart from the mast-house, the dock resembled in construction and purpose the much earlier Howland-Greenland dock at Rotherhithe. It was purchased in 1803 by the East India Dock Co. and incorporated in their Export Dock (see section 6). The mast-house remained until 1862.

2. WEST INDIA DOCKS, Phase 1 (1800–1803)

In response to a requirement by the Customs and Excise that the Import and Export docks must have separate entrances, a new plan with the docks in parallel (not in series as originally intended) was drawn in August 1799.[35] The docks were also increased in area to 29 and 23 acres respectively, the entrance basin enlarged to 6 acres and a 2 acre basin introduced at the west end. The line of the ship canal then had necessarily to be moved about 800 feet further south. These radical changes were worked out by Jessop, Dance and Walker, and in November 1799 Walker reported in some detail on the docks plan.[36]

The idea of an impounded (non-tidal) entrance basin had not yet been put into practice and uncertainty existed as to whether the river entrance should have a single pair of gates or a lock. The November 1799 plan adopted a pair of gates, but the final plan which emerged in February 1800, in consultation with Jessop,[37] shows locks at the Blackwall and Limehouse entrances; evidently after a decision that the extra capital cost was justified by greater flexibility in operation. The docks were also enlarged to 30 acres (2,600 by 510 feet) and 24 acres (2,600 by 400 feet). The final layout and design (by George Gwilt) of the warehouses on the north quay followed in April. Fig. 2 shows the docks as built.

At the time of his appointment as engineer in February 1800 Jessop was already preparing sketch designs and estimates, and immediately took full responsibility for construction. Up to October 1802 his salary was £500 p.a. During that period he worked 437 days on dock business, mostly on site: an average of 160 days per year. This far exceeded the duty normally expected from a chief engineer, so from November 1802 he received his usual consulting fee of 5 guineas per day plus expenses, and on completion of the works in July 1803 the directors awarded him a gratuity of 1,000 guineas. For two years he had as an assistant his sons Josias, who left in February 1803 to work on the Croydon–Merstham extension of the Surrey Iron Railway.

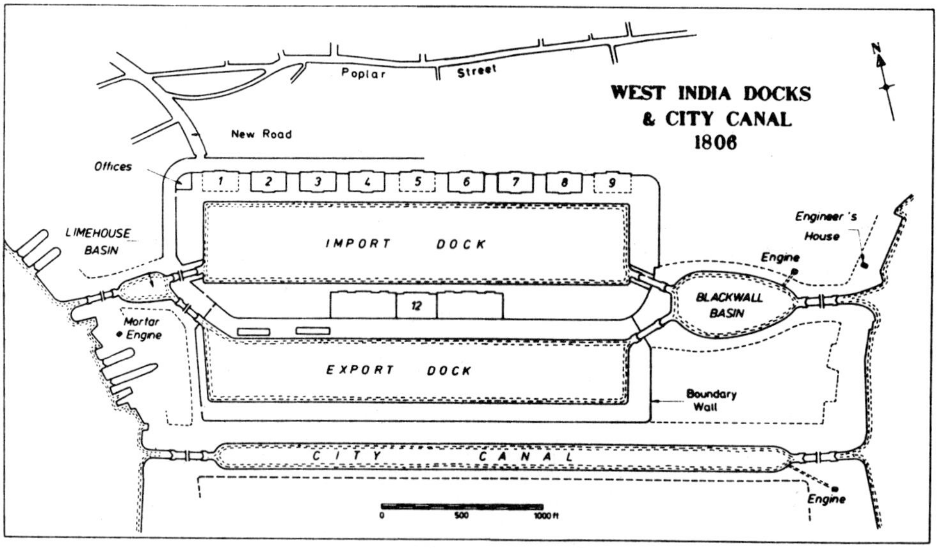

Fig. 2. West India Docks as built.

Ralph Walker was appointed resident engineer for the docks in August 1799. He had a salary of £600 p.a. with a rent-free house at Blackwall, a clerk and a draughtsman. His responsibilities included checking the quality and quantities of work done by the contractors and by direct labour, issuing working drawings, and keeping accounts in the engineering department. In addition there were superintendents of brickwork, masonry, carpentry and excavation. Walker left in October 1802 and his place was taken by Daniel Vaux, acting as deputy engineer to Jessop, until 1803.

Work began in February 1800 with the excavation for the Import dock, the firm of John Holmes & William Bough being the prinipal contractor. The top 5 or 6 feet of sandy clay was used for brick-making, though very large quantities of bricks had to be bought in addition. As digging proceeded into the underlying gravel, the excavations were drained through iron pipes laid beneath the river bank to a sluice at low water. At still greater depths drainage was achieved by a 28 h.p. Boulton & Watt engine established with a deep well near Blackwall entrance. Water from the excavations flowed along small trenches and finally through buried pipes to the well. Another engine, of 20 h.p., near Limehouse entrance worked the mortar mills, and limekilns were sited nearby. Most of the materials (bricks, limestone and sand, timber and building stone) were brought in by water to jetties at Limehouse and transported on wooden railways to the site.

The first brick of the dock wall was laid in June 1801 by the contractors Adam & Robertson, who also built the lock structures and warehouses. The walls, 6 feet thick with counterforts at 15 feet centres, were of brickwork throughout, set in hydraulic lime mortar and built to a curved batter (Fig. 3) with a coping of Dundee grit-stone. Iron reinforcement strips strengthened the bond between wall and counterfort. Puddle clay, covering the dock bottom and backing the walls, greatly reduced seepage from the dock. The walls have an economical section. The gravel admittedly provides a good foundation, but the concave profile (already used by Jessop in Ringsend Docks at Dublin[38]) by setting back the centre of gravity of the walls adds much to their stability, and they remain today in good condition after 175 years.

Fig. 3.

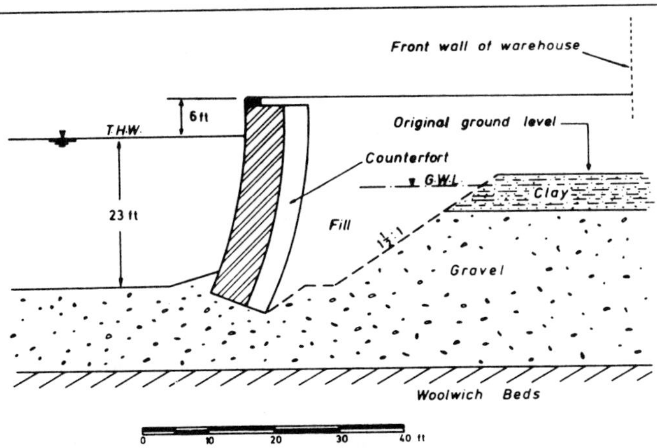

WALL SECTION: WEST INDIA IMPORT DOCK 1802

Blackwall basin, also excavated by Holmes & Bough at the same time as the dock, had sloping earth sides. John Dyson, excavation contractor for Limehouse basin, started work in February 1801. The walls of this basin were similar in design to those of the dock. Gravel from the excavations was placed as a general fill behind the walls and the embankment around Blackwall basin in order to raise ground level from its original height of 6 or 7 feet below, to the level of the coping 6 feet above, high-water of ordinary spring tides, or Trinity High Water (T.H.W.): see Appendix I.

A section through the chamber of Blackwall entrance lock (Fig. 4) shows the brick walls with counterforts and masonry facing, and the inverted arch construction of the floor.[39] Rows of grooved timber sheet piling driven beneath the floor at each pair of gates increased safety against piping under the considerable differential head (22 feet of water at exceptional spring tides) between the basin and low tide in the river. The pointing gates, 28 feet high and slightly curved in plan, were framed in oak; they moved on iron rollers running on iron *sweeps* and were operated by capstans and chains. Piling for the lock began in April 1801 (Thomas Clark, contractor) and the whole structure, including a double-leaf timber swing bridge over the lock, reached completion in August 1802. It had a length of 192 feet between gates, a clear width of 45 feet and 24 feet of water on the sills at T.H.W. The communication lock between Blackwall basin and the dock was 162 feet long and 38 feet wide, with the same depth of 24 feet, and could admit 650 ton ships fully laden.

Fig. 4.

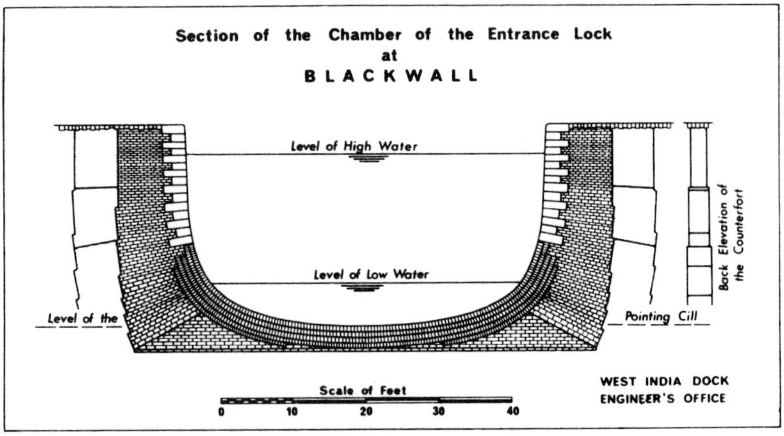

Limehouse entrance lock, also crossed by a timber swing bridge, had a length of 155 feet, a width of 36 feet and sills 22 feet below T.H.W. Intended chiefly for use by lighters, two of these small vessels could be admitted simultaneously at almost any state of the tide. The communication lock was 5 feet shorter but otherwise of the same dimensions.

Wing walls, forming a wide bell-mouth entrance to the Blackwall lock, had their foundations 7 feet below low tide level; so a cofferdam was necessary. Bough & Clark won the contract and started preparing timber and horse-operated pile engines in November 1801. After removing mud down to about low water, pile driving began in December and the work was completed in March 1802. Arched in plan, the cofferdam consisted of two rows of sheet piling 10 feet apart driven 7 feet into the gravel, supported by 14 inch by 14 inch gauge piles and horizontal walings; with additional support from rakers carried to a shore pile which itself was tied back to the inner gauge piles by two 12 inch by 4 inch planks (Fig. 5). Compacted earth filled the space between the sheet piles.[40]

All went well until Jessop, under pressure to achieve a deadline for opening the dock, allowed excavation behind the dam in order to minimise the amount of dredging after its removal. A close watch was kept, but without warning a failure occurred at high tide on 22 July. By then the excavation, 3 feet into the gravel, had been opened along 40 or 50 feet of the southern part of the dam. Theoretically a small factor of safety existed against piping, but this could easily be reduced in practice, e.g. by a pile not having reached its design depth or being broken in driving. Six of the men working there were trapped by the sudden inrush of water and lost their lives. A *preventer* dam of stop logs in front of the outer lock-gates saved the lock from damage, and with only a short delay the Export Dock was formally opened on 27 August 1802.[41] The opening of the Limehouse entrance followed in October. Soon afterwards part of a not quite completed wall in

ENGINEERING IN THE PORT OF LONDON, 1789–1808

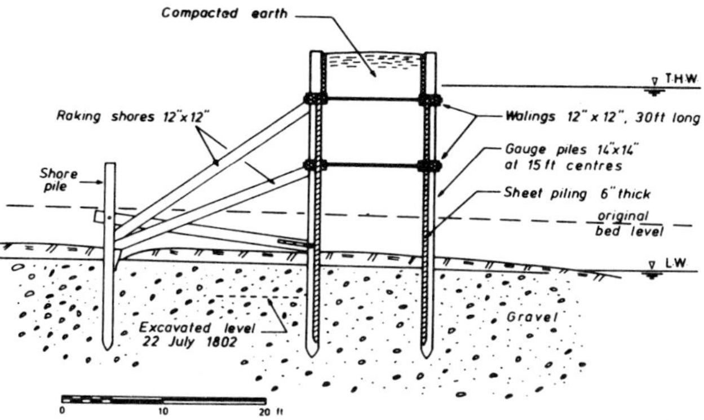

Fig. 5. Sketch Section of Blackwall Cofferdam: West India Docks.

Limehouse basin suffered a *rapid drawdown* failure, due to mistaken operation of the lock-gates following an exceptional high tide; but this was an accident, not a fault in design. Rebuilding the wall, completing the wing walls (in tide-work on piled foundations without a coffer dam) and several minor works, took until July 1803. By that date all engineering work and the north quay warehouses had been completed.

The West India Import dock, with a water area of 30 acres and 23 feet depth, and its basins, locks and warehouses, created a new era in dock construction and long remained the world's largest structure of its kind. The cost of the engineering works alone came to £310,000.[42]

3. WEST INDIA DOCKS, Phase 2 (1803–1806)

Jessop produced drawings, estimates and specifications for the Export dock in August 1803.[43] Work began next month with the contractor John Dyson excavating trenches (21 feet bottom width and 1½:1 slopes) for the wall foundations. Thomas Morris, who had been dock engineer at Liverpool, arrived in December to take up his post as resident engineer, at £840 p.a., and a little earlier William Pilgrim was appointed assistant engineer at £157 p.a. One of Morris' first jobs involved building, by direct labour, a swing bridge over the communication lock between Limehouse basin and the Import dock.

A contract for cofferdam construction went to Thomas Clark in January 1804. The dock walls and locks communicating with the entrance basin followed closely in design and dimensions those of the Import dock (Section 2); contracts for the brickwork (Aslat & Stewart) and masonry (T. & W. Crawford) were signed in February 1804.

Drainage of the excavation was at first achieved by using the 30 h.p. engine already installed for the City Canal (see Section 4) and this began pumping, for the dock, in December 1803; from which date the Dock Co. paid the Corporation of London £350 per month for use of the canal engine or £175 per month when it was employed for both jobs. Later, from September 1804, the Blackwall engine was used, pumping through iron pipes which had been laid beneath Blackwall Basin (see Discussion, page 106).

William Bough and John Hughes contracted for the main dock excavation, i.e. the *dumpling* left after digging the wall trenches, and began work in August 1804. They soon added to their work force a 6 h.p. steam engine (previously employed at London Docks, see Section 5), for hauling waggons on an inclined plane.[44] In October 1805 the job was finished; 490,000 cubic yards of earth having been removed at an average rate of 8,000 cubic yards per week charged at 11d. per cubic yard. It is worth noting that the total excavation for the West India Docks, between 1800 and 1805, came to about 1.5 million cubic yards.

By June 1805, with the work well under control, Jessop's services were no longer needed and, in any case, Bristol Floating Harbour now demanded much of his attention.[45] Everything proceeded according to plan, and the official opening of the dock took place on 12 July 1806. By this time, also, three large single storey warehouses had been erected on the south quay of the Import Dock. The engineering cost of the Export Dock with its two communication locks came to £205,000; and the total expenditure on the West India Docks to 1806 is as follows:[42]

Act of 1799	£12,000
Purchase of land, etc.	£94,000
Interest payments	£34,000
Engineering works	£515,000
Warehouses, quays, roads	£557,000
Boundary wall, offices	£27,000
	£1,239,000

This is a very large figure for the period, but it represents nevertheless a most economical piece of construction in view of the scale of the works and buildings. Moreover the docks proved to be an immediate, and lasting, financial success.

4. CITY CANAL (1800–1805)

As mentioned in Section 2, the line of the ship canal across the Isle of Dogs was settled in August 1799. Its dimensions remained unchanged from the first plan of 1797: top water width 176 feet at the level of ordinary spring tides and a depth of 23 feet (Fig. 6). Embankments 12 feet high along both sides, backed by a general fill, brought adjacent ground level up to a height of 6 feet above T.H.W. Locks at each end were practically identical to the Blackwall entrance lock of the West India Docks, and double-leaf timber swing bridges carried traffic over them.[46]

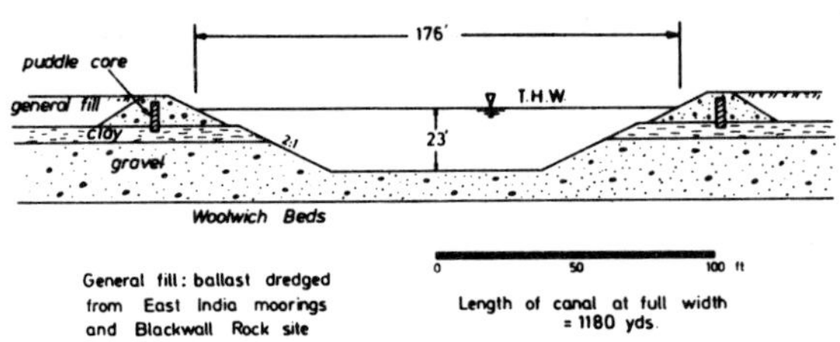

Fig 6. Sketch Section of City Canal.

The canal was built under the Act of July 1799 at the City's expense. Jessop held the position of engineer from that date until December 1805 at a fee of 5 guineas per day. John Foulds acted as assistant engineer to 1802 followed by Daniel Vaux from 1803. Accounts of work done were kept by James Mountague, a member, like Foulds and Vaux, of the City Surveyor's Office. Work began in January 1800, with John Clark & Thomas Thatcher removing the top 6 feet of clay. John Dyson contracted for excavating the underlying 11 feet of gravel and also the lock pits. He started in March 1801, and two months later a 15 h.p. Boulton & Watt engine came into operation for drainage. During the next year, excavations at the canal and at the West India Docks mutually benefited from pumping at both sites. But after 1802 the canal engine had to carry on alone, and as the excavations (particularly in the lock pits) reached deeper levels, more power was required. The engine was therefore converted from single to double acting and a larger boiler installed.

John Fentiman built the eastern lock, starting in June 1803, and Dyson followed soon afterwards with the western one. Both locks were finished around April–May 1805. Construction of the cofferdams, by James Spedding, began in March 1805 and when these were closed in May, William Stewart started the wing walls. Finally the river banks between the locks and cofferdams were removed by John Hughes and, after delays caused by some trouble at the eastern cofferdam (but without any injury to workmen), the opening of the canal took place on 9 December 1805.

The engineering works cost £126,000 out of £194,000 total expenditure. The canal was soon being used as much for laying-up as by shipping in transit,[47] and gradually the original function fell almost into abeyance. In 1829 the West India Dock Co. bought the canal and without structural change it became their South Dock, providing a useful 14 acres of water. A timber pond was added on the south side in 1832–33.

5. LONDON DOCKS, Phase 1 (1801–1805)

The Court of Directors of the London Dock Co. first met in July 1800 and appointed Daniel Alexander as Surveyor (and architect). This was virtually a full-time job carrying a salary of £1,200 p.a. Negotiations began at once for purchasing property within the boundaries laid down in the Act.[48] For advice on planning the main dock and Wapping entrance the Directors consulted Robert Mylne, John Rennie and Capt. Joseph Huddart. They reported in March 1801, approving the proposed scheme in general but recommending a smaller, rectangular dock with a water area of 20 acres (1,260 by 690 feet) and a depth of 19 feet; a single (not double) communication lock of 160 feet between gates, 38 feet wide and 19 feet deep; a basin of 3 acres, 23 feet deep; and an entrance lock also 160 by 38 feet, with 23 feet of water on the sills at T.H.W.

This report was accepted, Alexander drew up a detailed plan in May,[49] and submitted his designs for the warehouses in September. Further discussions with Rennie resulted in two changes: the depth of water in the dock and communication lock increased to 20 feet, and the entrance lock enlarged to 170 by 40 feet, keeping the 23 feet depth. The dock as built is shown in Fig. 7.

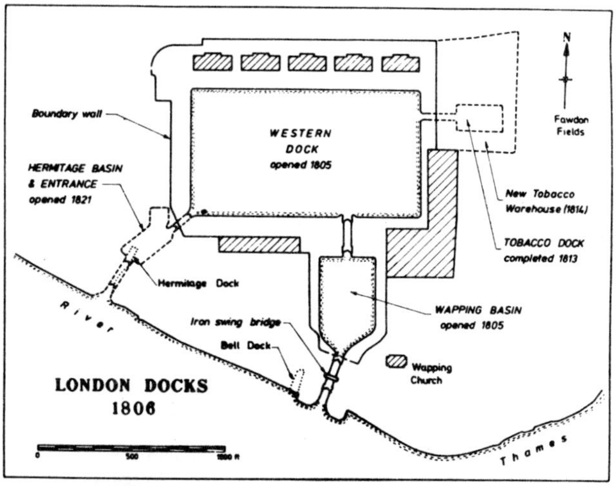

Fig. 7. London Docks as Built.

Rennie's appointment as Engineer dates from May 1801, his salary being £500 p.a. for 90 days service annually. He chose James Murray as Resident Engineer, who arrived in September and held this post for exactly three years at a salary of £500 p.a. full-time. The very able James Hollinsworth served as Assistant Engineer from June 1801 to February 1805, initially at £250 p.a. and from December 1802 at £350 p.a. Rennie also selected superintendents of masonry and carpentry who had previously worked under his direction. By October 1801 houses for a pair of Boulton & Watt 28 h.p. pumping engines and a 20 h.p. mortar engine were being erected, and the drainage well had been sunk; it was located beside Wapping basin; a jetty at Bell Dock near Wapping entrance had also been built for an iron railway leading to the construction site, the rails for which were being laid.

The ground having been cleared by the early months of 1802, Bough & Holmes began excavating in March. Meanwhile large quantities of timber, stone and bricks had been arriving, and John Hiscock started work on the cofferdam across Wapping entrance in June. As early as July 1801 Rennie decided to order an 8 h.p. engine for driving the coffer dam piles, and this came into operation, successfully, in June 1802: the first recorded example of its kind.[51]

For the dock walls and locks Rennie generally followed the designs of Jessop at the West India Docks, the differences being, for the walls, a row of sheet-piles at the toe and three courses of stone at top water level and, for the locks, masonry instead of brickwork for the inverted arch floor and walls. Another variation was the use of pozzolana mortar for the outer brick and stone courses. By November 1802 good progress could be reported on the dock walls (contractor William Hobson), the cofferdam had been closed, and a start made on the lock foundations. The locks, incidentally, were built by direct labour under Murray's direction. Drawings[51] are shown in Fig. 8.

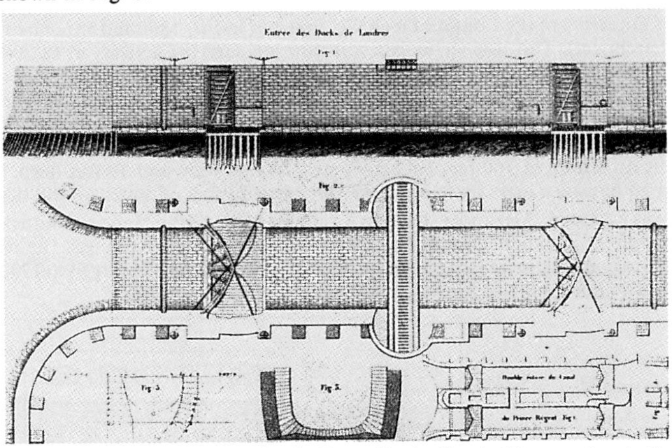

Fig. 8. A London Dock Entrance Lock. From Charles Dupin, *Voyages dans la Grande Bretagne*, (Travels in Great Britain), Part III, Paris, 1824.

Combining information from three sources, much can be learnt of the methods and rate of excavation.[52] Digging, of course, was done by spadework. The total amount of earth removed amounted to 938,000 cubic yards. About 40 per cent of the total was placed as fill behind the walls and to raise the general site level. Charged at 10d. per cubic yard, this material would have been brought out of the excavations on barrow and horse runs located close to the places where it was to be deposited. Another 40 per cent was barged away from jetties at Bell Dock and Hermitage Dock (Fig. 7) at 2/6d. per cubic yard. Horse-drawn waggons, each holding a cubic yard of earth, were conveyed on railways from the excavations, up inclined planes and along to the jetties. On the Hermitage line horses did all the work, but on the Bell Dock line the contractors, in June 1803, installed a 6 h.p. steam engine[53] to haul waggons up the plane, which rose 20 feet in a length of 120 yards. On this line Bough & Holmes agreed to carry 3,500 cubic yards per week, with a 400 yard haul from the dock to the jetty. To do so they employed 25 men filling waggons, 24 horses each with a boy driver, and 7 men unloading. In addition there were men shifting and re-laying rails, maintaining waggons (30 on this line), looking after the horses and operating the engine. Waggons on a third railway conveyed earth to be deposited on open ground east of the site in Fawdon Fields; with a shorter haul the cost was 1/2d. per cubic yard. From July 1803 to the end of the excavation contract in October 1804 the rate of excavation on the whole job, and including the winter months, averaged 8,500 cubic yards per week.

The railways had two tracks with cross-over points and lead-outs, and turntables at the jetties so that the waggons could be rotated through 90° to be end-tipped into barges. The plate rails, laid on wooden sleepers to facilitate shifting to new positions, weighed about 44 lbs. per yard and were supplied by Messrs. Outram of Butterley and Aydon & Elwell of Bradford. On the 'materials' line, which ran more or less on a level to all parts of the site, there were ordinary waggons for bricks, sand, mortar, etc., and trucks for stone or timber; and one horse could readily draw three waggons each holding a ton of bricks or one cubic yard of sand. The use of iron railways on a construction job was an innovation, the cost (nearly £2 per yard of double track) being justified by increased efficiency on works of massive scale, and the steam engine of 1803 can be regarded as the origin of mechanised excavation. It saved the labour of about 16 horses.

Another 'first' at London Dock is the double-leaf iron swing bridge across Wapping entrance lock. Rennie produced the drawings in April 1803 and the bridge was completed in October 1804.[54] It had a span of 40 feet and a roadway 15 feet wide; supported by 7 ribs. I have not found out who supplied the castings.

There is little more to report except that by December 1804 the first warehouse on the north quay and the tobacco warehouse at the south-east corner of the dock had been completed, and the dock was opened on 31 January 1805. All five north quay warehouses reached completion by July 1805 and the large south quay warehouse followed in December 1806, by which time the London Docks had almost exactly the appearance depicted by Daniell in his beautifully clear engraving of 1808 (Fig. 9).[55]

Murray left in September 1804 to take up an important post in Ireland as engineer (with Rennie as consultant) of the Royal Canal. Hollinsworth left in February 1805, just after the dock opening, but worked part-time until May on some minor jobs and settling accounts. He then became resident engineer on the Crinan Canal and, later, on Waterloo Bridge. Tidying up operations were continued by John Timperley. Rennie's appointment ended in February 1806, when he received a gratuity of £500 and a new appointment as consultant, at a reduced salary of £200 p.a. Every last detail in the engineering department seems to have been settled by September 1806, when Timperley left. He, too, subsequently had a distinguished career as a resident engineer.

London Docks show Rennie at the height of his powers, as an innovator, designer and director of a great undertaking in which excellence ranked as the main criterion. The engineering cost can be reckoned approximately at £320,000 and the total expenditure to 1806 was about £2.1 million of which more than half is due to the purchase of property.[56] This heavy outlay meant that the docks, though always very busy, returned only a modest 4 or 5 per cent on capital; and many years elapsed before the whole scheme, including the eastern dock and Shadwell entrance, could be finished.

Fig. 9. A View of the London Dock. From an engraving by William Daniel. 1808.

6. EAST INDIA DOCKS (1803–1806)

In August 1803 the East India Dock Co. appointed John Rennie and Ralph Walker to be joint Engineers. Rennie received a salary of £500 p.a. for at least 90 days attendance a year and, by an unusual arrangement, Walker was to be Resident Engineer; his salary being £1,100 p.a., out of which he paid two assistants (for measurement and accounts) and a draughtsman. Additionally he had the use of a rent-free house. Thomas Swithin was appointed Surveyor.[57] James Walker, who in 1800, at the age of 19, had become a pupil to his uncle on the West India Docks, now acted as his deputy on the East India job, and wrote progress reports to the Directors in the occasional absence of the two senior engineers. He was also engineer for the Commercial Road and its extension, the East India Dock Road.

The original plan of March 1803 involved transforming Brunswick western dock into an export dock, and the eastern dock into an entrance basin, with a new lock from the river, and building a new dock on the north side 13 acres in area, with a communication lock leading into the basin. Asked to submit detailed particulars of the works, Rennie and Walker reported on 18 August advising (i) an enlargement of the north (import) dock to 16 acres, (ii) the entrance and communication locks to be 210 feet between gates and 50 feet wide, and (iii) the docks, basin and locks to have a depth of 25 feet below T.H.W. A contract for excavating the import dock went to Hugh McIntosh, and he began work in September 1803. At the same time orders were placed for a 28 h.p. Boulton & Watt pumping engine, for bricks, timber and stone (from Dundee), and a second-hand 12 h.p. engine from Hull for the mortar mill. Also, Walker made arrangements for the Company to purchase a lime quarry at Guildford.

In November the import dock dimensions were finally settled as 1,410 by 560 feet, or 18 acres. But in February 1804 Rennie and Walker submitted a major revision of the rest of the scheme: namely, to convert the entire Brunswick Dock into a 9 acre export dock and to construct a new entrance basin of 3 acres on the east side (see Fig. 10). The Directors immediately agreed, and a few days later McIntosh signed a contract for excavating the basin and the two lock pits. As built, the Export dock had a slightly smaller width, the dimensions being 760 by 463 feet (8 acres), but otherwise the plan of February 1804 was followed exactly.[58]

However, after further discussion it was decided to make the Import dock and basin 22 feet deep below T.H.W. and to keep the 25 feet depth for the lock sills. The lock width was later fixed at 48 feet, and a decision to build the dock coping 8 feet above T.H.W. followed in July 1805. With these dimensions accommodation could be provided for fully laden ships up to 800 tons. The locks had sufficient length and width to admit the largest East Indiamen of 1,500 tons, but the restricted depth meant that the practice had to be continued of taking out their stores and part of their cargoes at Long Reach before coming up to Blackwall.

Fig. 10.

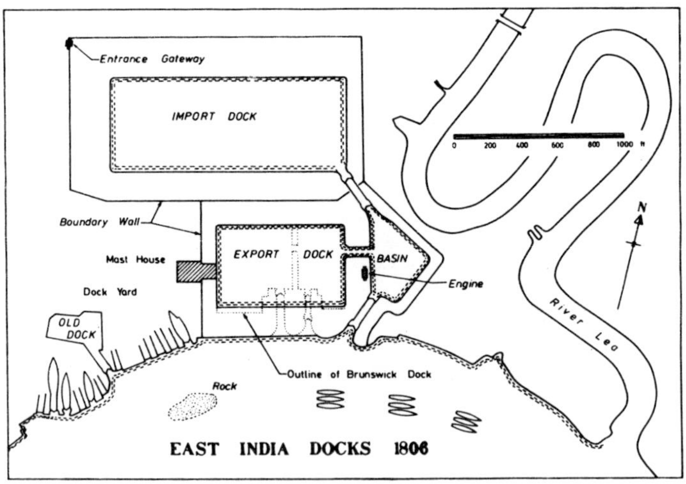

Meanwhile, work proceeded. John Hiscock made eight pile engines; construction of the engine house and well was finished in August 1804; pumping began in September, and Philip Richards & Thomas Crawford started building the Import dock walls in February 1805. The basin had open slopes on its north and east sides; work on the west wall of the basin began about March 1805.

The cofferdam was of novel design.[59] Built by James Spedding, and completed in November 1804, it comprised the following elements: a single row of 12 by 12 inch piles driven in contact with each other and bolted to two strong walings; shore piles at 16 feet centres driven 18 feet behind the outer piles and connected by 12 by 12 inch walings; diagonal and horizontal bracings between the shore piles and the outer sheeting; the whole structure forming a horizontal arch. Initially it proved to be satisfactory, as well as economical, and Walker noted that the City thought of adopting a similar design for the canal. But after just four weeks some the shore piles yielded inwards, leading to a compression failure in the outer waling and the collapse of a short length of the dam. Rennie and Walker proposed to remedy the defect by driving additional shore piles midway between each of the existing ones. Whether this or some other solution was adopted is not clear, but at any rate the dam as rebuilt by June 1805 proved to be stable.

More trouble followed, when part of the south wall of the dock failed during construction in June 1805. The walls were similar in design to those of London Dock but here they had clay, not gravel, beneath their foundations; and from a report by Rennie and Walker we learn that a bearing capacity failure occurred, the clay being heaved upwards outside the toe of the wall. The wall itself was 22 feet high above excavated level at the time. Remedial measures, which took no more than a month to complete, consisted of removing the broken portion, driving piles into the clay and rebuilding. Just before reaching its full height another portion of the dock wall, on the east side, failed in September; from which it is evident that the engineers had over-estimated the strength of the clay, and the earlier failure was not due to some exceptional foundation condition.

The locks were of similar design to those at Blackwall and Wapping entrances but here we fortunately have copies of drawings by Walker giving, among other details, sections through the recesses for the gates where the floor has to be horizontal, not arched.[60] Various structural features of this part of the entrance lock, including built-up timber beams 2 feet deep spanning 54 feet between the walls, are shown in Fig. 11. Philip Cayley, known to Walker from the West India job, contracted for all the carpentry of the locks. He started work on the communication lock in March 1805, and on the entrance lock in July. The walls were probably built by Crawford, and both locks had been completed by April 1806.

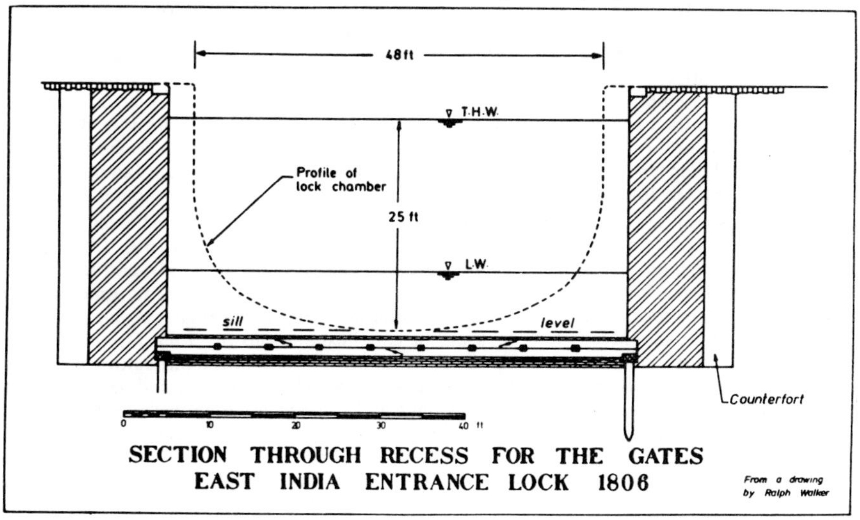

Fig. 11.

A horse-operated bucket dredger, previously employed in Brunswick Dock,[61] was purchased by the Company, and McIntosh contracted in September 1804 to use it to remove some 8,000 tons of mud before he began excavating what was to be the Export dock, in February 1806. Building the south wall of this dock began in March but the original timber walls in the north, east and west sides were retained; excavation alongside was presumably not taken to the full depth. Finally, having removed the cofferdam and dredged the entrance, the official opening took place on 4 August 1806. As usual, some minor works still had to be completed; parts of the wall coping, transit sheds on the west quay, timber wharfing at the entrance and along the riverside. But very little remained to be done in December 1806 when a statement of total expenditure showed £392,000. Of this the purchase of Brunswick Dock and other property accounted for £67,000 and the engineering works cost about £250,000.

Four comments can be made. First, the entrance lock, 210 feet between gates, 48 feet wide, with the sills laid 25 feet below T.H.W., could still be designated the largest in the world in 1838;[25] nor was it exceeded in size by any lock in London until the 1850s.[29] Second, there is evidence that the engineers were kept to a tight budget, which may help to explain their (unsuccessful) attempts to economise in cofferdam and dock wall design. Third, the choice of simple pile engines may imply a lack of economy in the pioneer steam pile-driving at London Docks. Fourthly, by using traditional methods Hugh McIntosh excavated a total volume of 857,000 cubic yards at the competitive price of 9¼d. per cubic yard, and with his large labour force (as many as 400 men and 100 horses) achieved an average rate of 8,000 cubic yards a week during the summer months of 1805 and 1806, and an average of 6,000 cubic yards a week throughout the whole year 1805. Finally, it may be noted that it was Walker who designed (in December 1805) the well-known gateway to the docks with its clock and bell-tower, and that he remained as consulting engineer to the Dock Company for many years. The first iron swing bridge at the East India Docks, over the link between basin and Export dock, was cast in 1815 to his designs by the Horsley Iron Co., and erected by Hunter & English in 1816.

7. GRAND SURREY DOCKS, Phase 1 (1804–1807)

In sharp contrast to their original intentions the Grand Surrey Canal Co. found themselves in 1807, after five years work, in possession of a barge canal only 3 miles long and an 8 acre dock at Rotherhithe.[62] The idea of building a 'Grand Bason for Shipping' at the canal entrance originated with one of the Directors, John (later Sir John) Hall, in July 1802 a few months after excavation of the canal had started. Asked to consider this proposal John Rowe, the company's engineer, submitted a less ambitious plan for a basin about 3 acres in area with a ship lock 140 feet in length, 32 feet wide and 18 feet deep below T.H.W. This was accepted in November, and by March 1803 the riverside property occupying the proposed lock site had been purchased. The management committee then vacillated, deciding at almost every meeting on some other, less costly, schemes; but a Special Assembly of the company in October again settled for Rowe's plan, and next month John Varley signed a contract to execute the dock works. Varley's sureties being considered inadequate, Hall's supporters called another meeting but lost the vote by a narrow margin. Next, Jessop was consulted and reported in December in favour of Rowe. A contract with Samuel Jones followed in January 1804, Rowe ordered a steam engine from John Curr of Sheffield, and work on the dock at last got under way in March.

Undaunted, Hall submitted what appears to have been a new plan, for an 'oblong Bason' with a central island: an arrangement providing maximum wharf length for a given water area (Fig. 12). This gained approval in April. For the time being it was decided to excavate the eastern branch only, and Jones agreed to work to the enlarged dimensions (about 4 acres) without changing his unit costs. In June, acting as a consultant, Ralph Walker reported favourably on the cofferdam design; the engine began pumping, and Jones now had 50 men on site. Meanwhile work on the canal excavations (by John Dyson) and bridges continued and more land still had to be purchased. The committee saw financial difficulties ahead, so they considered building the ship lock and also the canal lock (for which no contract had yet been made) of timber. Nothing came of this, and the ship lock as built had stone-faced brick walls; but its length was reduced to 130 feet by a decision in July, only to have the depth increased to 19 feet below T.H.W. in October. No further alterations in design followed, and the ceremony of laying the first stone of the lock took place in November 1804. By January 1805 work had temporarily to stop for lack of funds. Rowe's salary was reduced from £315 to £150. John Timperley, Assistant Engineer since August 1802, had his salary of £157 slashed to £70 (the company later made it up by a gratuity) and Jones' men were laid off. As a result Timperley left in March to take a post at the London Docks and Rowe, now in bad health, resigned in August.

Further subscriptions having been raised and a large bank loan obtained, work could be resumed in the autumn. Jones carried on with the interrupted work and signed a new contract for excavating the south and west branches of the dock, and Hall managed the job until July 1806 when Ralph Walker became 'Visiting Engineer'. Almost at once he dismissed Jones, for unsatisfactory masonry workmanship, though Jones subsequently won an arbitration against the company, and John Hooper contracted to finish the ship lock and to build the canal lock at the south end of the dock. In August, Nathaniel Simmons became Clerk of the Works at £109 p.a. and Hugh McIntosh contracted for opening the entrance.

Fig. 12.

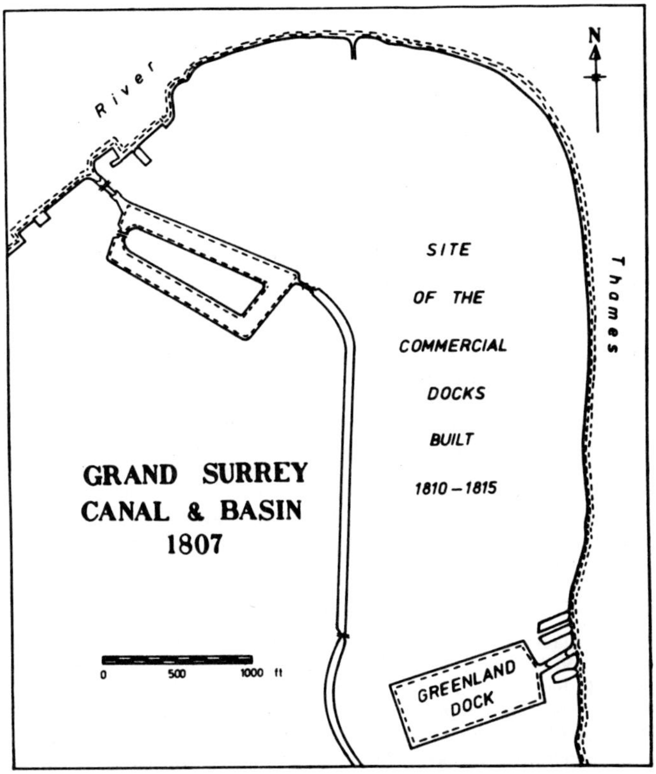

The first ship entered the dock on 13 March 1807, by which time the canal had been completed from Rotherhithe past the Croydon Canal junction to the Kent Road. The dock had timber wharfing backed by wide embankments formed from the excavated earth. A wooden swing bridge crossed the entrance lock and a drawbridge provided access from the north wharf to the island. Building plots for warehouses, to be let on 21 year leases, were set out on the island and west wharf; a towpath ran along the east side of the dock and canal.

Walker left in July with £380 in fees. Simmons remained for a further seven years, as engineer at an increased salary of £150. By 1810 he had extended the canal for one mile to Camberwell and in 1811–12 he enlarged the dock from 8 to 10 acres as well as opening a new dock of about the same area immediately to the south. Accounts drawn up in March 1811 show that the company had paid off most of their debts and were able to issue the first dividend. Up to that time capital expenditure amounted to £121,000 of which £40,000 was due to land purchase. The cost of the dock, as built in 1807, probably came to about £65,000.

8. EAST INDIA MOORINGS (1803–1807)

The dredging of ballast (i.e. gravel) from the Thames was carried out by Trinity House, using lighters equipped with a 'bag and spoon' dredge and a winch and operated by a crew of five men. The lighters worked mostly in the shallower parts of the river where ballast could more easily be obtained and navigation would be improved by deepening the shoals. Moorings, on the other hand, were maintained by the Corporation of London who levied a tonnage duty to pay for upkeep of the chains, salaries of harbour masters at the five stations, and occasional deepening of the river at moorings or elsewhere. These routine matters call for no further comment here, except to mention that the first steam dredger used in the Thames was built by Trevithick, in 1806, for raising ballast.[63]

However, in 1803 the City undertook a quite special operation of removing an outcrop of rock in the river bed at Blackwall, and at the same time embarked on a massive programme of dredging nearby for new East India moorings (Fig. 10). To finance the work an Act was obtained in July, granting powers to borrow £100,000 from the Consolidated Fund on the security of tonnage duties. Both jobs were controlled by the City's Port Committee in co-operation with Trinity House.[64]

At the East India moorings James Mountague took charge initially under the Surveyor but from December 1805 with full responsibility.[65] Preparations started already in September 1802 and by August 1803 dredging had begun. Using several lighters, perhaps as many as six at a time, about 800 tons a week were raised at a cost of 2/6d. per ton. Regular payments to Trinity House for this work continued until July 1806. Exactly what depth had then been achieved is not recorded, but the Port Committee and Trinity House jointly decided in October that dredging must be carried to 20 feet below low water springs; a figure which may have been without precedent.

Now, as it happens, the contractors Bough & Hughes in August 1806 applied for permission to bring into West India docks a 60 ton barge, the *Brunswick*, to be fitted out with bucket-ladder dredging machinery powered by the 6 h.p. engine they had recently used in the Export dock excavations[63] (see Section 3). And when operations at the moorings were resumed the work was being executed by the *Brunswick* steam dredger, on a contract signed by Bough & Hughes (with the lighterman John Mills as third partner) in December. This specified that the 20 feet depth had to be obtained, raising 200 tons per day at 3/8½d. per ton. John Hughes personally supervised the contract. After some teething troubles, which required modification to the machinery, the job was successfully completed between February and November 1807. The accounts show £9,346 paid during the period, corresponding to 50,300 tons at an average rate of 1,700 tons per week or 195 tons per day. The total cost of four years work at the moorings came to £52,000 plus about £2,000 for interest payments and parliamentary expenses.

9. BLACKWALL ROCK (1804–1808)

The rock outcrop at Blackwall (Fig. 10) was a serious navigational hazard, but as it occupied an area about 300 feet long by 150 feet wide, and lay only a few feet under low water, its removal presented formidable problems.[64] Robert Mylne first visited the site for the Port Committee in November 1802. He submitted a preliminary report in March 1803 and a detailed survey, with soundings by John Foulds, in October. Meanwhile the Act for borrowing £100,000 from the Consolidated Fund had been obtained in July (see Section 8), and in September it was agreed to make a trial with underwater blasting. The experiments, conducted by J. M. Warren, lasted several months but proved unsatisfactory. Urgent requests for advice resulted in reports by Mylne, Jessop, Walker and Chapman submitted in March and April 1804. The reports have not survived, but in June the Port Committee appointed Jessop as 'consulting Engineer', at his standard fee of 5 guineas per day with Warren as resident engineer, and by November his proposals had been accepted by the Committee and by Trinity House.

Jessop's scheme was to break up the rock by heavy iron chisels, operated like pile-drivers from a barge, and then, from a floating stage or pontoon, to lower into position a large diameter cylinder and pump it out so that miners could remove the fractured rock. The process of course had to be repeated at very many positions and at increasing depths, working two shifts every 24 hours at and around low tide. The cylinder had a 'skirt' at its lower edge, probably of leather, to effect a seal against inflowing water.

Trinity House lighters began dredging adjacent to the rock in August 1804. In October booms were placed around the working site, specially equipped barges and the floating stage arrived in November, and so did the miners, who came from Cornwall. What happened in the first phase of operations is not clear, for the iron cylinder (made at Butterley) appears not to have been delivered before March 1805. However, work thereafter proceeded satisfactorily until the winter. So far the job was carried out on a direct labour basis. Jessop made frequent visits; Warren, who received £5 a week, virtually lived on site as he had to be on duty at night and the day tides, and Mountague handled the quantities and payments. But, the method having been proved in practice, a contract could now be let and this went to James Spedding in January 1806.

Spedding had the use of the 'machine barge' with three rock-breaking chisels, a crane, and a spoon for dredging shingle; the floating stage with lifting tackle and pumps for the cylinder; a barge fitted out with living accommodation for the miners; and the necessary anchors, booms, etc. He undertook to achieve a depth of 18 feet below low water, at 10/– per ton of rock and 3/9d. per ton of gravel raised and removed. Moreover, he agreed to complete the work in two years, starting in March 1806, and this included, as well as the rock, dredging gravel between the rock and the East India moorings. From February 1806 at the express wish of the Surveyor, George Dance, Mountague took full charge[65] though doubtless he was pleased to have Jessop's support on several occasions during the early months of the contract. Spedding completed the work according to specifications and precisely on schedule in March 1808. The total cost came to £44,000 of which Spedding's contract accounted for £26,000 and interest charges, etc. about £2,000.

10. CONCLUSION

With the removal of Blackwall Rock all seven projects initiated in the years 1799–1803 had been completed, and as no new ones were started until 1808 this date is a convenient terminus for the present study. Capital expenditure from 1799 to 1808 totalled £4.1 million (equivalent to at least £100 million in modern terms) and the engineering work alone accounted for nearly £1.5 million. West India and London Docks, as built in 1806, rank among the half dozen largest schemes carried out in Britain at any time before 1830. The concentration of effort also can scarcely be rivalled, and the main objective was certainly achieved: London had become the world's best equipped port, as well as handling the greatest volume of shipping.

ACKNOWLEDGEMENTS

I am indebted to Mr. Edward Sargent and to Mr. J. G. James for finding, and sending me copies of, the Blackwall Lock drawing and William Vaughan's letter on London Dock excavations. I have received help from them in other ways, and from Mr. Richard Brown at London Dock House where the P.L.A. generously provided facilities for research in their archives. Grateful acknowledgement is due to Miss Lombard and Mrs. Cahalane for their secretarial assistance, to Miss Schuler and Miss Goodhall for expert preparation of drawings, and to my wife for her continued encouragement and interest. Fig. 4 is reproduced by permission of David & Charles Ltd.

NOTES

1. Data from references 7 and 25.
2. William Vaughan, *On Wet Docks, Quays, and Warehouses for the Port of London.* (London, 1793.)
3. [John Powsey] *Plan of the River with the Proposed Docks,* engraved plan published in William Vaughan, *Plan of the London-Dock, with some Observations.* (London, 1794.)
4. Daniel Alexander, *The London Docks,* plan engraved by J. Cary, 1794.
5. The Corporation of London set up its Port Committee in January 1794; Journal of the Court of Common Council (C.C.J.) 29 January 1794, Corporation of London Record Office (C.L.R.O.). Proposals for improving the Legal Quays were reported on by William Jessop, John Rennie and John Foulds, 10 April 1794. (Rennie papers, Box 13 F. 2 National Library of Scotland.)
6. C.C.J., 18 February 1796.
7. *Report from the Committee to enquire into the best Mode of providing sufficient Accommodation for . . . the Port of London, 13 May 1796.* Parliamentary Papers.
8. Daniel Alexander and John Rennie, evidence and estimates for the London Docks scheme, March and April 1796, with a new edition of plan 4: in reference 7.
9. Ralph Walker, evidence and estimates for alternative scheme for London Docks, April 1796, with *A Proposed Plan for Wet Docks at Wapping:* in reference 7.
10. George Dance and James Peacock, evidence and estimates for improved Legal Quays and for warehouses on south bank, April 1796, with engraved plans: in reference 7.
11. William Jessop, evidence and estimates, April 1796, with a *Plan of the River Thames with the Proposed Docks at Rotherhithe & in the Isle of Dogs,* engraved by R. Metcalf: in reference 7.
12. Daniel Alexander, *The London Docks,* November 1796; plan of a new scheme, engraved by J. Cary.

13. William Jessop, George Dance and Ralph Walker, report on proposed West India Docks and ship canal, 29 August 1797 (transcribed in C.C.J., 16 January 1798).
14. Dance, Jessop & Walker, 29 August 1797, *A Plan of the Proposed Canal and Wet Docks for the West India Trade in the Isle of Dogs*, engraved by R. Metcalf: in *Report of a Committee of West India Planters and Merchants... November 16, 1797.* (London, 1797.) Shows, from west to east, dock A (21 acres), dock B (10 acres) and basin C (5 acres). The plan was re-issued, unsigned, from the Surveyors Office with a different title.
15. William Jessop and Ralph Walker, estimates (November 1798) and evidence (February and March 1799) on revised plan for West India docks: in reference 18.
16. Second plan by Dance, Jessop & Walker, engraved by R. Metcalf (1798): in reference 18. Same title as plan 14 but larger docks in reversed order.
17. Daniel Alexander and John Rennie, evidence and estimates for London Docks scheme, April 1799, accompanied by a reprint of plan 12: in reference 19.
18. *Minutes of Evidence taken at the Committee on the Bill for rendering more commodious and for better regulating the Port of London (City Plan)*, 7 May 1799. Parliamentary Papers.
19. *Minutes of Evidence taken at the Committee on the Bill for making Wet Docks... within the Port of London (Merchants Plan)*, 7 May 1799. Parliamentary Papers.
20. John Rennie and Ralph Walker, estimates for the East India Docks, 8 March 1803 (transcribed in E.I. Minute Book) and plan engraved by B. Baker.
21. For comparision: Liverpool's docks totalled 26 acres in 1796, and 64 acres by 1832. Hull had 17 acres in 1809 and 23 acres by 1829.
22. *Report from the Select Committee on the Port of London, 1836* gives numbers and tonnage of ships entering each of the eight docks in 1830 and 1835. Parliamentary Papers, 557 of 1836.
23. Sir Joseph Broodbank, *History of the Port of London.* (London, 1921.)
24. John Pudney, *London's Docks.* (London, 1975.)
25. James Elmes, *A Scientific, Historical and Commercial Survey of the Port of London*, in Simms *Public Works of Great Britain.* (London, 1838.)
26. *Life of Telford... edited by John Rickman*, (London, 1838) and Atlas of plates.
27. Sir John Rennie, *The Theory, Formation and Construction of British and Foreign Harbours.* (London, 1854.)
28. W. P. Shepherd-Barron, Presidential Address *Proc. Inst. Civ. Eng.* **3**, Pt. 1 (1954), 12–40.
29. Charles Hadfield & A. W. Skempton, *William Jessop, Engineer.* (Newton Abbot, 1979.)
30. John Powsey, evidence to Parliamentary Committee, March 1799: in reference 18.
31. See the plan by Rennie & Walker, 1804: reference 58.
32. Engraving by and after William Daniell, published October 1803: reproduced in reference 23.
33. John Rennie & Ralph Walker, report to East India Dock Co., 18 August 1803: in reference 57.
34. H. Green & R. Wigram, *Chronicles of Blackwall Yard.* (London, 1881.)
35. Unless noted otherwise, information in Section 2 is based on Minute Books of the West India Dock Co., vols. **1–5** (P.L.A. library). These include transcriptions of many reports, specifications, etc., by Jessop and Walker.
36. The plan accompanying Walker's report of 15 November 1799 is probably Sheet 14 in the West India Docks portfolio (P.L.A.).
37. Jessop spent 8 days in the first two weeks of February 1800 'Settling the General Plan with Mr. Walker, making Sketches of Designs and making Estimates'. The resulting plan was published by John Fairburn in May 1800. Sheet 10 in the W.I. portfolio, signed by Walker, is slightly later and shows the final lay-out of the north quay warehouses.
38. Ringsend Docks, water area 25 acres, were built 1792–96 at a cost of £113,000. The main entrance lock had a length of 149 feet between gates, a clear width of 35 feet and a depth of 18 feet on the sills at high water springs; it was constructed with an inverted arch floor: reference 29.
39. Copied from P.L.A. Engineers Office drawing 6760108.
40. Dimensions of timbers from Jessop's specification (25 September 1801): depth of piles, etc., from report by Jessop and Walker (29 July 1802) on the failure; some construction details from drawings by Walker of an almost identical coffer dam, in reference 25.
41. The opening ceremony amounted almost to a national celebration, see *European Mag.* **42** (1802), 213–17 and an engraving by P. W. Tomkins. It is important to note that Daniell's engraving published October 1802 shows the docks as intended to be completed, not as built at the time. Both engravings are reproduced in reference 23.
42. *Annual Accounts of the West India Dock Co. from July 1799 to February 1822.* (P.L.A.)
43. Information in Section 3 is taken from West India Dock Co. Minute Books, vols. **6–15**, (P.L.A.).
44. Report by John Hughes quoted in *The Civil Engineer and Architect's Journal*, **2** (1839), 9–10.

45. Bristol Floating Harbour, built 1804–9; engineering cost £457,000, total expenditure £611,000. Jessop's main entrance lock to Cumberland Basin, had a length of 200 feet between gates, a width of 45 feet and a depth of 25 feet over the sills at normal high water. During 1804–5 the rate of excavation averaged 15,000 cubic yards per week: reference 29.

46. There are no surviving Minute Books for the City Canal, but detailed information is given in the Port Committee Accounts 1799–1829 and Contracts Book 1799–1811. (C.L.R.O.)

47. See the engraving by Henry Moses published 1824: reproduced in reference 29.

48. Primary sources for the London Docks are: Court of Directors Minutes, vols. **1 & 2**, and Committee of Treasury Minutes, vol. **1**, (P.L.A.), and Rennie's Letter and Report Book (Inst. Civ. Eng. library).

49. Plan dated 4 May 1801, Sheet 65 in Vaughan's portfolio (P.L.A.). The final and parliamentary plans are shown in an engraved map published by John Fairburn, August 1802.

50. This first steam pile-engine was a Boulton & Watt crank engine. After use at the coffer dam it was sold to a rope-maker at Limehouse in 1803. (Boulton & Watt Engine Book.)

51. Drawings of a London Dock entrance lock, engraved from Rennie's originals, are shown in *Voyages dans la Grande-Bretagne*, Part 3 (Paris, 1824), by Charles Dupin. In the text he says they refer to Hermitage entrance, but there the lock had a length of 150 feet between gates whereas the drawings show 170 feet, and a width of 40 feet, exactly as for Wapping entrance lock (dimensions given in *Royal Commission on the Port of London*, 1902). The section shows three rows of sheet piling under each gate.

52. Report by Murray of 10 June 1803, transcribed in the Committee of Treasury Minutes; letter from William Vaughan, dated 14 June 1804, addressed to the Minister of Marine, St. Petersburg, printed in *Archives of History of Science & Technology*, **4** (1934), 400–409; and the detailed returns of excavation progress in Rennie's account book (I.C.E. Library).

53. Probably the 6 h.p. bell-crank engine ordered by Bough & Holmes: drawings in Boulton & Watt portfolio 723, dated 1802–3. (Birmingham Reference Library.)

54. Rennie mentions the swing bridge drawings in his Notebook No. 54, and the date of completion in a letter to Capt. Huddart. (Rennie Collection, Box 26 and 13, Nat. Lib. Scotland.)

55. William Daniell, engraving of London Docks published October 1808: reproduced in reference 23.

56. These costs are deduced from various entries in the Minute Books. Regrettably I can find no summarised accounts.

57. Section 6 is based on Minutes of the East India Dock Co., vols. **1–3** and vol. **6** (P.L.A. library), and Rennie's report book (I.C.E. Library).

58. The plan was engraved and published by Robert Wilkinson in December 1804, by order of the Directors. It is entitled *Plan of the East India Docks. John Rennie & Ralph Walker, Engineers.*

59. In a report on its failure, Rennie and Walker give details of the cofferdam (E.I. Minutes, 28 December 1804). Elmes in reference 25, shows a double-wall earth filled cofferdam said to be from a drawing by Walker of the East India structure. Perhaps this was originally intended; it closely resembles the West India cofferdam of 1802.

60. From a drawing by Walker, in reference 25.

61. See reference 63 for an illustration of the dredger.

62. Information in Section 7 is based on Minute Books of the Grand Surrey Canal Co., vols. **1–6** (P.L.A. library).

63. A. W. Skempton 'A History of the Steam Dredger, 1797–1830', *Trans. Newcomen Soc.*, **47** (1975), 97–115.

64. The principal sources for Sections 8 and 9 are: Account Book of the £100,000 Fund 1803–9 and Port Committee Contracts Book 1799–1811 (C.L.R.O.). Supplementary information is given in Trinity House By-Minutes (Trinity House archives). The Port Committee Minutes have been lost, except a rough copy for the period April 1803 to June 1804 (P.L.A. library).

65. Report of the Port Committee, 7 April 1815 (C.L.R.O.).

66. R. W. Horner, 'Current proposals for the Thames barrier' *Phil. Trans. Roy. Soc.* A. **272** (1972), 178–185.

General note: It is not always possible to find definitive dimensions of locks, maximum tonnage of ships admitted to the docks, and so on. The figures quoted represent my best judgement from available evidence.

APPENDIX I

Trinity High Water (T.H.W.) was established as a datum for the Port of London by Trinity House in 1800. It is at +11.4 feet O.D. Newlyn. Intended to represent the mean height of ordinary spring tides, it actually lies a few inches below this level in the upper docks area. Trinity standard low water is 18 feet below T.H.W. The 1953 tidal surge rose 6 feet above T.H.W. but, for reasons not fully understood, surge height has increased steadily during the past 150 years, and the highest known tide in London before 1830 was only 2½ feet above T.H.W.[66]

APPENDIX II
DOCKS AND SHIP CANAL IN THE PORT OF LONDON, 1807

	Construction Dates	Water Area Dock	Water Area Basin	Water Depth	Lock Length	Lock Width	Lock Depth	Approximate Total Cost	Engineer
Greenland Dock	1697–99	10		16	150	44	16		John Wells
West India: Import Dock	1800–02	30		23	162	38	24		
Blackwall Basin	1800–02		6	23	192	45	24	£1,239,000	William Jessop
Limehouse Basin	1801–03		2	21	155	36	22		
Export Dock	1803–06	24		23	160	38	22		Wm. Jessop 1803–05 Thos. Morris 1805–06
City Canal	1800–05	14		23	192	45	24	£194,000	William Jessop
London: Western Dock	1801–05	20		20	160	38	20	£2,100,000	John Rennie
Wapping Basin	1801–05		3	23	170	40	23		
East India: Import Dock	1803–06	18		22	210	48	22		John Rennie
Entrance Basin	1804–06		3	22	210	48	25	£392,000	&
Export Dock	1805–06	8		22					Ralph Walker
Grand Surrey: Outer Dock	1804–07	8		18	130	32	19	£65,000	John Rowe 1804–05 Ralph Walker 1806–07

Water area to nearest acre. Lock dimensions, to nearest foot: length between gates, clear width, depth of sills below T.H.W.

DISCUSSION

Mr. G. M. Binnie opened the discussion by asking for more details of the pumping operations. **Professor Skempton** said that for the West India Import Dock the 28 h.p. engine at Blackwall pumped continuously from a well, sunk to a depth of about 12 feet below low tide level, into a small reservoir which was discharged twice a day at low water. This was an early application of steam power to the de-watering of deep excavations, but not the first. It was followed on the other dock works. He added that Jessop wisely ordered a second, stand-by engine, but in the event this did not have to be used and it was sold in 1802 to the Kennet & Avon Canal Co. On the question of the City Canal engine being employed for pumping out the Export Dock excavations, Professor Skempton, when drafting the paper, had thought this might be an instance of 'ground-water lowering', but he now rejected the idea. Water from the dock works must have been led into the canal excavations and pumped from a sump adjacent to the canal engine.

Mr. E. H. Sargent, referring to the last point, said there was evidence that the dock engine at Blackwall was used for de-watering the Export Dock, iron pipes having been laid across the bottom of Blackwall Basin before filling, in 1802. **Professor Skempton** said he would check this (and his reply is given at the end of this discussion).

In response to further questions from Mr. Sargent, Professor Skempton replied that the engineer's house at Blackwall was occupied by Ralph Walker and probably, later, by Thomas Morris; and, regarding the City Canal coffer dams, no details were available of their construction though the failure of the west cofferdam in July 1805 apparently had been due to underseepage.

Miss Julia Elton asked why the City Canal was used so little by ships in transit. **Professor Skempton** said that by 1815 the transit trade had fallen to half its volume in the first few years, and he considered this was due in part to opening of the docks, particularly the Commercial Docks at Rotherhithe, which reduced congestion in the river round the Isle of Dogs.

Mr. J. W. Butler queried whether the Napoleonic Wars had any effect on the finance and labour required in building the docks, and on trade. **Professor Skempton** replied that apart from some temporary trouble from press-gangs the wars seemed to have had no influence. Capital was raised without difficulty, there is no evidence of labour shortage, and from the moment of opening the docks they were used almost to full capacity. He added, in reply to **Mr. J. G. James,** that the merchants got their money back by dock dues and by avoiding the heavy losses which had arisen from plunder in the pre-dock days.

Mr. D. B. Hague asked about the mast house, believing there to be no comparable example in the naval dockyards, and also about the use of sheet-piling. In reply **Professor Skempton** agreed that the Brunswick Dock mast house was most unusual (at which point a visitor interjected that one had existed at Copenhagen and there were other continental examples), and he said that interlocking sheet piles went back certainly to the early 18th century, when they were used by John Perry in closing Dagenham Breach, and possibly still earlier in France and Italy. As used by Jessop and Rennie at the docks the sheet piles were generally 6 by 12 inch planks with a V-shaped joint on the narrow faces.

Professor J. N. Hutchinson raised a question on the geology of Blackwall Rock, and **Professor Skempton** said it was thought to be a cemented part of the Woolwich Beds.

In reference to a further query from **Mr. Sargent, Professor Skempton** replied that some of the cargoes unloaded at the docks were taken upstream by barges and others were carried by waggons. Commercial Road and East India Dock Road had been built, with the young James Walker as engineer, largely to cope with traffic to and from the docks.

Mr. R. J. M. Sutherland asked if the wall reinforcement took the form of hoop iron and if any earlier examples are known. **Professor Skempton** replied that it was hoop iron, probably two strips at each counterfort, every 3 feet apart vertically, and he knew of no example of such reinforced brickwork before its use at the West India Docks. Mr. Sutherland also queried the absence of sheet piles in the East India lock section, and Professor Skempton said this may have been due to the foundations of the lock being on clay, not gravel as at Wapping and Blackwall.

After some discussion concerning engineers' fees, a subject dealt with rather thoroughly in the paper, **Mr. Sargent** showed several slides of early dock plans from which he concluded that the layout of the West India Docks owed much to ideas by Ralph Walker. **Professor Skempton** did not dissent from this and said that Walker, as well as being an experienced ship's captain, was evidently a highly intelligent man, but at the start of the West India job knew little about practical engineering construction. This goes some way to explaining why Jessop had to spend so much time on site, though the terrific pace and scale of the work were other factors.

Mr. J. G. James, in conclusion, said that is gave him great pleasure to propose the vote of thanks. Professor Skempton had done wonders in condensing, and presenting in a logical manner, the enormous volume of material in the Minute Books that had formed the basis of this paper. The vote of thanks was passed by acclamation.

CORRESPONDENCE

In a later reply to the point raised by Mr. Sargent, **Professor Skempton** wrote:

"I have looked again at the problems of de-watering the West India Export Dock excavations. There can be no doubt the canal engine was used, as the Dock Co. during 1804 paid the City a total of £1,575 for the use of this engine for the purpose. But I agree with Mr. Sargent that it was intended, in 1802, to lay pipes under Blackwall Basin. In July 1803 repairs were made to the Blackwall engine and Jessop was authorised 'to proceed with the Caisson and laying down the Pipe, he having undertaken that it shall not in the least endanger the Bank of the Bason'. This implies that the pipe beneath the Basin was about to be extended below the embankment and along to the site of the Export Dock. However, these operations seem to have taken about a year to complete, and only in September 1804 could Morris report that 'the steam engine is at work and keeps down the water'. He presumably refers to the Blackwall engine; and a further argument that this engine was being used in the later stages can be based on the fact that the canal excavations were completed in April 1805, six months before those of the Export Dock. I am grateful to Mr. Sargent for prompting me to make a more thorough examination of this matter, which has now been briefly covered in the final text of the paper".

In a further contribution **Mr. J. G. James** wrote as follows:

"The iron swing bridge over the London Dock entrance at Wapping is described in *The Star*, 31 January and 1 February 1805. Each leaf had seven ribs at about 2 feet 6 inch centres and each rib weighed about 3 tons. Over 5 tons of pig iron was used as kentledge on each leaf to balance the cantilever. The leaves revolved on rollers working between cast iron plates. To open the bridge, hinged locking-plates were raised and eight men then winched it round. Peter Nicholson stated that the ribs were 1½ inches thick, that the whole bridge weighed 85 tons, and that it could be opened or closed in under 3 minutes. It is almost certainly this bridge which was illustrated by Dupin in 1824 (see Fig. 13).

The engine bought from Curr for Surrey Docks in 1804 was not very successful and another (second-hand?) one was bought a few months later from Wm. Crawshay. I mention this primarily to record that it was repaired and installed by the Lambeth engineer John Collinge, a now little-remembered figure whose son Charles was one of the founders of the Institution of Civil Engineers.

An alleged 'first' at the East India Docks may be added. George Godwin the concrete historian wrote in 1838 that McIntosh the contractor told him (5 October 1835 when they were 'directing the composition of concrete for the foundation of the arch over Grange Road, Bermondsey') that 'I (McIntosh) executed concrete 36 years ago at the East India Docks. The ground was a mud bank with here and there gravel and sand affording, therefore, most unequal bearing; but the concrete has nevertheless answered admirably. I think this was the first time concrete was used in England; indeed I know it was'. It would be interesting to know if Professor Skempton found any supporting evidence for this statement which, despite the error in dating, sounds authentic. Godwin's statement is in *Civ. Eng. & Archit. Journ.*, January 1838, vol. 2, 18–19".

In reply to the last point **Professor Skempton** wrote that he had found no supporting evidence from the East India Minute Books, but he agreed with Mr. James that the account by McIntosh seemed to be authentic; it possibly applied to the foundations of the south wall of the Export Dock in 1806. He welcomed the additional information on the iron swing bridge at Wapping, which there is every reason to suppose was the first of its kind, and on the Crawshay engine for the Grand Surrey Canal Dock excavations.

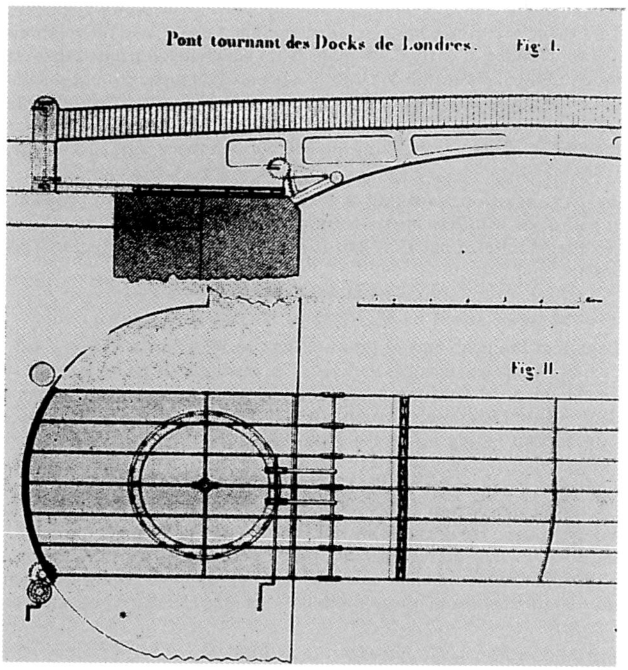

Fig. 13. Cast Iron Swing Bridge over an Entrance Lock, London Docks. From Charles Dupin, *Voyages dans la Grande Bretagne* (Travels in Great Britain), Part III, Paris, 1824.

XII

Engineering in the Port of London, 1808 – 1834

In a paper read to this Society on 14 March 1979, (*Trans.* Vol **50**, pp87 – 108), I traced the construction history of the first great docks in the Port of London: the West India, the London, and the East India Docks. All three were in full working order by 1806, having been built on a scale without precedent in this or any other country and within the amazingly short time of six years. The smaller Grand Surrey Dock (opened 1807) and the City Canal in the Isle of Dogs (opened 1805) were also described. In the present paper I shall consider the next two stages of development. The period from 1808 to 1821 saw the construction of the Commercial Docks in Rotherhithe, the small East Country Dock nearby and Regent's Canal Dock at Limehouse, and the provision of a second entrance (at Hermitage) for the London Docks. There were also some extensions at the Grand Surrey, London, and East India.

Nothing of particular note happened during the next four years, but a new phase of activity started in 1825 with construction of St. Katharine Docks (completed 1830), a new dock and entrance for the London Docks at Shadwell (opened 1832), an addition to the West India, and finally (1833 – 34) the Brunswick Wharf for steam ships at East India Docks. By this time more than 4000 ships were entering the docks annually, with a tonnage not far short of 1 million tons (see Appendix I). Most of the colliers and coasters still moored in the river or alongside the quays, but practically all ships from foreign ports now entered the docks. The vision of those far-sighted individuals who battled for the dock system in the 1790s had been amply fulfilled.

No more than passing reference is made to the warehouses and other buildings; a proper study of them would far exceed the time and space at my disposal. Another problem arises from the fact that during the period under consideration most of the docks were in business, not simply being built. The Minute Books, which remain the chief (often the only) primary source, are therefore full of matters having nothing to do with construction. Consequently it is easy to miss a brief entry that might be the sole clue to a completion date of a bridge, the appointment of an assistant engineer, the cost of some item, and so on. I can only hope there are not too many of such omissions.

The Port of London Authority again very generously provided facilities for research at London Dock House. Mrs. Anne Langford made the tracings from my drawings and Miss Joyce Gurr was responsible for the photography.

1. EAST COUNTRY DOCK

Few records seem to be available on the construction of this dock which was built by the East Country Dock Co. on a site adjacent to the Greenland Dock in Rotherhithe. The entrance lock had been finished by February 1808.[1] A few months earlier the Commercial Dock Co. contemplated purchase of the premises, but nothing came of the negotiations and work continued (apparently with somewhat limited funds) until the dock was eventually opened in 1811 at a cost of the order £40,000.[2]

David Mathews was the engineer[3] but I have found no reference to the contractors.

As shown in a plan by James Walker published in 1820 (Fig. 1) the dock had an area of 4½ acres and the entrance lock was about 160 ft long with a width of approximately 40 ft.[4] The sills were at a depth of 23½ ft below Trinity High Water (T.H.W.).[5]

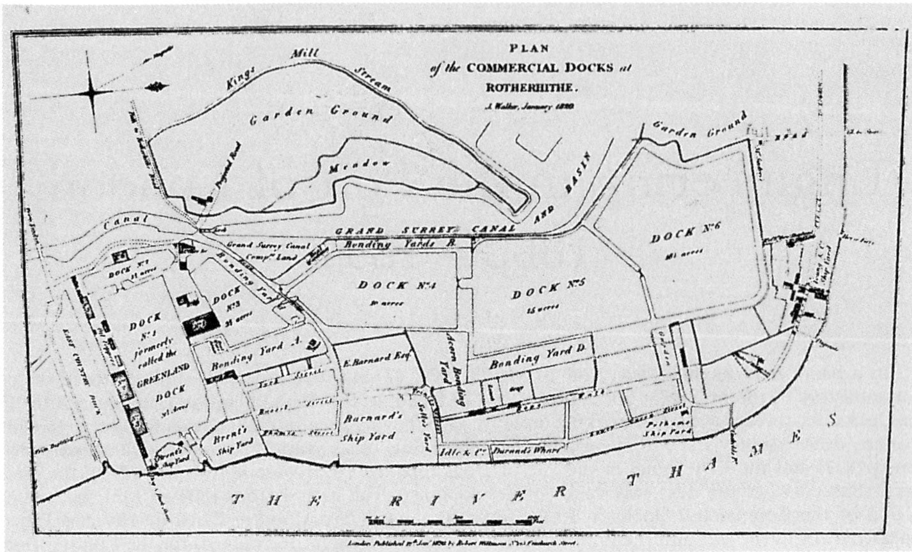

Fig. 1. Commercial Docks, plan dated 1820.

The Company dealt principally in trade from the Baltic ports. Returns for the number of ships entering the dock and their tonnage in the years 1820, 1830 and 1835 are given in Appendix I.

The dock and premises were finally purchased by the Commercial Dock Co. in 1850. They proceeded to enlarge the dock and its entrance lock, and built a communication passage into the Greenland Dock. Opened in 1855, it then became the 'South Dock' of the Commercial system.

2. COMMERCIAL DOCKS

The nucleus of this establishment (see Fig. 1) was the old Greenland Dock of 9¾ acres, built at the end of the 17th century, to which had been added about 100 years later a small extension (No. 2) of 1¾ acres and a timber pond (No. 3) of 3¾ acres. These docks and the adjacent premises were purchased in November 1807 from Mr. William Ritchie by the newly formed Commercial Dock Company for £35,000.[6]

Repairs to the lock gates were immediately put in hand and soon afterwards Bough & Hughes began removing a thick deposit of mud from the dock.[7] Meanwhile, as mentioned above, negotiations to purchase the East Country Dock premises were also under way, but in May 1808 it was decided not to pursue the matter and, instead, to deepen and enlarge the Greenland entrance lock. Ralph Walker was appointed engineer and in June Messrs Adam & Robertson contracted to carry out the work in accordance with his plans. A week later James Walker joined his uncle as assistant engineer.

Work proceeded very efficiently. The foundation stone of the new lock was laid on 16 November 1808 and the re-opening of the dock took place about July 1809. By the end of the year rather more than 200 ships had entered the dock, with a total tonnage of nearly 50,000 tons.[8]

As rebuilt, the lock had a width of 42 ft, a depth of 18½ ft of water on the sills at T.H.W., and a length of 204 ft between the gates.[9] Its construction (Fig. 2) embodied the inverted arch floor and curved-profile walls[10] which had become almost standard features since their introduction by Jessop at Ringsend (Dublin) and the West India Docks.

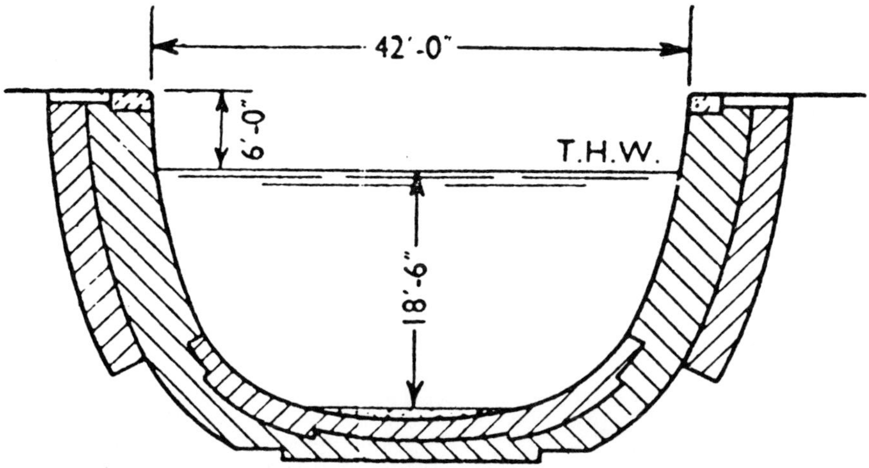

Fig. 2. Greenland Dock, section of entrance lock, 1809.

Ralph Walker submitted his fees, amounting to £525, in December. By then most of the accounts had been settled, including £18,000 to Adam & Robertson and about £7000 to William Bough (now without his former partner John Hughes) for excavation and mud removal.

After completion of these works James Walker became engineer to the Company, a position he retained for many years. His formal appointment dated from October 1810, but he received 1000 gns for the period June 1808 to September 1810 including the amount (about £200) he earned while working on the lock. His salary on appointment was fixed at £800 p.a.

The Company entered into an agreement in 1809 to purchase some 40 acres of land lying to the north of Dock No. 3. This enabled a carefully planned extension to be carried out,[11] under an Act obtained in June 1810 which authorised the raising of £130,000 extra capital. Before the end of 1811 Dock No. 3 (later known as Norway Dock) had been deepened and Dock No. 4 (later called Lady Dock) of 10 acres was opened; as also a large warehouse (No. 1 on the plan in Fig. 1, page 74). William Bough contracted for the excavations and Adam & Robertson built the warehouse, the iron columns of which were supplied by Hunter & English.

These works were followed in 1812 by the 15 acre timber pond marked as Dock 5 in the plan, and later called Acorn Pond. In the same year Hunter & English erected an iron swing bridge, to Walker's design, over the communication passage between Dock 3 and 4, and Warehouse No. 2 was completed by the builders Barker & Howkins. The masonry walls of the passages were built by direct labour, not by contract.

The timber pond had sloping earth sides topped by a wide embankment rising well above high water mark. A note of 4 March 1812 records 'Water in No. 5 up to level of high water & bank safe and watertight'. The depth of water was about 15 ft below T.H.W.

The final accounts of Bough and Adam & Robertson were settled in the early months of 1813, and the Company commissioned a view of the docks from the topographical artist William Daniell. His large aquatint engraving (Fig. 3) gives an accurate representation of the scene.[12] The entrance lock, the docks and timber pond and the two warehouses can be seen, and also the old Greenland warehouse (No. 3 in the plan) which had been re-slated in 1812. A glimpse is provided of ships lying the the East Country Dock and, in the background, of ships in the new basin of the Grand Surrey Docks.

Though Walker attended closely to the execution of the works, as well as their design, he must have had an assistant or resident engineer and superintendents of excavations, building and so on, but their names were not found in the surviving minute books. With his agreement, his salary was reduced to £400 in January 1815 and to £300 in July 1816, as little new work was now in hand.

Completion of the first phase in the history of the Commercial Docks can be dated to January 1822, when water was admitted into timber pond No. 6 (later called Lavender Pond). A plan and estimate for this pond, 18½ acres in area, had been submitted by Walker in July 1815 and it seems that excavation started soon afterwards; there are payments to Hugh McIntosh 'on account of the new timber pond' in August and September. But the bottom proved to be too permeable and would have led to trouble in neighbouring property due to leakage, for it must be remembered that natural ground level was several feet below high water mark and protected from flooding by the river embankment.

Fig. 3. Commercial Docks, view in 1813.

Surprisingly the Company delayed doing anything to remedy this state of affairs until July 1821, when a contract was let to Hugh McIntosh to complete the excavation and puddle the sides and bottom with 3 ft of clay. He employed about 200 men and finished the job in December. Water was then admitted gently from No. 5, through a sluice in the cofferdam at the communication passage, and by 12 January 1822 the pond had been filled. Observations showed no leakage; the cofferdam was removed, and before the end of the month 6000 loads of timber were floating in No. 6.

3. GRAND SURREY DOCKS: PHASE 2

During the years 1811 – 12 the dock of the Grand Surrey Canal Co. was enlarged from 8 to 10 acres and a new inner dock or basin formed by widening and deepening a portion of the canal about 2000 ft in length between the outer dock and a bridge on the road leading into the Commercial Docks (Fig. 1, page 74). The basin had an area of 9 acres and sloping earth sides, with mooring posts for the ships.[13]

Originally there was a lock leading from the dock into the canal. This had to be removed, and a new lock with three pairs of gates was built adjacent to the bridge at the south end of the basin. Plans drawn up jointly by the Company's engineer Nathaniel Simmons and a land surveyor C. A. Weir were approved in July 1811. The Company raised extra capital of £30,000 and work began in September.

Few details are available, but we can assume that most of the basin was excavated by spadework, leaving a strip of land as an embankment between the excavation and the canal. The canal would then have been closed while the old lock was removed and the new one built, and while removing the temporary embankment. The Minutes record that works on the new dock were in progress night and day (presumably during canal closure) and that water was let in on 6 October 1812. What I take to be an earth-fill cofferdam for the lock was then removed by dredging and, at a meeting held on 11 November, canal navigation is reported as having been restored.

It is not possible to interpret the accounts with certainty as other works were in hand at the same time on the canal further inland. However, it seems that the basin and lock were constructed partly by direct labour and partly by the contractor Thomas Cooper. Simmons, who had been engineer to the Company since 1807 at a salary of £150 p.a., retired in April 1813. The basin was later transformed into the Russia Dock of the Surrey Commercial Dock system.

4. LONDON DOCKS: PHASE 2

The second phase of construction at London Docks begins with the erection of a double-leaf cast iron swing bridge over the communication lock at Wapping Basin. The contractor Aydon & Elwell ran into financial difficulties and in June 1811 G. Mawson agreed to complete the job.[13] Then came the Tobacco dock and warehouse, at the north-east corner of the main dock (Fig. 4),[15] opened in 1813 and 1814 respectively. Finally, a second entrance to the docks was built at Hermitage, into the south-west angle of the main dock. Constructed in two stages, the inner lock reached completion in 1815 and the basin and entrance lock in 1821.

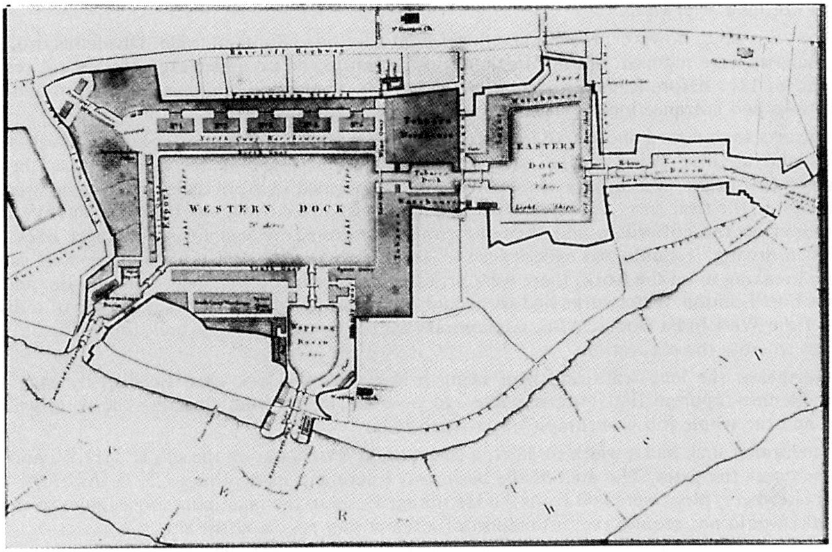

Fig. 4. London Docks, plan dated 1831.

John Rennie was chief engineer throughout this period. From about November 1810, until he left in 1814 to take charge of the works at Sheerness Dockyard, John Thomas was resident engineer with his younger brother Richard as assistant from the latter part of 1811. James Hollinsworth came as resident engineer on the Hermitage basin and entrance lock, and saw these works to completion. Daniel Alexander continued as Surveyor, responsible for all building work, and in the later stages of the period Richard Douthwaite was his chief assistant.

Tobacco Dock. The general lay out of the Tobacco Dock and Warehouse was settled by May 1811. When building the main dock an opening had been left in the wall for future extension to the east, and this opening was used as part of the communication passage to the Tobacco Dock. Work began in August, the principal contractors being William Bough for excavation and William Hobson for building the walls. John Hiscock drove the piles for a bridge over the communication passage but there seems to be no record of who built the bridge itself.

The dock (area 0.8 acre, depth 21 ft) and the passage and bridge were completed by October 1813 at a cost slightly below Rennie's 1811 estimate of £32,240. The Tobacco Warehouse, a single storey structure with vaults beneath, covering an area of 210,000 sq. ft., was opened about February 1814.

Hermitage Entrance. From the earliest days the intention had always been to provide a second entrance, primarily for small ships and lighters, at the south-west angle of the dock. Consequently when building the dock walls an opening was made for connecting with a future lock. Shortage of funds prevented any further action until 1811 when plans for the entrance were again considered and most of the required land was purchased. The original idea had been for a tidal basin with a single lock leading into the dock, but in August 1811 Rennie proposed that there should be two locks impounding a non-tidal basin, and submitted estimates totalling nearly £140,000. After further discussions his scheme was approved in December 1812.

Meanwhile work had started earlier that year on the lock adjacent to the dock, a feature common to either plan. The 20 h.p. mortar-mill engine was adapted to pumping out the excavation, the mill being moved elsewhere and provided with a new engine. Bough and Hobson extended their operations (at the same unit prices) to this opposite corner of the site, and contracts were signed for the lock gates (Thomas Clark) in October 1814 and for the ironwork of the gates in January 1815. Doubtless the work was completed not long afterwards.

About this time, however, the Company suffered a serious fall-off in trade. Dividends dropped to 3 per cent, salaries were reduced, and no further construction could be undertaken for several years. So it was the end of 1817 before Rennie produced estimates for completing the basin excavation and building the basin walls and entrance lock.

Contracts were signed shortly afterwards by Joliffe & Banks for the cofferdam, excavation, piling, lock gates and an iron swing bridge, and by William Hobson for all the masonry and brickwork. Preliminary work began in February 1818 but had to be stopped in April as legal problems were arising over property on the river front. Eventually these were resolved, but it was not until March 1819 that work could proceed on the cofferdam and then the contractor found he had to use stronger piles to avoid difficulties in driving. Trouble was experienced in keeping a sewer, which ran along the west side of the lock, from breaking in on the work, there were problems in shoring up buildings on each side, and a water main from East London Waterworks had to be laid across the entrance, requiring the use of a diving bell borrowed from West India Docks. Also, a temporary timber bridge had to be built for traffic on Wapping High Street to cross the excavation.

Nevertheless the lock walls and iron swing bridge over the lock were finished by May 1820. In December Rennie reported that little remained to be done, apart from finishing the wing walls at the entrance, and the whole job was completed in April 1821.

The entrance lock had a width of 38 ft, a depth of 20 ft of water on the sills at T.H.W. and a length of 150 ft between the gates. The area of the basin was 1 acre and its depth also 20 ft. As at the Tobacco Dock and elsewhere, pipes were laid from the Hermitage Basin to the main pumping engines so that future repair works would not require the installation of a temporary engine at the site.

Hollinsworth stayed on until September 1821, Rennie died in October and Alexander left in December, Douthwaite then remaining in sole charge. The year 1821 indeed marks the end of an era at London Docks. In very approximate figures £500,000 had been spent on engineering works since 1801 and £1.4 million on warehouses and other buildings. Trade was beginning to increase again; the annual tonnage of ships entering the docks now totalled rather more than 200,000 tons, the largest amount for any establishment in the Port of London (see Appendix I).

5. EAST INDIA DOCKS

The docks and entrance basin at Blackwall having been opened in 1806 little more engineering work was required for several years. However, as trade increased the basin tended to become over-crowded and in February 1815 the Company decided to enlarge it and, at the same time, to provide proper wharves along its eastern side instead of open earth slopes as in the original basin.[16]

Ralph Walker, consulting engineer to the East India Docks, at first thought of excavating in the dry down to the water table and dredging below that level, but judging from the accounts it seems likely that the area behind a strip of land about 60 ft wide (see Fig. 5) was excavated to or near the full depth and then about a 90 ft length of the strip was removed by dredging, to create a passage into the new, eastern basin, leaving the rest of the strip as a jetty. In this way the water area increased from 2¾ to 4½ acres.

ENGINEERING IN THE PORT OF LONDON, 1808 – 1834

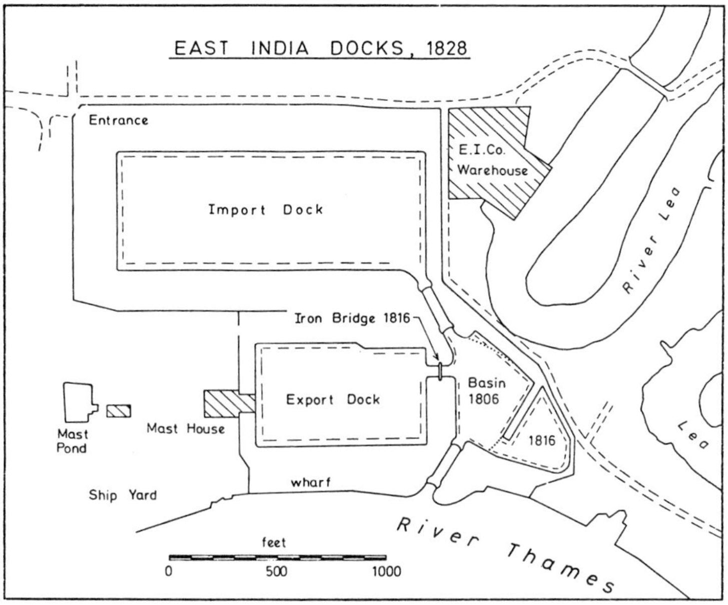

Fig. 5.

Under Walker's direction works began in May 1815 with William Bough as contractor. Thomas Gladstone started the timber wharfing in October. Meanwhile Bough had died but the firm continued the contract, with payments being made to his executors. Gladstone's accounts were settled in July 1816, for a total of about £1700 excluding the cost of timber which had been supplied by the Company. Bough & Co. finished around November, and their bills totalled about £6800. During the last two months of 1816 McIntosh had a small contract for dredging (just over £1000); so the engineering work cost roughly £10,000.

While this job was in hand Hunter & English erected a cast iron swing bridge over the communication passage between the basin and Export Dock. The castings, to Walker's design, had been tendered for by the Horseley Iron Works at £12.10s per ton delivered on site, together with the rollers, wrought iron railings etc. Quite extensive building works were in progress at this time, also under Walker's direction, but I have not obtained details.

James Walker succeeded his uncle as consulting engineer to the East India Docks in 1824.[17] The plan in Fig. 5 is based on a drawing by him dated 1828.[18]

6. REGENT'S CANAL DOCK

Regent's Canal was built, under an Act of 1812, from Paddington Basin (of the Grand Junction Canal) around the north of London and down to the Thames at Limehouse.[19] John Nash, the architect, played a leading part in its promotion and his former assistant James Morgan became engineer to the canal. The original plans envisaged a small basin at Limehouse for barges only. Later the Company contemplated a more ambitious scheme of providing accommodation for ships, and purchased extra land accordingly, but then reversed their opinion on grounds of cost. However, when the canal was nearing completion they finally decided to build a ship dock and obtained an Act in June 1819 granting powers to enlarge and deepen the barge basin and channel from the Thames.[20]

Work proceeded very vigorously, with the engineer James Tate as assistant to Morgan, and the dock was opened on the same day as the canal in August 1820. Hugh McIntosh carried out the excavations.

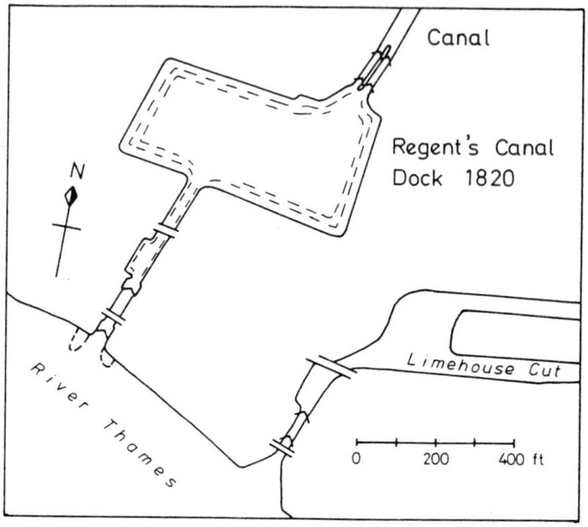

Fig. 6.

The dock (Fig. 6) had an area of 4½ acres. An engraving published in 1827 from a drawing by T. H. Shepherd[21] shows open slopes on the north and east sides and what appear to be timber wharfing wing walls at the canal lock. The ship entrance lock was about 32 ft wide and 130 ft in length between gates, with 19 ft of water on the sills at T.H.W.[22] There were two iron swing bridges, over the lock and over the passage leading to the dock. Another engraving after T. H. Shepherd, published in 1828, gives a view of the entrance (Fig. 7).[23]

At a later date the open-work timber wing walls were extended further into the river, as shown by the dotted lines in Fig. 6. This modification, in effect, removed the outer gates further from the run of the tide, creating slack water in front of them, and gave more room for vessels to manoeuvre in entering or leaving the lock.[24]

Morgan remained engineer to the Regent's Canal until 1835. The dock was used chiefly by colliers and lighters bringing coal and building materials for distribution in London. It soon became one of the busiest in the Port.

7. WEST INDIA DOCKS: PHASE 3

In this phase of construction the wing walls at Limehouse entrance lock were rebuilt and the timber swing bridges replaced by iron.[25] Much activity also took place in building operations. Those, like the engineering works, came under the direction of John Rennie who was appointed engineer in March 1809 with an annual retainer of 100 gns and fees for individual works. Thomas Morris continued as resident engineer until retirement in March 1811. William Pilgrim succeeded him, to be followed (after he died in April 1816) by Thomas Shadrake. The latter remained in office for many years, first under Rennie and from 1821 under Sir John Rennie.

Fig. 7. Regent's Canal Dock, entrance 1828.

Limehouse Wing Walls. Most of the works carried out to Jessop's design in the early stages of construction were on such a vast scale and so solidly built that no extensions and few modifications were needed. But, to save time, the wing walls at Limehouse entrance had been built in tide-work on piled foundations. This proved to be most unsatisfactory, and one of Rennie's first jobs at West India was to rebuild the walls with much deeper foundations, requiring the use of cofferdams.[26]

Unfortunately the contractor, William Bough, failed to drive some of the piles to a sufficient depth. As a consequence at high tide on 3 October 1810, with men excavating for the wall foundations, water burst under part of the dam. The men got out safely, but in a few minutes the piles were forced bodily upwards and the shoring collapsed.[27] The difficult task of reconstruction was given to Joliffe & Banks at their own estimate of cost which, of course, Rennie checked as being reasonable in the circumstances.

Where the foundations were to be deepest the cofferdam piles were driven 16 ft into the ground, as compared to about 5 ft originally. All went well and both walls had been finished by October 1811, within the contract time and cost (£14,300). Additionally, the accident involved a loss of about £5000.

Iron Bridges. In this report of June 1809, where he advises on re-building the wing walls, Rennie remarks that the Directors already had in mind replacing the old (timber) bridge over the Limehouse entrance lock by an iron one, for which Aydon & Elwell had submitted an estimate. I have not found any reference to this in the Minutes, but in a report of March 1811 Rennie mentions that gravel from the wing wall excavations 'has been used for making up the road to the new Iron Bridge'.[28]

The decision to provide an iron bridge over Blackwall entrance lock and another over the communication lock between Limehouse Basin and the Import Dock dates from October 1811. Aydon & Elwell were again involved. There seems to have been some delay, but in May 1813 Rennie reports that the bridge at Blackwall is being erected and iron for the communication lock bridge is on site.[29] It may be noted that the ironwork for these bridges cost £1700 and £1400 respectively. At about this time Rennie also designed the West India diving bell.

The Butterley Company submitted a tender in 1818 for replacing the last of the timber bridges; over the communication lock between Limehouse Basin and the Export Dock. Presumably this was built, but I have not found definite evidence for a date of completion.

Finally, the Butterley Company also supplied two iron swing foot-bridges in 1823 and 1825 (at about £460 each): for Harrow Lane on the north side of the dock premises, and the communication lock from Blackwall Basin to the Import Dock.

8. St. KATHARINE DOCKS

The Act for building St. Katharine Docks dates from 10 June 1825. Based on outline plans by Thomas Telford and Philip Hardwick, made in 1824,[30] and a valuation of the property which would have to be purchased before clearing the site, the estimated expenditure came to £1,353,000. The whole of this sum had to be subscribed before the Act could be put into force, and the Company was empowered to raise a further £500,000.

At the first meeting of the Board,[31] on 16 June 1825, Telford and Hardwick were appointed Engineer and Architect respectively, both at salaries of £500 p.a. Shortly afterwards (the exact dates are not given in the Minutes) Peter Logan, who had been working on Dundee harbour, came on the staff as Resident Engineer (salary £500 p.a.) and Richard Douthwaite left the London Docks to become Chief Clerk of the Works (at £300 p.a.) under Hardwick. Three interesting early items in the accounts may be noted: £291 to Telford and £259 to Hardwick for the period preceding their appointment, and £125 to John Goode 'for boring and proving the Stratum for the site of the Docks and warehouses'. The borings were made to a depth of 40 ft.

The plans of 1824 correspond broadly to the docks as built, but Telford made several changes (notably in the design of the entrance) before October 1825 when, accompanied by Logan, he submitted to the Board 'plans, working drawings and specifications of the whole of the work in the Engineer's department', the estimated cost of which came to £242,000. At the same meeting Hardwick 'attended by

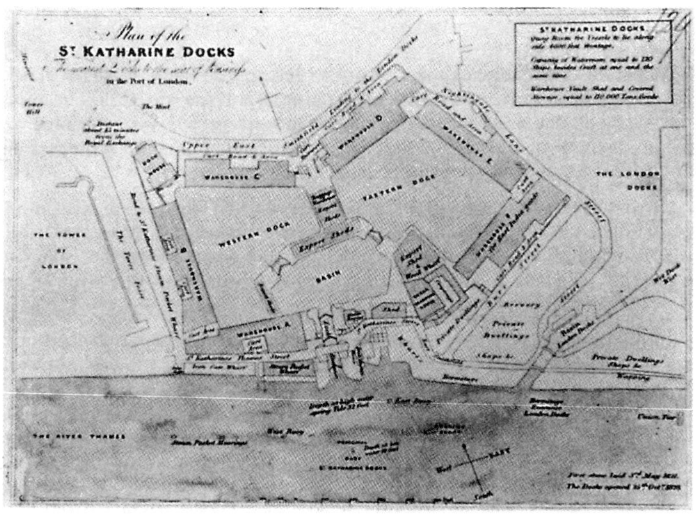

Fig. 8. St. Katharines Docks, plan 1830.

Mr. Douthwaite, presented plans, working drawings and specifications of the Vaults and foundation walls ... and also an Elevation of a stack of Warehouses'. These plans were accepted and in essentials represent the docks as finally constructed (Fig. 8): an entrance lock with three pairs of gates, a basin of 1¾ acres water area and two docks, of 4 acres (west) and 4¼ acres (east). Each dock had a single pair of gates, to maintain water level, and there were to be lofty warehouses surrounding the docks on three sides, the warehouses having their front wall supported on massive iron columns on the edge of the quay.[32]

Western Dock, Basin and Entrance Lock. Contracts having been prepared in November, four tenders were received in January 1826 and that of George Burge proved to be the most acceptable: about £190,000 for all items in the engineering department apart from the lock gates, dock gates and bridges. He was instructed to be ready to commence work in March, but in the event delays in clearing the ground prevented a start from being made until the end of May. On 13 May 1826 the Minutes record that an 'inspecting foreman' is to be appointed under Mr. Logan at 2½ gns per week (£136.10s per year). This I believe signals the arrival of Thomas Rhodes, one of Telford's most trusted assistants, who had been working on the Menai suspension bridge.

Some progress is reported during the next few months, though Burge remained much hampered by lack of room, and it was only in October that he began to achieve a reasonable rate of progress. By that time 8 pile engines were at work on the cofferdam, with four more almost ready, and a steam engine for pumping the excavations was nearing completion.[33] Each of the pile engines drove at least one pile per day; not bad going when it is realised that the piles were 12 inches square and 44 ft long, set precisely into position and driven 10 ft into the blue London Clay.[34]

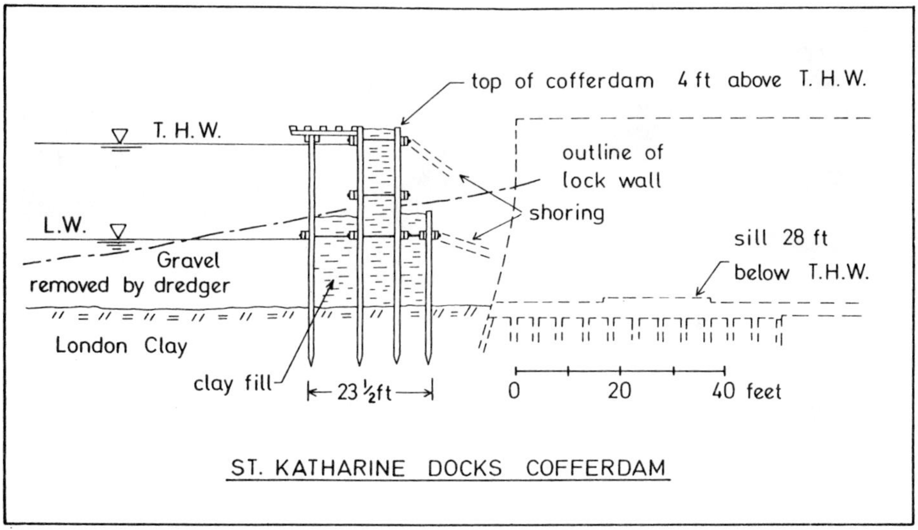

Fig. 9.

Referring to Fig. 9, gravel of the river bed was first dredged out to the top of the clay 12 ft beneath low tide level. Gauge piles were then driven 10 ft apart in three rows, with a row of shorter piles on the inside. Next a bay between the gauge piles was filled in with nine 12 ins piles, driven between temporary walings fixed to the gauge piles, great care being taken to make the piling as watertight as possible. After completing the pile rows the temporary walings were replaced by permanent ones, at the levels shown, together with 2 ins diameter iron tie bolts at 5 ft centres. Finally, the spaces between the rows having been pumped out, clay was compacted between the piles, the topmost part of the main dam being finished with bricks laid in sand to prevent shrinkage of the clay below. During excavation behind the cofferdam shoring would have been provided (no details of this are available) and as soon as the walls of the lock began rising above their foundations the dam was braced against the structure.

By mid-January 1827 payments to Burge totalled £14,000; the cofferdam had probably been finished and excavation could now proceed below tide level. Indeed this is very likely, as the first stone of the walls was laid on 3 May 1827. A drawing of unique interest made somewhat later that year by Clarkson Stanfield, and published as an engraving by George Cooke, see Fig. 10, gives a view of the basin excavations.[35] Spoil is being carried away in horse-drawn waggons on three railways, with numerous lead-outs into various parts of the works. There are two barrow runs on the north side, and the contractor's engine house can be seen near a temporary timber bridge over the railways. These lead to a jetty near the cofferdam (out of sight in Stanfield's drawing) whence the earth was carried away in barges. A Swedish visitor, Captain Carlsund, adds the information that the railways were of iron and that steam engines were used, presumably to haul the waggons up inclined planes to the jetty[36] as at London and West India Docks in 1803 – 5.

Fig. 10. St. Katharines Docks, excavations in 1827.

Meanwhile tenders had been accepted from Messrs. Bennet & Hunt for building the three warehouses, for £228,000, and from Joseph Bramah & Sons for making the lock and dock gates with their associated machinery, for £13,690. Both these contracts were signed in May 1827. A month later Messrs. Seaward of the Canal Ironworks, Limehouse contracted to build the swing bridge over the entrance lock, their tender being £4010 for castings etc. and erection. In the architect's department we note that George Aitchison took Douthwaite's place in July.

Captain Carlsund says he frequently witnessed a thousand men employed in the excavations. He may have been exaggerating but there is no doubt that Burge was now proceeding rapidly, with the west dock as well as the basin; and this despite a catastrophe on the night of 31 October. An exceptionally high tide, rising 3 ft above T.H.W., breached the bank of St. Katharine Thames Street with which the west wing of the cofferdam was connected.[37] The inrush of water swept away the timber bridge and filled the excavation in less than a quarter of an hour. Fortunately no workmen were on site at the time, and the cofferdam itself sustained no injury. Working night and day the breach was soon repaired and the engine (which had been flooded) re-started, so that the whole work-force resumed their labours after about two weeks.

By January 1828, as shown in an engraving after W. Ranwell (Fig. 11), the walls of the basin and dock were well advanced and of the three warehouses under construction one was already nearing completion.[38] John Hall now took over from Logan as resident engineer. Progress in both departments continued to be made at a headlong pace and the formal opening of the basin, west dock, warehouses, and a large shed facing the basin, took place on 25 October 1828. The fine painting by W. J. Huggins (Fig. 12) marks the occasion.[39]

Fig. 11. St. Katharines Docks, progress of construction in January 1828.

Fig. 12. St. Katharines Docks, the first ship entering on 25 October 1828.
By Courtesy of the Port of London Authority.

A feature of particular interest is the arrangement for filling the lock to allow entry of ships even at fairly low states of the tide, and for topping up the basin to dock level (i.e. to T.H.W.), a most necessary requirement as lack of space dictated the use of a single pair of gates (not a lock) at the entrance to each dock. The scheme, for which James Watt, Jr. was chiefly responsible, involved filling the lock partly by pumping from the river and partly by gravity flow through sluices from the basin, and replenishing the basin by pumping while ships were worked through the lock. Calculations showed that two 80 h.p. engines would be required. Boulton & Watt delivered the engine parts in May-June 1828; by the end of June three boilers (one for each engine and a spare) were in place, and after intensive work during four months under the direction of Watt's assistant James Brown everything was ready just before the opening ceremony in

October. When the engines had been properly adjusted and experience gained in the operation, tests in December 1828 showed that the lock could be filled in 5½ minutes from a level of 12 ft below T.H.W.[40] The cost of the engines, pumps, etc. came to £14,775 to which must be added the cost of the engine house, probably about £3000.

The lock had a length of 180 ft between gates, and a middle pair of gates could be used for smaller ships. The sills of the outer and middle gates were 28 ft below T.H.W., thus ensuring a depth of not less than 20 ft of water from 2 hours before to 1½ hours after high tide neaps and for about 2 hours longer at springs.[41] The inner sill, though 4 ft higher, provided ample depth for entry into the basin at all times as water level in the basin rarely fell more than a foot or so below T.H.W.

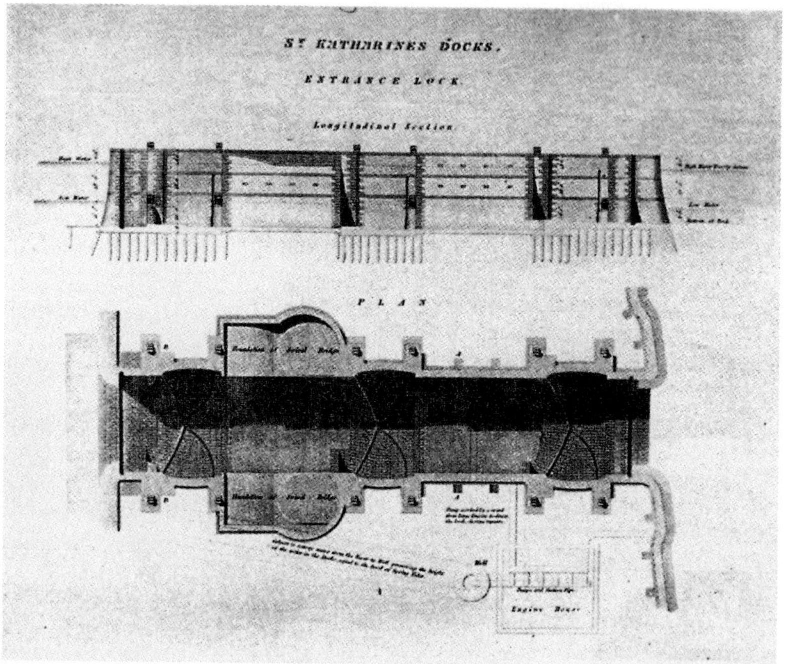

Fig. 13. St. Katharines Docks, entrance lock.

A plan and longitudinal section of the lock are shown in Fig. 13. Details of the gates have been published, as also of the swing bridge.[42] With a span of 45 ft and a width of 24 ft, this was the largest structure of its kind yet built. It had 8 ribs and contained 144 tons of cast iron with 10 tons of iron kentledge counterbalancing each of the cantilevers.

As ground water level outside the docks was 8 ft below T.H.W. the dock (and basin) bottom was lined with puddle clay to prevent loss of water by seepage under the walls. The walls rose 33 ft above dock bottom (25 ft of water plus 8 ft of freeboard). A line of 9 inch sheet-piling 14 ft deep ran under the toe of the walls. They were founded on a 12 inch layer of lime concrete, had a base width of 10 ft, a curved front batter and a top width of 5 ft 3 inches, counterforts 3 ft 9 inches square at 18 ft centres, and a backing of puddle clay.[43]

Throughout the period from May 1827 the resident engineer had on his staff two superintendents, of masonry and brickwork, and similarly there were three superintendents in the architect's department. It only remains to add that a footbridge over the west dock entrance was built by Bramah & Sons in 1829 for £610.

Eastern Dock. John Hall died in November 1828, a week or two after the opening ceremony, and Thomas Rhodes was promoted to the post of resident engineer in his place. Excavation of the east dock had already started, it being a continuation of Burge's contract, and work proceeded so efficiently that everything was ready for water to be admitted on 27 October 1829. The dock walls were also part of Burge's contract. On the same day John Lloyd, who had supplied many of the warehouse cranes, contracted to build a foot-bridge over the east dock entrance for £450. The present bridge is probably of later date. The warehouses around the dock, again to Hardwick's designs, were declared to be completed a year later in 1830.

Steam Packet Wharf. Thomas Carpe contracted in February 1829 to build a wharf wall on the river front, west of the entrance lock, for landing passengers from steam vessels. This he finished in November of that year, and had then only to dredge the gravel in front to an agreed profile. But there was a ready sale of the gravel for the foundations of King's College, at Somerset House, and despite warnings from the resident engineer the contractor dredged well below the proper depth; as a result part of the wall cracked and settled 2 or 3 ft.[44] Carpe was discharged and Burge satisfactorily completed the repairs. The wharf had a length of 170 ft.

By about May 1830, four years after the start of construction, virutally all work had been finished in the engineer's department, at a total cost around £250,000. Telford's salary was paid to June. Rhodes stayed until the end of the year,[45] to be followed as Civil Engineer by James Jones. Hardwick also left in December 1830, Aitchison remaining in charge of the architect's department. But little more than maintenance and minor improvements had to be carried out for several years. Total expenditure by the Company, including the heavy cost of purchasing property, was of the order £2 million by 1830.

9. LONDON DOCKS, PHASE 3

The Act of June 1800 empowered the London Dock Co. to build an eastern dock and entrance at Shadwell. Some land in this area had been purchased early on, but it was not until July 1824 that they decided to go ahead with a new dock.[46]

Eastern Dock. The Company called on Daniel Alexander to draw up an outline scheme and then brought in William Chapman, from Newcastle, to produce more detailed plans and estimates. Chapman visited the site, accompanied by Douthwaite, and after three weeks study reported with an estimated cost of about £280,000 divided almost equally between engineering and building works. It is to be noted that the dock plan now resembled the first proposal of 1794 far more than the revised scheme of 1796 on which the Act had been based.

After receiving Chapman's report, in September 1824 Alexander was appointed Surveyor in charge of both departments at a salary of £600 p.a., together with J. W. Hemingway as resident engineer at £300 p.a. John Arnold later joined the staff as Hemingway's assistant.

Work began in April 1825 on the communication passage to the new dock (see Fig. 4, page 77). This involved taking down part of the Tobacco warehouse, placing a temporary dam across the Tobacco dock entrance, pumping (from the Wapping engines), excavating, building walls and an access bridge for traffic on Old Gravel Lane.

While these operations were in hand work could start on the dock excavations. Alexander prepared specifications in May. A 10 h.p. engine was ordered and Hugh McIntosh assembled his workforce in June. By December the Wapping engines could be used for supplementary pumping and by February 1826 about half the excavation had been completed. Alexander's specifications for the dock walls followed in March and next month Thomas Smith signed a contract to build them for £49,200.

During the following year two important staff changes occurred. Hemingway having died in June 1826, Henry Robinson Palmer, hitherto one of Telford's assistants, came to London Docks as resident engineer at a salary recently increased to £500 p.a. This proved a most fortunate move for him as Alexander retired in April 1827, finally leaving the Company's service, and two months later Palmer was appointed in his place at a salary, also recently increased, of £800 p.a.

By February 1827 the excavation had nearly been finished and the north and west walls were well advanced. Hunter & English contracted in May to build two iron swing bridges of 50 ft span over the communication passage, to Palmer's design, and in October Palmer produced drawings for a floating dam to be used at the passage. This was built by James Jones who, until 1831, was contractor for maintenance of the machinery at the docks. The dock walls were completed in February 1828, the two bridges a month later (cost £5180), and the opening of the dock took place on 3 May 1828. It had an area of 7 acres of water and a depth of 21 ft. There were vaults for wine and spirits on the north and east sides, a warehouse on the west, and the usual sheds with cranes along the quays, as well as a boundary wall enclosing the premises.

The total expenditure since 1825 amounted to £303,000 (including some building work at the western dock) plus £19,000 for purchase of property.

Shadwell Entrance. Although an instruction had been issued in November 1826 to leave an opening in the east wall of the dock for a possible future entrance from the river at Shadwell, the decision actually to build such an entrance (broadly on the line projected in the original plan) dates from July 1827. Palmer submitted an outline scheme for discussion later that year, followed by detailed plans in October 1828. His estimates came to £85,000 with a further £65,000 for purchase of property.

Referring to Fig. 4 (page 77) it will be seen that the works involved an entrance lock, a basin of $1\frac{3}{4}$ acres, and a communication lock leading into the eastern dock. The entrance lock was designed, and built, with a width of 45 ft, a length of 180 ft between gates, and a depth of 26 ft of water on the sills at T.H.W. The basin also had a depth of 26 ft.

Richard Walker joined the staff as Palmer's assistant in January 1830 at 4 gns per week. Working drawings for the locks etc. were produced in March and for two swing bridges in May. William Hazledine contracted to build these for £1936 each. A contract with Robert Streather to carry out the main works for £55,700 was signed in September. Palmer in February 1830 prepared specifications for the lock gates, to be made in cast iron. Tenders were received from Hunter & English, Bramah & Sons, Horseley Iron and William Hazledine, but the successful bid came from John Sturgess & Co. of Bradford: £10,400 for the four pairs of gates.

After the customary delays at the beginning of works of this kind, Streather got well under way in the early months of 1830. Hazledine delivered the bridge castings on site in August, and the excavations were in an advanced state by October. Building of the lock walls must have started about that time, and by April 1831 Hazledine was erecting the bridges, one at each lock. Details of the bridge at Shadwell entrance have been published.[47] It had a span of 46 ft and a width of 16 ft.

Construction of the entrance lock proceeded behind the river embankment. A temporary timber dam now had to be placed across the lock mouth; the embankment would then be removed by spadework and dredging, and finally the wing walls were to be built in tide-work. This was the state of affairs in April 1831 when Streather was declared bankrupt. His payments to date totalled £51,200. Palmer estimated £12,500 to be required for finishing the job. Hugh McIntosh agreed to complete the work 'eastward of the inverted arch of Shadwell bridge' for £9300 and the Company undertook completion of the passage from the communication lock to the dock, using direct labour.

Sturgess & Co. completed their contract for the lock gates in July 1831. John Duffus built an iron footbridge over the entrance to the dock (for which he was paid £260 in October). Edward Turrell engraved a plan of the docks (see Fig. 4, page 77). Trinity House dredgers deepened the river. A borrowed diving bell enabled Palmer to inspect works at the entrance underwater, and it was probably used also for extending the front apron of the lock. Thomas Smith built the dock master's house at the entrance, Hunter & English supplied the bollards and the first ship passed through the lock on 1 March 1832.

The new works having been completed, Palmer's salary was reduced first to £300 and then in February 1833, after maintenance had been placed under the superintendence of an 'Engineers' Committee', to £150 p.a. Two years later he left, possibly without regret on either side, as considerable trouble was being experienced at the Shadwell entrance. Much effort had to be expended on mud removal, the north wing wall (like those at the West India Limehouse entrance) was in danger of collapse, and ships found difficulty in manoeuvring in or out of the lock.

Plans drawn up by George Rennie, as consulting engineer, for rebuilding the wing wall were approved and he prepared specifications which led to a contract with Plews & Slater in August 1835. John Smeaton (no relation, so far as I am aware, to his famous namesake) acted as site engineer. Though details are lacking it seems that the operations were extended to include the south wall and timber piers to improve the entrance[48] (Fig. 14). The works, completed in June 1837, cost £21,600. Rennie received £1050 for his services plus £445, 'being the charge for Mr. Smeaton'. The latter had clearly proved himself to be an able man and the Company immediately appointed him as their Resident Engineer at £300 p.a., a post he held until his death in 1842.

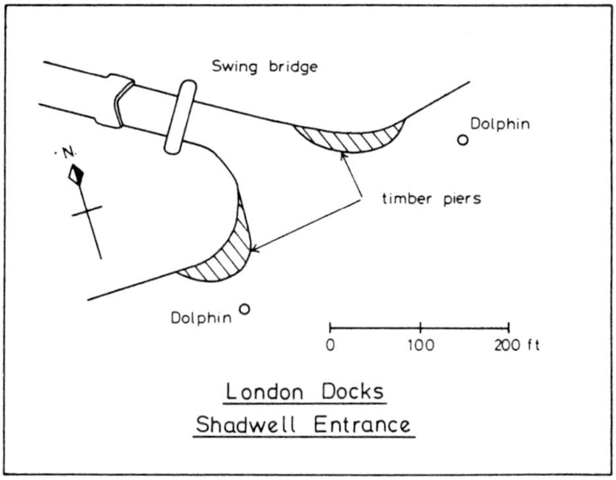

Fig. 14.

10. WEST INDIA DOCKS: PHASE 4.

In 1829 the West India Dock Company purchased the City Canal for £120,000. This provided an extra 14 acres of water, ready made, with entrance locks having the same dimensions as the Blackwall lock (192 × 45 × 24 ft). Numerous short jetties gave access to the ships, from the north bank.

Three years later it was decided to add a timber pond parallel to and south of the South Dock (as the Canal was now called) and connected to it by a short passage.[49] The pond had an area of 16 acres (approximately 2000 ft long by 350 ft wide)[50] with mooring posts and sloping earth sides topped by an embankment. Wave protection for the east bank was provided in the form of wooden hurdles. Hugh McIntosh carried out the work, under Sir John Rennie's direction, at a cost of about £4000.

Filling took place in April 1833 and preparations were at once made for admitting timber previously kept in the South Dock and Blackwall Basin.

11. BRUNSWICK WHARF

This review of engineering in the Port of London ends with an account of an exceptionally interesting and well designed work: Brunswick Wharf, built in iron sheet piling on the river front at Blackwall for the use of steam vessels.

Iron sheet piling appears to have been introduced about 1820 by David Mathews (previously engineer of the East Country Dock) when building the foundations of the north pier at Bridlington harbour, and its first use on the Thames, also for foundations, dates from 1824 in James Walker's reconstruction of part of the quay wall at Downs Wharf near St. Katharine Docks.[51] At Brunswick Wharf, however, it was employed on a large scale as an essential feature of the design.[52]

Detailed plans and a model were submitted to the East India Dock Co. by James Walker in November 1832. They met with approval and, after sanction to build the wharf had been obtained from the Thames Navigation Committee, Walker recommended acceptance of tenders from the following firms: Birtley Iron Co. of Newcastle for the cast iron piling, Messrs. Gordons for the wrought iron tie bars, and Hugh McIntosh for all site work including the granite coping, concrete, etc.

Work began in March, with G. P. Bidder as assistant or resident engineer. Timber piles were first driven accurately in position in a trench dredged out to low water level along the line of the wharf. Two rows of walings, fixed to these piles, acted as guides for the main iron piles, which were placed at exactly 7 ft centres. For ease of handling the main piles were cast in two parts. The lower, 25 ft in length and weighing 28 cwt, was the part actually driven into the river bed and the upper part was bolted on, making a total length of 37 ft (see Fig. 15). The sheet piles, each 22 ft long, were then driven in the bays and secured

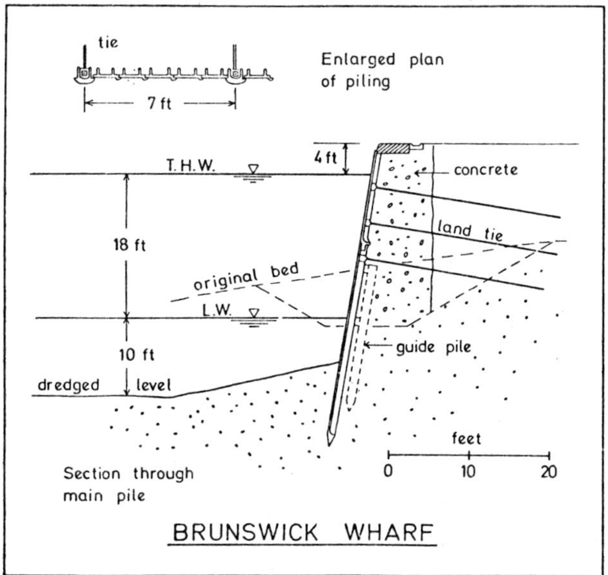

Fig. 15.

to the upper waling by bolts. Three iron plates in each bay, spanning between the main piles, together formed the top 14 ft of the wall. Next came the land ties and mass concrete cast behind the sheeting, and finally the bed was dredged to a depth of 10 ft below low tide level.

The pile engines had a 15 cwt ram, falling 3 ft 6 ins as a maximum, operated by a winch and release mechanism. Driving was very hard, through coarse gravel and a stratum of the Blackwall rock in places, but after adopting a slightly thicker section, following some trouble in the early stages, only 5 piles out of a total of about 600 were broken in driving.

To avoid delays, orders for the piles were soon placed with the Horseley Iron Co. as well as with Birtley. From June 1833 all went well. Dredging in front of the wall started in December and the wharf was completed, and ready for use, in April 1834. It had a length of 720 ft, contained upwards of 900 tons of cast iron, and the construction cost amounted to approximately £16,000. Fig. 16 shows a general view of the wharf with two steamers lying alongside.[53]

Fig. 16. Brunswick Wharf, view in c.1836.

APPENDIX I

Return of the number of Vessels, and their Tonnage, that entered the Docks.

Docks	1820		1830		1835	
	Ships	Tons	Ships	Tons	Ships	Tons
West India	618	170,456	930	206,971	944	202,443
London	1,255	205,388	1,197	207,708	1,559	273,508
East India	140	92,981	98	72,646	72	47,897
Grand Surrey	209	47,397	158	35,504	183	46,177
Commercial	270	69,631	283	71,231	172	49,620
East Country	74	19,925	58	17,366	65	22,716
Regent's Canal	31	6,868	633	116,324	721	151,771
St. Katharine	—	—	893	141,771	804	151,970
Total	2,597	612,646	4,250	869,521	4,520	946,102

These figures have been compiled from data given in the *Report from the Select Committee on the Port of London*. House of Commons, 12 August 1836 (No. 557).

APPENDIX 2

Summary of Water Areas and Entrance Lock Dimensions, 1833.

	Total Area (acres)			Entrance Lock			
	Docks	Entrance Basins	Timber Ponds	Length (feet)	Width (feet)	Depth (feet)	Opened
West India	68	8	16	192	45	24	1802
London	28	6	0	170	40	23	1805
,,				180	45	26	1832
East India	26	4	0	210	48	25	1806
Grand Surrey	19	0	0	130	32	19	1807
Commercial	25	0	33	204	42	19	1809
East Country	4	0	0	160	40	23	1811
Regent's Canal	4	0	0	130	32	19	1820
St. Katharine	8	2	0	180	45	28	1828
	182	20	49				

REFERENCES

1. On a site visit to the East Country Dock in February 1808 the City of London Navigation Committee report that the river bank had been cut through for a length of 50 ft or more, and gates and piers erected across for a dock of large extent. Minute Book of the Committee, 3 March 1808 (P.L.A. archives).
2. In February 1811 the East Country Dock Co. petitioned Parliament for powers to raise £40,000 to pay debts already incurred in making the dock and to complete the works (Journal of the House of Commons). Vessels were entering the dock by August 1811, if not earlier. Navigation Committee Minute Book, 30 August 1811.
3. David Mathews is mentioned several times as engineer of the East Country Dock, between 1808 and 1813, in Minute Books of the Navigation Committee and the Grand Surrey and Commercial Dock Companies. He obtained the post of resident engineer to Bridlington Harbour in 1815 (personal communication from Mr. John James).

4. *Plan of the Commercial Docks at Rotherhithe* J. Walker. London, published 12 Jan. 1820 by Robert Wilkinson (scale 1 inch to 400 ft).
5. Elmes, James. *A Scientific, Historical, and Commercial Survey of the Harbour and Port of London.* (London, 1838).
6. Griffin, Josiah. *History of the Surrey Commercial Docks.* (London, 1877).

7. Except where other references are given, the following account is taken from Minute Books Nos. 1 – 4 (1807 – 1824) of the Commercial Dock Co. and the Dock Committee Minutes Nos. 19 – 22 (1811 – 1823). P.L.A.

8. *Report from the Select Committee on the Port of London,* 12 Aug. 1836 (557). Appendix No. 24.
9. Rennie, Sir John. *The Theory, Formation and Construction of British and Foreign Harbours.* (London, 1854).
10. Shepherd-Baron, W. P. Presidential Address. *Proc. I.C.E. Part I,* vol. **3** (1954) 12 – 40.
11. *Plan of the Commercial Docks at Rotherhithe with the Intended Improvements.* J. Walker, Engineer. Published 1 May 1810 by Robert Wilkinson.
12. *A View of the Commercial Docks at Rotherhithe.* James Walker Engineer, B. Gilbeck Superintendent. Published 1 June 1813 by William Daniell.

13. This account is based on Minute Books Nos. 7 and 8 (1810 – 14) of the Grand Surrey Canal Co. (P.L.A.)
14. The relevant Minute Books for Phase 2 of the London Docks are: Court of Directors, Books 3 and 4 (1811 – 22) and Committee of Treasury, Books 2 – 4 (1810 – 23), in P.L.A. The dimensions of Hermitage entrance lock are given by Sir John Rennie op. cit (9).

15. *Plan of the London Docks as completed in 1831.* Henry R. Palmer, Engineer. Engraved by E. Turrell. (scale 1 inch to 400 ft).
16. East India Dock Co. Minute Books 6 and 7 (1814 – 19). P.L.A.
17. East India Minute Book No. 9
18. Walker's plan of 1828 (scale 1 inch to 250 ft) is in the East India portfolio (P.L.A.)
19. For a general account see Spencer, H. *London's Canal. The History of Regent's Canal* (London, 1961) and also Hadfield, Charles *The Canals of the East Midlands* (Newton Abbot, 1970) who gives a plan of the dock.
20. From the petition submitted to Parliament in February 1819 it is clear that work was then in progress on the barge basin and channel.
21. *The Limehouse Dock, Regent's Canal.* Drawn by Thos. H. Shepherd. Engraved by F. J. Havell. Published 23 August 1827 by Jones & Co.
22. The depth of water on the sills is given by Elmes op. cit. (5).
23. *Entrance to the Regent's Canal, Limehouse.* Drawn by Thos. H. Shepherd. Engraved by F. J. Havell. Published 26 January 1828 by Jones & Co.
24. Redman, J. B. 'Remarks on the formation of entrances to wet and dry docks, situated upon a tideway; illustrated by the principal examples in the Port of London'. *Min. Proc. I.C.E.* 7 (1848) 159 – 174.
25. West India Dock Co. Minute Books (2nd Series) Nos. 5 – 33 (1809 – 23) P.L.A. No. 6 of this set is missing.
26. Rennie's reports 26 April and 12 June 1809. Rennie Letter Books (Inst. Civ. Eng. library).
27. Rennie's report 5 October 1810 (Letter Books) and contemporary newspaper cutting (West India portfolio, P.L.A.).
28. Rennie's report 5 March 1811 (Letter Books).
29. Rennie's report 8 June 1813 (Letter Books).
30. *Plan of the Proposed St. Katharine's Docks* Designed by Thomas Telford, Civil Engineer, F.R.S.E. Philip Hardwick, Architect. C. Hullmandel's Lithography. (scale 1 inch to 100 ft). A copy in the St. Katharine portfolio (P.L.A.) is endorsed 'For Jas. Mountague Esq. 1 Nov^{r}, 1824 With Mr. W. Tooke's Comp^{s}'
31. Most of the information on building St. Katharine Docks is taken from the Board of Directors Minute Books Nos. 1 and 2 (1825 – 34). P.L.A.
32. For a description of the warehouses see Tucker, M. 'St. Katharine Docks' *Arup Journal* 5 (1970) 10 – 18.
33. Letter to the Secretary (John Hall) from Telford 28 October 1826. In Letter File on the Steam Engines (P.L.A.)
34. 'Specifications of entrance lock and coffer dam. St. Katharine's Docks. Engineer, Thomas Telford'. *Civil Engineer and Architect's Journal* 2 (1839) 430 – 433. Text of the specifications with two plans and a section of the cofferdam.
35. *The Tower of London. From the Excavations for the St. Katharine's Docks.* Drawn by Clarkson Stanfield. Engraved by George Gooke 1827. Printed by S. H. Hawkins. London, Published by Longman & Co. ... and J. & A. Arch ... and G. Cooke.
36. Captain Carlsund's account of St. Katharine Docks is quoted in part by Elmes op. cit (5).
37. Contemporary newspaper cuttings. (West India portfolio, P.L.A.).
38. *To the Chairman ... of the St. Katharine's Dock Company this Plate representing the Works in their progress as they appeared in the month of January 1828 is inscribed* ... Painted by W. Ranwell. Engraved by J. Phelps. Published 25 October 1828 by G. R. Phelps.
39. The painting by Huggins is reproduced by courtesy of the Port of London Authority. It was the subject of a fine engraving: *This Plate representing the Opening of the St. Katharine Docks on Saturday the 25th of October 1828, is Most Respectfully Dedicated to the Board of Directors* ... Painted by W. J. Huggins, Marine Painter to his Majesty. Engraved by E. Duncan. Published 1 January 1829 by W. J. Huggins. The ship entering the basin in this picture is the *Elizabeth*, an East India free trader of 516 tons burden. More than 10,000 people were present at the opening.
40. Letter to the Secretary from Boulton, Watt & Co. 14 February 1829. Steam Engines Letter File. Further information in this source includes letters from James Brown covering the period June 1828 to July 1830. See also 'Description of the Pumping Apparatus and Culvert' by Thomas Rhodes in *Life of Thomas Telford* (London, 1838) p. 337 where the date of the December trials is misprinted 1827.
41. From hourly tidal records at springs and neaps taken at St. Katharine's Dock entrance, quoted by Elmes op. cit (5).
42. Published drawings can be found in the following sources. *Atlas to the Life of Thomas Telford* (London, 1838). Pl. 40 plan of the docks, scale 1 inch to 100 ft; Pl. 41 swing bridge; Pl. 42 entrance lock and dock wall. Elmes op. cit. (5). Pls. 5 – 7 lock gates; Pl. 8 swing bridge. Brees, S.C. *Second Series of Railway Practice* (London, 1840) Pls. 37 – 43 swing bridge.

43. Details of the dock walls are given by Telford in *Life of Telford* pp. 156 – 157 and *Atlas* Pl. 42.
44. Contemporary newspaper cutting (St. Katharine portfolio, P.L.A.).
45. In 1830 Rhodes produced a set of record drawings. They came in the Telford Bequest to the Inst. Civ. Eng. and were catalogued as 'Drawings of St. Katharine Docks, including the Coffre Dam, Locks, Dock Gates, Cranes and other Machinery connected therewith designed by Thos. Telford Esq., F.R.S.L. & E., P.C.E., by T. Rhodes Resident Engineer 1830. 29 large Drawings, Index & Title in Portfolio on guards'. Inst. Civ. Eng. Basis Catalogue of Maps, Drawings and Papers 1836. Most of these, disbound, together with some original plans, etc., were sent to the London & India Dock Co. in 1905.
46. The Minute Books covering Phase 3 of the London Docks are: Court of Directors Nos. 6 – 9 (1826 – 41), Committee of Treasury Nos. 5 – 9 (1823 – 50) and Engineers Committee Nos. 2 – 4 (1835 – 45). Court No. 5 and Engineers No. 1 were not handed over to the P.L.A. and at the time of writing Court Nos. 6 and 8 and Engineers No. 2 were missing.
47. Details of the Shadwell swing bridge are given in Brees op. cit. (42) Pls. 12 and 13. These plates are undoubtedly based on Palmer's drawings of which eleven, relating to the eastern entrance lock etc., were given to the Inst. Civ. Eng. and in 1906 transferred to the London & India Dock Co.
48. Redman, op cit. (24).
49. West India Dock Co. Minute Books Nos. 51 – 53 (1832 – 33) P.L.A.
50. Dimensions from *Plan of the East and West India Docks,* 15 April 1841. Engraved by J. & E. Walker. (scale 1 inch to 300 ft).
51. Borthwick, M. A. 'Memoir on the use of cast iron in piling, particularly at Brunswick Wharf, Blackwall'. *Trans. Inst. Civ. Eng.* **1** (1836) 195 – 205.
52. This account is based on East India Dock Co. Minute Books Nos. 11 and 12 (1828 – 35), P.L.A., with technical details from Borthwick, op. cit. (51).
53. *Brunswick Steam Wharf East India Docks.* Coloured lithograph by C. Hullmandel. Undated, but c. 1836.

NAME INDEX

SUBJECT INDEX